Lecture Notes
in Business Information Processing   115

Series Editors

Wil van der Aalst
Eindhoven Technical University, The Netherlands

John Mylopoulos
University of Trento, Italy

Michael Rosemann
Queensland University of Technology, Brisbane, Qld, Australia

Michael J. Shaw
University of Illinois, Urbana-Champaign, IL, USA

Clemens Szyperski
Microsoft Research, Redmond, WA, USA

Kurt J. Engemann
Anna M. Gil-Lafuente
José M. Merigó (Eds.)

Modeling and Simulation in Engineering, Economics, and Management

International Conference, MS 2012
New Rochelle, NY, USA, May 30 - June 1, 2012
Proceedings

 Springer

Volume Editors

Kurt J. Engemann
Iona College
Hagan School of Business
New Rochelle, NY, USA
E-mail: kengemann@iona.edu

Anna M. Gil-Lafuente
University of Barcelona
Department of Business Administration
Barcelona, Spain
E-mail: amgil@ub.edu

José M. Merigó
University of Barcelona
Department of Business Administration
Barcelona, Spain
E-mail: jmerigo@ub.edu

ISSN 1865-1348   e-ISSN 1865-1356
ISBN 978-3-642-30432-3   e-ISBN 978-3-642-30433-0
DOI 10.1007/978-3-642-30433-0
Springer Heidelberg Dordrecht London New York

Library of Congress Control Number: 2012937597

ACM Computing Classification (1998): I.6, J.1, G.1, I.2

Typesetting: Camera-ready by author, data conversion by Scientific Publishing Services, Chennai, India

Printed on acid-free paper

Springer is part of Springer Science+Business Media (www.springer.com)

Preface

The Association for the Advancement of Modeling and Simulation Techniques in Enterprises (AMSE) and Iona College are pleased to present the main results of the International Conference of Modeling and Simulation in Engineering, Economics, and Management, held in New Rochelle, New York, May 30 to June 1, 2012, through this book of proceedings published with Springer in the Series *Lecture Notes in Business Information Processing.*

MS 2012 New York was co-organized by the AMSE Association and Iona College through the Hagan School of Business, New Rochelle, New York, USA. It offered a unique opportunity for students, researchers and professionals to present and exchange ideas concerning modeling and simulation and related topics and see how they can be implemented in the real world.

In this edition of the MS International Conference, we gave special attention to the use of intelligent systems in science. Especially, we focused on the use of these techniques in business administration. The title of the book is Modeling and Simulation in Engineering, Economics, and Management. The importance of having the most efficient techniques in economic sciences in order to maximize benefits or minimize costs is clear when dealing with all the complexities of our uncertain world. Every day the world is changing and we need to adapt our economic models so we can efficiently adapt to the necessities of the market. Therefore, it is necessary to improve the available models with new technologies such as the general field of intelligent systems so we can efficiently manage the economy.

The MS 2012 proceedings are constituted by 27 papers from 17 countries. We have also included a summary of the plenary presentation given by Ronald R. Yager, an IEEE Fellow. The book mainly comprises papers close to the field of intelligent systems and its applications in economics and business administration. It also covers some papers with a stronger orientation to modeling and simulation in general fields of research.

We would like to thank all the contributors, referees and the Scientific and Honorary Committees for their kind co-operation with MS 2012 New York; Joseph Nyre, President of Iona College, Brian Nickerson, Interim Provost of Iona College, Vin Calluzo, Dean of the Hagan School of Business, for his support; Jaime Gil Aluja for his role as the President of AMSE; the whole team of the Organizing Committee, including Warren Adis, Shoshanna Altschuller, Jaime Gil Lafuente, Anna Klimova, Salvador Linares, Donald Moscato, Robert Richardson, Ore Soluade, Heechang Shin, Emilio Vizuete and Ronald Yager; Megan Droge, Dawn Insanalli, Joanne Laughlin Steele, Cindy Zapata and the rest of the Iona team for their support regarding the logistics of the conference;

and to Ralf Gerstner, Christine Reiss and Viktoria Meyer (Springer) for their
kind advise and help in publishing this volume. Finally, we would like to express
our gratitude to Springer and in particular to Wil van der Aalst, John Mylopoulos, Michael Rosemann, Michael J. Shaw and Clemens Szyperski (editors of the
book series *Lecture Notes in Business Information Processing*) for their support
in the preparation of this book.

March 2012

Kurt J. Engemann
Anna M. Gil-Lafuente
José M. Merigó

Organization

Honorary Committee

Special thanks to the members of the Honorary Committee for their support in the organization of the MS 2012 New York International Conference.

Jaime Gil-Aluja   President of *AMSE* and President of the
                                     Spanish Royal Academy of Financial and
                                     Economic Sciences
Lotfi A. Zadeh   University of California at Berkeley

Scientific Committee

Special thanks to all the members of the Scientific Committee for their kind support in the organization of the MS 2012 New York International Conference.

Christian Berger-Vachon   Anna M. Gil-Lafuente
(Co-President)   (Co-President)

Jihad M. Alja'am, Qatar	Yuriy Kondratenko, Ukraine
Said Allaki, Morocco	Gang Kou, China
Noura Al-Khaabi, UAE	Viktor Krasnoproshin, Belarus
Mercedes Ayuso, Spain	Prabhat Kumar Mahanti, India
Bernard De Baets, Belgium	José M. Merigó Lindahl, Spain
Houcine Chafouk, France	Radko Mesiar, Slovakia
Nashaat El-Khameesy, Egypt	Ramir Mirsalinov, Azerbaijan
Kurt Engemann, USA	Francesco C. Morabito, Italy
Joan Carles Ferrer, Spain	Vladimir S. Neronov, Kazakhstan
Fares Fraij, Jordan	Ahmad Nuseirat, Jordan
Hamido Fujita, Japan	Ahmed Oulad Said, Morocco
Jaime Gil Lafuente, Spain	Viji Pai, India
Robert Golan, USA	Witold Pedrycz, Canada
Marta Gómez-Puig, Spain	Emmanuel Perrin, France
Federico González Santoyo, Mexico	Mohamed Saighi, Algeria
Montserrat Guillén, Spain	Wilfred I. Ukpere, South Africa
Rainer Hampel, Germany	Emili Vizuete, Spain
Kedi Huang, P.R. China	Iskander Yaacob, Malaysia
Korkmaz Imanov, Azerbaijan	Ronald R. Yager, USA
Janusz Kacprzyk, Poland	Constantin Zopounidis, Greece
Uzay Kaymak, The Netherlands	

Organizing Committee

Special thanks to all the members of the Organizing Committee for their support during the preparation of the MS 2012 New York International Conference.

Organizing Committee Chairs

Kurt J. Engemann
Anna M. Gil-Lafuente

Organizing Committee Co-chair

José M. Merigó Lindahl

Organizing Committee

Warren Adis, USA
Shoshanna Altschuller, USA
Vin Calluzo, USA
Jaime Gil Lafuente, Spain
Anna Klimova, Russia
Salvador Linares, Spain

Donald Moscato, USA
Robert Richardson, USA
Ore Soluade, USA
Heechang Shin, USA
Emilio Vizuete, Spain
Ronald R. Yager, USA

External Reviewers

We would like to thank to all the referees for their advice in the revision process of the papers submitted to the MS 2012 New York International Conference.

Moses A. Akanbi
Igor P. Atamanyuk
Amir Azaron
Luciano Barcelos
Abdelhakim Begar
Joel Bonales Valencia
Sefa Boria
Kurt J. Engemann
Mark Evans
María Ángeles Fernandez-Izquierdo
Idoya Ferrero-Ferrero
Antoni Ferri
Anna M. Gil-Lafuente

Jaime Gil Lafuente
Montserrat Guillen
Aras Keropyan
Anna Klimova
Yuriy P. Kondratenko
Salvador Linares
Carolina Luis Bassa
José M. Merigó
Donald Moscato
María Jesús Muñoz-Torres
Robert Richardson
Wilfred I. Ukpere
Emilio Vizuete

Table of Contents

Social Networks and Social Information Sharing Using Fuzzy Methods

Ronald R. Yager

Machine Intelligence Institute, Iona College
New Rochelle, NY, USA
ryager@iona.edu

Abstract. Web 2.0 has provided for a rapid growth of computer mediated social networks. Social relational networks are becoming an important technology in human behavioral modeling. Our goal here is to enrich the domain of social network modeling by introducing ideas from fuzzy sets and related granular computing technologies. We approach this extension in a number of ways. One is with the introduction of fuzzy graphs representing the networks. This allows a generalization of the types of connection between nodes in a network from simply connected or not to weighted or fuzzy connections. A second and perhaps more interesting extension is the use of the fuzzy set based paradigm of computing with words to provide a bridge between a human network analyst's linguistic description of social network concepts and the formal model of the network. We also will describe some methods for sharing information obtained in these types of networks. In particular we discuss linguistic summarization and tagging methods.

K.J. Engemann, A.M. Gil Lafuente, and J.M. Merigó (Eds.): MS 2012, LNBIP 115, p. 1, 2012.
© Springer-Verlag Berlin Heidelberg 2012

Using Analytical Methods
in Business Continuity Planning

Holmes E. Miller[1] and Kurt J. Engemann[2]

[1] Muhlenberg College, Allentown, PA USA
[2] Hagan School of Business, Iona College, New Rochelle, NY USA

Abstract. Business continuity focuses on ensuring an organization can continue to provide services when faced with various crisis events. Part of the business continuity planning process involves: Business Impact Analysis; Risk Assessment; and Strategy Development. In practice, these activities often rely on ad hoc methods for collecting and analyzing data necessary for developing the business continuity plan. In this paper we discuss how various analytical methods that have been successfully used for addressing other problems, may be applied to the three phases of business continuity planning mentioned above.

Keywords: Business Continuity, Business Impact Analysis, Risk Assessment, Strategy Development, Analytical Methods, MS/OR methods.

1   Introduction

Business continuity focuses on ensuring an organization can continue to provide services when faced with various crisis events. These events can directly impact an organization, resulting in immediate losses (primary effects) or impact the organization indirectly, such as when a crisis event affects a supply chain partner (secondary effects). Often natural disasters cause crisis events. Primary effects from the 2011 Japanese tsunami involved organizations such as the Fukushima nuclear plant. Secondary effects were borne by organizations indirectly affected by the event, such as closing automobile assembly plants in the United States via disruption of supply chains with suppliers located in the impacted area in Japan [1].

Ensuring business continuity involves actions to sustain or resume operations when faced with crisis events. Business continuity planning (BCP) is the ongoing process that ensures the plans an organization has in place can effectively deal with all the events that may confront an organization. Since ensuring that an organization can function at optimal levels when faced with all possible events is not feasible, BCP involves analyzing possible threats, crisis events, the risks, and the resulting business impact, and from this analysis, developing strategies to keep the critical components of an organization functioning at acceptable levels. This "sub optimality" involves prioritizing and making decisions involving resource allocation subject to various

K.J. Engemann, A.M. Gil Lafuente, and J.M. Merigó (Eds.): MS 2012, LNBIP 115, pp. 2–12, 2012.

constraints. It also involves the ability to understand increasingly complex supply chain networks to facilitate decision-making.

1.1 The Current State of Analytics in BCP

Although BCP's visibility has increased in recent years, many organizations still view it as an auxiliary, albeit necessary, activity. Often the resources allocated to BCP may not fully support many required steps in the process. How does this affect the process itself? In some cases it may mean skipping certain steps entirely. BCP plan maintenance and testing for example, is a critical step but may receive insufficient resources or lack of organizational support. Another deficiency occurs when a step is done partially, or does not fully exploit the tools available. An example of this occurs when in the risk analysis [2], when certain relevant threats are ignored.

Using available analytical tools and information is the subject of our paper. Cursory interviews with managers and back-of-the-envelope calculations may generate numbers that look credible, but that lead to plans and strategies that are inappropriate for an organization. Conversely, a more analytical approach might yield data suggesting a plan using a completely different approach that focuses on different events and addresses different weaknesses. Today, when many organizations depend on complex supply chains and combinations of infrastructures that may pose unseen risks, firms miss opportunities and create hidden exposures when they use ad-hoc rather than analytical methods.

1.2 Business Impact Analysis, Risk Assessment, and Strategy Development

Given these observations, our paper will focus on how analytical methods may be applied to three phases of the BCP process: Business Impact Analysis; Risk Assessment; and Strategy Development. We will discuss how using simulation, algorithms and data analysis techniques enhance the process. The discussion is introductory and to point out directions rather than to fully explain the details of the methods apropos for the specific step. Our goal is to provide a road map for further research to enable practitioners to develop more robust business continuity plans.

2 Business Impact Analysis

Business Impact Analysis (BIA) helps develop business recovery objectives by determining how disruptions affect various organizational activities. BIA seeks to quantify the impact of possible events and provides the foundation for developing continuity and recovery strategies. An underlying assumption is that not all activities need recover to their pre-disaster levels at the same time. Given resource restrictions, some activities will use more resources than others. Specific BIA steps include [2]:

- Determine and prioritize the objectives of an organization
- Determine the critical deliverables of an organization
- Identify critical resources required by the deliverables
- Determine the impact over time of a disruption

- Determine resumption timeframes for critical operations
- Provide information from which appropriate recovery strategies can be formulated

In practice many of these steps are conducted informally. Procedures may involve interviews and cursory questionnaires, which overlook the interdependencies and complexities present in many organizations. Other approaches are discussed in [3].

The specific BIA steps mentioned above can be consolidated into a series of questions amenable to more in-depth, analytical treatment. These questions include:

- What are the objectives of the organization and how are they prioritized?
- What are the critical organizational deliverables, when are they required to resume operation and at what levels of pre-disaster performance?
- What is the impact of a disruption and what are the interdependencies?
- What resources are necessary for successfully implementing the BCP and how should these resources best be allocated?

Each of these questions will be discussed in the following sections.

2.1   Organizational Objectives and Priorities

In many organizations with multiple product lines and supporting infrastructures, individual objectives combine to form a network of objectives. In practice determining and prioritizing these objectives may involve interviewing managers, obtaining rankings and developing an organization-wide picture.

Keeney [4] discusses a more formal process of prioritizing objectives, using a structured, quantitative-based approach. This includes identifying an organization's fundamental objectives, its mean-ends objectives, and the relationships among these objectives. These can be analyzed in a hierarchical network framework that facilitates identifying interdependencies and reduces the likelihood that the final stated priorities will misstate the real priorities of the organization.

In some cases objectives can be quantified using a value model and mathematical programming methods can be applied, for example, for supply chains [5] or for energy supply applications [6]. To be effective, methods to prioritize objectives should have some analytical rigor while at the same time, not being so complex that decision makers cannot use them. The increasing power of placing complex models in the hands of practitioners along with user-friendly interfaces, creates a number of opportunities along these dimensions.

2.2   Deliverables, Resumption Requirements, and Levels Performance

After a crisis event, questions arise what must be done the resume processing, at what levels, and when. Finding answers involves understanding how the operational system can respond in the new environment and what are the internal and external needs of relevant stakeholders and customers. This involves identifying each deliverable (e.g., being able to process new orders), timing requirements (e.g., one day after the event) and quality requirements (e.g., process all orders greater than a specified minimum amount). This step may also include identifying implications arising from networked relationships. For example, if A must resume operation on day 4 but if B and C depend on A and must resume operation on day 2, and if B and C are of high priority, this affects the deliverables and timing requirements for A.

2.3   Impact of a Disruption and Interdependencies

In complex networked organizations accurately determining these interdependencies goes beyond using check sheets and tables. One class of methods that may be applied involves network analysis. Here, the organizational entities form a network that may be analyzed mathematically via network means [7] or via a simulation method [8].

One classical network method is the PERT/CPM method where a critical path is identified for completing a project [9]. Delay of any activity lying on the critical path also delays the project completion date. Not only are the interdependencies identified, but specifying the critical path facilitates applying resources to expedite various completion times for activities, which may allow the project to be completed sooner.

In a simulation model the user may exercise the system under various infrastructure scenarios, or conversely, generate overall measures of vulnerability given a specified structure. For example, a complex supply chain can be modeled and then exercised to see how disruptions at various tiers of the supply chain affect delivering the products and services for the supply chain [10]. Simulation models can incorporate the complexities of the underlying systems and, although they do not yield optimal answers, can be used to develop output response distributions, which can facilitate finding "good" solutions.

2.4   Necessary Resources for Implementing the BCP and Their Allocation

Resources may be thought of in two ways. First, are the resources necessary for the individual BCP components. For example, these may include computer processing, human resources, and supplies. Each activity may have its unique set of resources and each of these may have various attributes such as how much of the resource is needed, when it is needed, and also certain quality characteristics such as employee skill certification levels. Fungible resources, such as financial resources, are another class. Given a fixed budget, the decisions here involve how to allocate those resources such that the overall objectives of the organization are optimized. While resources can be allocated using informal methods, mathematical programming models can yield solutions that are far better than those made by instinct alone. Optimal allocations can speed-up bringing certain activities back online and in complex environments, help managers avoid allocating resources that "miss" important activities that may subvert the overall system objective. Examples of how mathematical programming and other operation research methodologies have been applied to BCP appear in [11], and [12].

3   Risk Assessment

Risk management in BCM provides an analytical foundation for decision making. A key tenet of risk management is that risk cannot be eliminated but can be controlled. The appropriate controls to employ depends both on the likelihood of the risk occurring and the magnitude of the loss if the risk does occur. Often risk can be quantified; however, when risk cannot be quantified, either because the underlying information does not exist or because it is too expensive to collect, principles of risk management can still be applied. These principles include:

- Identifying what can go wrong;
- Identifying what controls are currently in place;
- Evaluating the current exposure to the organization;
- Identifying new controls that can be implemented to reduce this exposure;
- Evaluating whether these controls should be implemented by investigating the costs and benefits.

Because business continuity management deals with events that are improbable, analyzing risks is challenging. Analysis begins with identifying threats. A threat (hazard) is a source of potential negative impact. A crisis (crisis event) is a manifestation of a threat. The core of a risk analysis involves specifying a set of encompassing crisis events, which represent what can go wrong. A minor crisis has limited impact and does not affect the overall functioning capacity of an organization, whereas a major crisis has the potential to seriously disrupt the overall operation of an organization. A disaster is a major crisis event that imperils an organization due to factors such as loss of life, environmental damage, asset damage and duration of disruption. A catastrophe is an extreme disaster.

Often in the context of business continuity planning, risk analysis is done in an impromptu manner, e.g., checklists. Using the arsenal of available analytical methods, the output of a risk analysis can go beyond the subjective. The following discusses some key steps involved in the risk analysis process: Identifying significant threats to critical operations; identifying and evaluating controls; estimating event probabilities and impacts; developing risk measures prioritizing risks.

3.1   Identify Significant Threats to Critical Operations

Assuming critical operations have been identified in the BIA, the first step in a Risk Assessment (RA) is to identify relevant internal and external threats. For each operational component this involves understanding the significant internal and external dependencies, redundancies, and single points of failure [13], [14]. Threats can be identified using historical records and in discussions with technical and business managers responsible for operational and support components. Augmenting these methods with more systematic analysis can uncover threats that may be missed by a cursory analysis. Two methodologies that can be used come from quality management and from reliability. In quality management process flow charts are used to identify opportunities for quality improvement. Process flows can also be used to identify dependencies and from these, threats. For example, using process flows in analyzing risks in an ATM network might reveal that data is sent to and retransmitted from a satellite. Threats affecting this satellite are relevant in analyzing threats that might affect the ATM network itself [15]. For a more analytical approach to identifying threats reliability theory can be used [16]. In reliability theory, specific linkages in a network can be specified and detailed probabilities can be determined.

3.2   Identify and Evaluate Controls

Controls can either prevent a threat from manifesting itself into a crisis event, or can mitigate the effects of an event when one does occur. For example, a levee may prevent a rising river from flooding a facility but given that a flood does occur, a backup facility may enable the organization to function, even at a degraded level of service.

Identifying controls involves examining the process, the documentation, and interviewing area experts. Evaluating controls is more difficult. A control is effective only to the extent that is reduces the likelihood of a crisis event occurring or mitigates the effects when it does occur. Individual controls might be effective, but when taken together still may be vulnerable to various threats. Analytical methods can be used in evaluating controls in two ways: First, they can provide information when individual controls' are exercised. Second, via means of network analysis and simulation, an existing configuration of controls effectiveness can be evaluated when confronted with various threats [17], [18]. Using this marginal benefit logic, other methodologies such as simulation can also be used to quantify a control's effectiveness. Moreover, using network models, areas where controls are missing also can be identified.

3.3   Estimate Event Probabilities and Impacts

Estimating event probabilities and their impacts are necessary steps in any quantitative analysis of risk. These may be estimated in various ways, ranging from informal estimates and tabular estimates to estimates derived from historical records, simulation models, and mathematical models.

Informal methods often involve managers and other "experts" using their experience and their knowledge of the underlying systems to estimate the probability and the impact of an event. While better than no estimates at all, especially if exercised via a sensitivity analysis, subjective probability estimates and impact estimates risk being disconnected from reality. Using probabilities to inform later quantitative analysis may result in solutions that appear to be precise but in fact, are anything but precise. The same may be said for estimates of impacts. Tabular estimates of probabilities in risk classes may be slightly more precise but suffer from similar deficiencies. For impacts, precision may be greater than for probabilities but still may include a large margin for error when indirect impacts are taken into account,.

Historical records often can provide better estimates. Engemann and Miller [19] used historical records of power outages obtained from the power utility to develop estimates of power outage probabilities in evaluating the purchase of a back-up generator for a large New York bank. A problem in using historical records to estimate probabilities arises when the future environment differs from the past. A second deficiency involves improbable events that have never occurred. In this case, absence of occurrence in the past does not preclude occurrence in the future. If only past experience is to be a guide, many threats may be essentially "invisible" since they have never manifested themselves in crises events.

Estimating the impact of events may be more precise, since specific departments have records of business activity and direct losses from various outages of specified durations can be determined. As mentioned above, however, indirect losses are more subjective and inexact. Simulation models and mathematical models, informed by the past and by the current system, can be used to generate more precise probability estimates. Grounded in logic based on the underlying system, these models can use methods such as fault tree analysis and reliability analysis to develop probability estimates [20]. Aven and Steen [21] developed a model for estimating probabilities in an environment when uncertainties are large: estimating the probability of pirate attacks the Gulf of Aden. Mathematical models, including Bayesian analysis, can also be applied to complex problems, such as supply chains [22], and catastrophe modeling in insurance [23].

3.4   Developing Risk Measures and Prioritizing Risks

Risk is the combination of the probability of occurrence of events and the impacts of events when they do occur. Risk may be expressed in several ways. When the information is qualitative in nature (e.g. probabilities being high, medium or low and impacts being high, medium or low), tabular measures may be developed. The 3x3 case just mentioned, would result in nine risk classes, which could be prioritized to provide guidance for further development of alternatives and strategies [2], [24].

More precise risk estimates result when probabilities and impacts are quantitative. In this case, expected losses may be developed and prioritized from high to low. Given the imprecision discussed above, sensitivity analysis can be developed which measures risk under various scenarios and the outcomes for a set of cases may allow risk to be prioritized over a broader spectrum of possible outcomes. Engemann and Miller [19] analyzed risk using this approach. Here, various alternatives for managing the risk of power outages were examined and, since both probabilities and impacts were quantitative, expected losses were used to analyze various cases. Engemann and Miller [25] also use a computational intelligence approach to analyze risk in using smart technologies for critical infrastructures.

Once the probabilities and impacts are determined, measuring and prioritizing risk follows. Specific methods may be drawn from the plethora of analytical methods available. The increasing volume and availability of information should only strengthen this trend and allow measuring and prioritizing risk to be more quantitative and more finely tuned in the future.

4   Developing Strategies

Developing strategies for BCP involves developing strategies for individual business and support units and ensuring that they also are integrated into a coherent organizational strategy. In many ways the process for developing BCP strategies mirrors the process for developing overall business strategy and many similar methods may be applied.

4.1   Strategies for Individual Business and Support Units

Although each unit may have a BCP strategy, this does not imply that these strategies should be developed independent of each other. For example, to deal with threats to physical structures, two units located in different states might combine to support each other in case of a crisis event. Alternatively, each might opt for its own individual strategy depending on its objectives, resources, and requirements regarding being back online.

Analytical methods may be used in two ways: First, in formulating the strategies the presence of data and statistical methodologies can lead to creating strategies to better serve the organization and its customers. Big data methods [26], which facilitate processing large amounts of data quickly, are examples. While normally the value is in ensuring data collection and processing can continue in a crisis event, the information of past history also is useful in identifying critical activities and priorities, which is at the core of strategy development. Many of the simulation and network methods discussed above also can be applied, especially when seeing how individual strategy alternatives interact with other unit strategies. Finally, when individual units have several alternative strategies to choose from, decision analysis [27] and mathematical programming-based methods [5] can be applied. Using these methods to choose among strategies can yield "optimal" solutions, although a downside is the methods themselves may be to complex and time-consuming to be practically applied. They are best used in large, resource rich environments where the benefits from higher cost selection methods outweigh the costs of doing the selecting.

4.2   Integrating Strategies

It is particularly important to integrate business unit and support unit strategies, and to integrate these strategies with of organizations throughout the supply chain. This is includes identifying what services (an levels of service) are delivered, and when they are delivered. Ensuring that strategies are coherent and consistent also is critical. At one level, analytical methods are not as important as basic information sharing, especially as requirements change across the supply chain. Strategies that are appropriate at one point in time may become obsolete and one must insure all members of a supply chain are in synch. Information sharing allows proper strategies that address current business needs to be formulated in the first place.

A second level is to identify the vulnerabilities when strategies are integrated and to test strategies to see if they are properly aligned. When strategies are improperly integrated, individually functioning units may not be able to deliver customer value because they may be achieving different and perhaps incompatible objectives. Testing strategies is important because it may be impossible to spot strategic misalignment in complex environments. For this simulation and network modeling are appropriate. Simulation models can capture the complexity of real environments and can be exercised to test how changes in parameters, variables and the underlying strategies affect performance. For years, simulation has been used successfully in examining the interactions in business strategies [28].

5   Conclusions and Discussion

Analytical models and methods can be successfully applied to support the BCP process. The three phases of the process – Business Impact Analysis, Risk Assessment, and Strategy Development – all depend on collecting and analyzing information, and generating from various alternatives, conclusions and courses for action. Ad hoc methods may be used to perform these activities but will miss many opportunities that analytical methods provide.

Many existing tools are amenable to the BCP process. These include more sophisticated statistical and predictive methods of analyzing large amounts of information, which are layered on data warehouses that contain massive collections of data generated by corporate activities. This is an area where BCP must keep up with ongoing business practices. Whereas ten or fifteen years ago, asking managers about the volume and timeliness of information suggests hoc answers may ignore underling realities. Using them also risks missing opportunities and ending up with suboptimal BCPs that place the organization itself at a competitive disadvantage.

The increasing complexity of supply chains also drives the need for employing more sophisticated analytical methods in developing BCPs. Years ago an organization could look inward and develop a BCP by considering risks that either were internal to its operation or external as they affected critical infrastructures on which the organization depended. Today, complex supply chains coupled with the reality that a chain is as strong as its weakest link, means that business continuity planners must expand their horizons and understand and incorporate the plans of their supply chain partners in their own plans. This is particularly challenging even within an organization, when documenting maintaining and testing BCPs.

Challenges to managers seeking to develop more analytically-based BCPs include: Cost efficiency of models; resource availability; and management support. While using models to support the BCP process makes sense for many reasons already mentioned, doing so can be expensive, in terms of dollars and time. Just because business-generated data exists, does not mean it is easily accessible or in the proper format to used to drive BCP analytics. The models themselves require using computer and human resources and suggest obtaining a level analytical sophistication that may not be present in the organization. Often, this means using expensive outside resources, such as consultants.

Even when individuals are available, training in both analytical methods (for BCP professionals) and business continuity (for analytical professionals) may be necessary. Since senior management is the final arbiter for allocating resources, management support is necessary for any increased application of analytics to BCP. In its infancy, many organizations did not want to employ BCP methods at all because they were expensive and managers often felt the risks – often unseen and thus ignored – did not justify taking financial resources away from ongoing operations. This view has changed and most managers agree that BCPs are necessity. Indeed, failure to properly plan for crisis events can get a manager fired, even before disaster strikes.

The quality of the plans is an emerging issue. As with a products, quality is not absolute. Rather, quality evolves relative to the environment. The same is true with BCPs. Failure to use analytical methods when the competition is doing so places firms at risk. Moreover, any member of a supply chain should realize that the "bigger fish" in the chain may require increased use of analytics in the analysis of risks and the development of plans. Given this reality, it behooves all organizations to plan ahead, not only for disasters but also for the evolving state-of–the art in order to meet the new challenges in developing business continuity plans.

References

1. Helft, M., Bunkley, N.: Disaster in Japan Batters Suppliers, New York Times (March 14, 2011)
2. Engemann, K.J., Henderson, D.M.: Business Continuity and Risk Management: Essentials of Organizational Resilience. Rothstein Associates, Brookfield (2012)
3. Sikdar, P.: Alternate Approaches to Business Impact Analysis. Information Security Journal: A Global Perspective 20(3), 128–134 (2011)
4. Keeney, R.A.: Value Focused Thinking. Harvard University Press, Cambridge (1992)
5. Mula, J., Peidro, D., Díaz-Madroñero, M., Vicens, E.: Mathematical Programming Models for Supply Chain Production and Transport Planning. European Journal of Operational Research 204(3), 377–390 (2010)
6. Karvetski, C.W., Lambert, J.H., Linkov, I.: Scenario and Multiple Criteria Decision Analysis for Energy and Environmental Security of Military and Industrial Installations. Integrated Environmental Assessment and Management (7), 228–236 (2011)
7. Zio, E., Sansavini, G.: Modeling Interdependent Network Systems for Identifying Cascade-Safe Operating Margins. IEEE Transactions on Reliability 60(1), 94–101 (2011)
8. Setola, R., Bologna, S., Casalicchio, E., Masucci, V.: An Integrated Approach For Simulating Interdependencies. In: Papa, M., Shenoi, S. (eds.) Critical Infrastructure Protection II. IFIP, vol. 290, pp. 229–239. Springer, Boston (2009)
9. Kerzner, H.: Project Management: A Systems Approach to Planning, Scheduling, and Controlling. John Wiley and Sons, New York (2008)
10. Miller, H.E., Engemann, K.J.: A Monte Carlo Simulation Model of Supply Chain Risk Due To Natural Disasters. International Journal of Technology, Policy and Management 8(4), 460–480 (2008)
11. Brysona, K., Millar, H., Josephc, A., Mambourin, A.: Using Formal MS/OR Modeling To Support Disaster Recovery Planning. European Journal of Operational Research 141(3), 679–688 (2002)
12. Altay, N., Green, W.G.: OR/MS Research in Disaster Operations Management. European Journal of Operational Research (175), 475–493 (2006)
13. Christopher, M., Peck, H.: Building the Resilient Supply Chain. International Journal of Logistics Management 15(2), 1–14 (2004)
14. Zsidisin, G.A., Ritchie, R.A.: Supply Chain Risk: A Handbook of Assessment, Management, and Performance. Springer (2008)
15. Mack, E.: Major Satellite Outage Affecting ISPs, ATMs, Flights; CNET (October 6, 2011), http://news.cnet.com/8301-1023_3-20116846-93/major-satellite-outage-affecting-isps-atms-flights/
16. Cardoso, J.M.P., Diniz, P.C.: Why Both Game Theory and Reliability Theory Are Important in Defending Infrastructure against Intelligent Attacks. In: Bier, V.M., Azaiez, N.M. (eds.) Game Theoretic Risk Analysis of Security Threats. International Series in Operations Research & Management Science. Springer, Boston (2009)

17. Bajgoric, N.: Server Operating Environment for Business Continuance: Framework for Selection. International Journal of Business Continuity and Risk Management 1(4), 317–338 (2010)
18. Escaleras, M.P., Register, C.A.: Mitigating Natural Disasters through Collective Action: The Effectiveness of Tsunami Early Warnings. Southern Economic Journal 74(4), 1018–1034 (2008)
19. Engemann, K.J., Miller, H.E.: Operations Risk Management at a Major Bank. Interfaces 22(6), 140–149 (1992)
20. Aven, T.: Risk Analysis. Wiley, Chichester (2008)
21. Aven, T., Steen, R.: On the Boundaries of Probabilistic Risk Assessment in the Face of Uncertainties, A Case of Piracy and Armed Robberies Against Ships in the Gulf of Aden. International Journal of Business Continuity and Risk Management 1(2), 113 (2010)
22. Paulsson, U., Nilsson, C., Wandel, S.: Estimation of Disruption Risk Exposure In Supply Chains. International Journal of Business Continuity and Risk Management 2(1), 1–19 (2011)
23. Grossi, P., Kunreuther, H.: Catastrophe Modeling: A New Approach to Managing Risk. Huebner International Series on Risk, Insurance and Economic Security (2005)
24. Aven, T., Aven, E.: On how to understand and express enterprise risk. International Journal of Business Continuity and Risk Management 2(1), 20–34 (2011)
25. Engemann, K.J., Miller, H.E.: Critical Infrastructure and Smart Technology Risk Modeling Using Computational Intelligence. International Journal of Business Continuity and Risk Management 1(1), 91–111 (2009)
26. Bughin, J., Chui, M., Manyika, J.: Clouds, Big Data, And Smart Assets: Ten Tech-Enabled Business Trends to Watch. McKinsey Quarterly, 1–14 (2010)
27. Hammond, J.S., Keeney, R.L., Raiffa, H.: The Hidden Traps in Decision Making. Harvard Business Review, 1–9 (September-October 1998)
28. Morecroft, J.D.W.: Strategy Support Models. Strategic Management Journal 5(3), 215–229 (1984)

Using Neural Networks to Model Sovereign Credit Ratings: Application to the European Union

Raúl León-Soriano and María Jesús Muñoz-Torres

University Jaume I,
12071 Castellón, Spain
{rleon,munoz}@uji.es

Abstract. Credit rating agencies are being widely criticized because the lack of transparency in their rating procedures and the huge impact of the ratings they disclose, mainly their sovereign credit ratings. However the rationale seems to be that although credit ratings have performed worse than their aim, they are still the best available solution to provide financial markets with the information that their participants base their decisions on. This research work proposes a neural network system that simulates the sovereign credit ratings provided by two of the most important international agencies. Results indicate that the proposed system, based on a three layers structure of feed-forward neural networks, can model the agencies' sovereign credit ratings with a high accuracy rate, using a reduced set of publicly available economic data. The proposed model can be further developed in order to extent the use of neural networks to model other ratings, create new ratings with specific purposes, or forecast future ratings of credit rating agencies.

Keywords: Neural networks, Sovereign credit rating, European Union, Credit rating modeling.

1 Introduction

During the very last years, credit ratings have become the most important tool for bankruptcy prediction of firms and countries. Credit ratings can be defined as evaluations of a potential borrower's ability to repay debt, prepared by a credit bureau at the request of the lender.

In spite of their increasing importance in financial markets, credit rating agencies do not disclose the precise factors and weights used in their ratings generation. Rating agencies claim to use complex processes considering non-linear interactions between different variables, which include quantitative and qualitative data on financial and non financial information [1]. As a result, modeling credit rating by means of quantitative or ruled based techniques is a challenging process.

Literature provides many research works aimed at modeling credit ratings by means of different statistical methods. Those include multiple discriminant analysis [2], regression analysis [3], ordered probit [4] and options-based approaches (e.g. KMV Corporation, see [5]), among others. However, the use of those econometric

K.J. Engemann, A.M. Gil Lafuente, and J.M. Merigó (Eds.): MS 2012, LNBIP 115, pp. 13–23, 2012.

methods involves introducing assumptions with respect to the underlying properties and relationships within the data. Those assumptions are particularly difficult to hypothesize in the context of credit rating because of the lack of transparency of credit rating agencies.

The use of neural networks for credit rating modeling is particularly appropriate because they do not require prior specification of theoretical models [6] and they are able to deal with incomplete information or noisy data [7,8]. They are usually used in economics and finance in researches where variables are in non-linear relations [9].

The aim of this work is to advance in the use of neural networks in credit ratings. To this end, a neural networks based structure is proposed to simulate sovereign credit ratings of the European Union countries. The model uses publicly available data on economic and financial issues of European countries, as well the ratings published by some of the most important agencies.

The rest of this work is structured as follows. Section 2 provides theoretical background on neural networks and their use in credit ratings modeling. In Section 3, it is presented a brief review on credit ratings, credit rating agencies and sovereign credit ratings. Section 4 presents the data and models employed to evaluate the feasibility of using neural networks for sovereign credit rating modeling. Main results are then presented in Section 5. Finally, Section 6 discusses on results and future research.

2   Neural Networks

Neural networks are powerful technique inspired by the way biological nervous systems process information. This technique consists on a large number of interconnected elements (neurons) working at the same time through a set of algorithms. Thus, neural networks can solve real problems if these can be mathematically represented.

Neural networks have the ability to learn from experience, adapting their behavior according to a specific environment. To that end, neural networks need to be trained to approximate an unknown function or process, based on available input and output data.

A significant group of researchers [7,8] argue that neural networks can be a more appropriate tool than traditional statistical tools, because: (a) the underlying functions controlling business data are generally unknown and neural networks do not require a prior specification function, only a sample of input-output data to learn and train the process; (b) neural networks are flexible – the brain adapts to new circumstances and can identify nonlinear trends by learning from the data; (c) neural networks can work with fuzzy data (very common in economics); and (d) they are able to deal with incomplete information or noisy data and can be very effective especially in situations where it is not possible to define the rules or steps that lead to the solution of a problem.

In this regard, numerous empirical research has compared neural network methodology with econometric techniques for modeling ratings, and consistent with theoretical arguments, the results clearly demonstrate that neural networks represent a superior methodology for calibrating and predicting ratings relative to linear regression analysis, multivariate discriminant analysis and logistic regression (e.g [1,10]). Leshno and Spector [11] even point out that the methods traditionally used in credit risk analysis, which include regression analysis, multivariate discriminant analysis, and logistic regression among others, represent special cases of neural networks, and therefore it is not surprising that they are appropriate for credit rating modeling.

Increasingly, academics are exploring neural networks methodology as a new tool for modeling decision process and predict in the area of finance, such us corporate financial distress and bankruptcy [12], loan evaluation and credit scoring [8,13], and bond rating, both from private corporations and sovereign issuers [1,14].

3   Sovereign Credit Ratings

Ratings are an invaluable instrument in globalized financial markets. Lenders and investors have nowadays a huge number of possible credits, but only a passing knowledge of most of borrowers. Furthermore, during the last few years, global securities markets have become an increasingly important source of external funding for many emerging economies, including countries and corporations. Within this context, credit rating agencies play a useful role in financial decision-making by providing market participants with information about the credit risk associated with different financial investments [15], and more precisely, about the probability that borrowers will default in its financial commitments.

A wide range of participants, including issuers, investors and regulators use the information provided by rating agencies in their decision-making [16,17]. The information generated by credit rating agencies is mainly delivered in the form of credit ratings, which are alphabetical indicators of the probability that borrowers will not fulfill the obligations in their debt issues. Typically, a credit rating has been used by investors to assess the probability of the subject being able to pay back a loan. However, in recent years, participants have started to give new uses to credit ratings, which include among others to adjust insurance premiums, to determine employment eligibility, and establish the amount of a utility or leasing deposit [18].

Rating agencies emphasize that the credit rating process involves the use of a wide range of variables considering financial and non-financial information about the object being assessed, and also variables related to industry and market-level factors. [19]. However, rating agencies do not specify neither what determines their ratings nor their rating procedure [20]. The precise factors they use in their ratings, and the aggregations processes and the related weights of these factors used in determining a credit rating, are not publicly disclosed by the rating agencies. Furthermore, the credit rating agencies underline that they do not employ a specific formula to combine their assessments of political and economic factors to derive the overall rating [15]. Instead, credit rating agencies use their experience and judgment, based on both objective and subjective criteria, in determining what public and private information – quantitative as well as qualitative- should be considered and how it has to be weighted in giving a rating to a particular company or government.

As a result, credit ratings issued by different rating agencies on a specific object do not have necessarily to be the same, what originates a lack of security in the decision making of financial markets participants.

During the last few decades, sovereign credit ratings have gained importance as more governments with greater default risk have borrowed in international bond markets. Sovereign credit ratings have important implications for international capital flows and for the linkages between company ratings and sovereign ratings [20]. They have a strong influence on the cost borrowers must face, and they can enhance the capability of countries' governments and private sectors to access global capital markets, attract foreign direct investment, encourage domestic financial sector development, and support governments' efforts on financial and economic improvements and transparency [21]. It is generally believed that improving a country's transparency, information control, financing costs and sovereign risk levels is expected to increase international capital inflows and improve the general level of development in financial markets and their financial integration with world capital markets [22].

There is also evidence that sovereign ratings strongly determine the pricing of sovereign bonds and that sovereign spreads incorporate market participants' views of expected credit rating changes [23]. A good country rating is a key success factor of the availability of international financing since it directly influences the interest rate at which countries can borrow on the international financial market, and it may also impact the rating of its banks and companies, and is reported to be correlated with national stock returns [24]. By opposite, as a direct consequence of being lowly rated, firms and countries may face higher interest rates, or even to the refusal of a loan by creditors [18].

While the ratings have proved useful to governments seeking market access, the difficulty of assessing sovereign risk has led to agency disagreements and public controversy over specific rating assignments [25]. Furthermore, during the last few years, and as a consequence closely linked to global crisis, sovereign credit ratings have become widely criticized for exacerbating the crisis when they downgraded the countries in the midst of the financial turmoil. As a result, the financial markets have shown some skepticism toward sovereign ratings when pricing issues. However, they are still considered as the best measure of a country's credit risk available nowadays as internal default data is missing [26].

In addition to the above mentioned problems, which are related to all credit ratings, sovereign credit ratings present one additional problem. Countries, unlike firms, rarely default because of the availability of international emergency credit (most of their problems being liquidity and not solvency related) and the high cost of future credit should they default. As a consequence, the rating agencies' objective of providing estimations of the probability that borrowers will not fulfill the obligations in their debt issues can be considered plausible at the level of corporations, but it is problematic at the sovereign level [20].

This paper focuses on sovereign ratings for several reasons. From a usefulness perspective, investors are increasingly focused on international diversification, and hence an understanding of sovereign credit risk is very important. Furthermore sovereign ratings have historically represented a ceiling for the ratings assigned to financial institutions, corporate and provincial governments, although the ceiling is no longer applied in an absolute sense by the largest three agencies [27]. Therefore, contributing on the understanding of sovereign credit rating agencies, and the factors and weights they use in their ratings is essential for the transparency of the markets. From a practical perspective, the amount of public available historic data on sovereign credit ratings is considerably higher than data on the other categories, which makes it easier to have at disposal enough data for training and testing the proposed models.

4   Models and Data

The development of a neural network system for facing classification problems, as well as others, starts with de definition of a sample. This study uses sovereign credit rating of countries in the European Union between years 1980 and 2011.

Once the sample is defined, it is necessary to define the neural networks input and output variables, a neural network model suitable for the problem being addresses and the selected input and output variables, and two datasets containing the necessary data to accomplish the training process of the neural network, which are a dataset of input data and a dataset with the corresponding target values. Those datasets are essential for the accuracy of the trained neural network, since they are the examples where the networks learn from.

4.1   Input Variables and Training Input Dataset

Literature has largely examined the variables that mainly determine the sovereign credit ratings, and findings may indicate that sovereign ratings are generally consistent with economic fundamental. Cantor and Packer [28] show that per capita income, inflation, external debt, an indicator of economic development and an indicator of default history all explain 90% of the variation in the ratings of 49 countries observed in 1995. They conclude that high ratings are assigned to countries with high per capita income, low inflation, more rapid growth, low ratio of foreign currency external debt to exports, absence of a history of default on foreign currency debt since 1970, and a high level of economic development. Those variables have been used by many authors in sovereign credit rating modeling [1,4,20,29] and it has to be noted that they are also consistent with those that the rating agencies claim being using in assessing sovereign credit worthiness [30,31].

In line with literature, following input variables have been chosen: gross domestic product (GDP) per capita, external balance relative to GDP, fiscal balance relative to GDP, gross debt / exports of goods and saves, inflation, and unemployment rate. All data used to populate the training input dataset have been sourced from the World Economic Outlook Database [32], and the AMECO database [33].

4.2   Output Variables and Training Target Dataset

There are a large number of credit rating agencies that publish their ratings, but only a few have risen to international prominence. The most popular credit rating agencies are Moody's Investor Service, Standard & Poor's and Fitch [28]. Table 1 summarizes the measurement scales used by the tree most important rating agencies.

While Fitch and Moody's agencies offer a whole history register of their sovereign credit rating in its website, Standard and Poor's only offers access to historic credit ratings initially determined on or after June 2007, which is very scarce for being used in neural networks training. Therefore, rating used in this study are Moody's and Fitch, and more concretely, their foreign currency sovereign credit ratings for countries in the European Union (EU27).

The ratings timelines provided by both agencies were used to create a dataset for each agency containing the distribution of the ratings at the end of years. A total of 615 and 450 samples were obtained for Moody's and Fitch respectively.

4.3   Neural Networks Model

Regarding the neural networks employed, there are some aspects that should be addressed in order to select the most suitable configuration for the problem being addressed. That includes selecting the type of neural network, the number of layers, the number of neurons used at each layer and the propagation functions.

In the case of the modeling a credit rating that uses a close set of categories to asses issuers repayment capability, literature and neural networks software providers recommend using a N-layered feed-forward network ($N>=2$) [34,35], with outputs as categories and with sigmoid transfer function in hidden layers and competitive transfer function at output layer [36].

Literature provides indications on how determine an approximation of the optimal number of neurons of hidden layer [37]. However, the optimal number of neurons must be found empirically because feed-forward networks use gradient method in learning, and it is therefore possible that the learning algorithm gets stuck in local minimum within the error function. That problem can be solved by adding additional neurons or changing other network parameters.

One single neural network can solve quite complex problems. However, neural networks can also be combined in complex structures specially designed to maximize the accuracy by exploiting the specific properties of different neural networks. In this paper it is used a structure similar to the one employed by [18], which is expected to minimize the potential error in classifications using neural networks with non dichotomous outputs.

That structure consists of n levels and 2^n dichotomous neural networks, being 2^{n-1} the number of networks in the level l. Each input data is processed by one single neural network at every level, and the output it generates determines which net will be used in the next level. The combination of the outputs of the three neural networks that process each specific case determines its rating.

As the above described structure allows classifying data in 2^n categories, the ratings being modeled have been reorganized in 8 categories. Table 1 shows the employed structure of neural networks. As groups G7 and G8 have a very limited representation, training Net 1.0.0 is not possible. Those categories are not considered in this research.

Table 1. Rating Groups linked to Moody's, Standard and Poor's and Fitch rating systems

Group	Characterization of debt(source: Moody's)		Moody's	S&P	Fitch	Net 1	1.1	1.0	1.1.1	1.1.0	1.0.1	1.0.0
G1	Highest quality		Aaa	AAA	AAA	1	1	-	1	-	-	-
G2	High quality	Investment grade	Aa1	AA+	AA+	1	1	-	0	-	-	-
			Aa2	AA	AA	1	1	-	0	-	-	-
			Aa3	AA-	AA-	1	1	-	0	-	-	-
G3	Strong payment capacity		A1	A+	A+	1	0	-	-	1	-	-
			A2	A	A	1	0	-	-	1	-	-
			A3	A-	A-	1	0	-	-	1	-	-
G4	Adequate payment capacity		Baa1	BBB+	BBB+	1	0	-	-	0	-	-
			Baa2	BBB	BBB	1	0	-	-	0	-	-
			Baa3	BBB-	BBB-	1	0	-	-	0	-	-
G5	Likely to fulfil obligations, ongoing uncertainty		Ba1	BB+	BB+	0	-	1	-	-	1	-
			Ba2	BB	BB	0	-	1	-	-	1	-
			Ba3	BB-	BB-	0	-	1	-	-	1	-
G6	High credit risk	Speculative grade	B1	B+	B+	0	-	1	-	-	0	-
			B2	B	B	0	-	1	-	-	0	-
			B3	B-	B-	0	-	1	-	-	0	-
G7	Very high credit risk		Caa1	CCC+	CCC+	0	-	0	-	-	-	1
			Caa2	CCC	CCC	0	-	0	-	-	-	1
			Caa3	CCC-	CCC-	0	-	0	-	-	-	1
	Near default with possibility of recovery		Ca	CC	CC	0	-	0	-	-	-	1
			C	C		0	-	0	-	-	-	1
G8	Default		C	SD	RD	0	-	0	-	-	-	0
			D	D		0	-	0	-	-	-	0

5   Results

The experiment was conducted separately for each one of the rating agencies –using in both cases the same training input dataset, and two scripts were codified. The first script generated the necessary training sub datasets for each of the 6 neural networks according to groups previously defined. It is necessary to clarify that during the training process, the training target dataset is used to determine what neural network will be trained at each level with a specific case, without considering the output this case would have generated in a trained system. The second script trained the networks and reorganized their outputs in a dataset containing, for each input case, a set with the outputs of the corresponding nets at each of the three levels.

Finally, a confusion matrix was generated to determine the accuracy of each system. Figure 1 shows confusion matrixes for Fitch and Moddy's. As it can be observed, both systems reach a quite high level of accuracy, and most misclassifications are one-group error. However, a higher sample and a more accurate selection on input variables and neural network parameters would be expected to increase the accuracy of the system. Furthermore, it has to be noted that errors are mainly related to cases which have bee better classified by the neural network systems than they were classified by rating agencies. This tendency may arise as a result of the training datasets, which present an unbalanced distribution where the highest classification is much more frequent than any other, and where negative ratings –those referred to speculative categories- are very infrequent.

Moody's

Outputs	G1	G2	G3	G4	G5	G6	
G1	223	58	13	6	2	0	73.8%
	36.3%	9.4%	2.1%	1.0%	0.3%	0.0%	26.2%
G2	10	50	2	2	6	0	71.4%
	1.6%	8.1%	0.3%	0.3%	1.0%	0.0%	28.6%
G3	10	22	118	36	9	4	59.3%
	1.6%	3.6%	19.2%	5.9%	1.5%	0.7%	40.7%
G4	0	0	1	24	5	3	72.7%
	0.0%	0.0%	0.2%	3.9%	0.8%	0.5%	27.3%
G5	0	0	0	0	4	0	100%
	0.0%	0.0%	0.0%	0.0%	0.7%	0.0%	0%
G6	0	0	0	2	0	5	71.4%
	0.0%	0.0%	0.0%	0.3%	0.0%	0.8%	28.6%
	91.8%	38.5%	88.1%	34.3%	15.4%	41.7%	68.9%
	8.2%	61.5%	11.9%	65.7%	84.6%	58.3%	31.1%
	G1	G2	G3	G4	G5	G6	

Targets

Fitch

	G1	G2	G3	G4	G5	G6	
G1	144	38	0	0	0	0	79.1%
	32.0%	8.4%	0.0%	0.0%	0.0%	0.0%	20.9%
G2	13	56	7	4	1	0	69.1%
	2.9%	12.4%	1.6%	0.9%	0.2%	0.0%	30.9%
G3	0	8	66	18	2	0	70.2%
	0.0%	1.8%	14.7%	4.0%	0.4%	0.0%	29.8%
G4	0	0	13	52	11	3	65.8%
	0.0%	0.0%	2.9%	11.6%	2.4%	0.7%	34.2%
G5	0	0	0	1	7	0	87.5%
	0.0%	0.0%	0.0%	0.2%	1.6%	0.0%	12.5%
G6	0	0	0	1	0	5	83.3%
	0.0%	0.0%	0.0%	0.2%	0.0%	1.1%	16.7%
	91.7%	54.9%	76.7%	68.4%	33.3%	62.5%	73.3%
	8.3%	45.1%	23.3%	31.6%	66.7%	37.5%	26.7%
	G1	G2	G3	G4	G5	G6	

Targets

Fig. 1. Confusion matrixes for Moddy's and Fitch models

6   Discussion and Final Remarks

The obtained results, whereas not surprising, confirm that, due to their usefulness for generalization and classification, neural networks can be used for credit ratings modeling. Using a comprehensive dataset comprising sovereign credit ratings of rating agencies and public information on countries over the period 1980–2011, this paper shows that neural networks can provide accurate classifications of sovereign credit ratings. Obviously, the problem of modeling credit rating agencies scores is very close to the variables introduced in the model, and a more accurate selection of input variables could significantly enhance results. Furthermore, it is also possible that using averages or tendencies of some variables during some previous years, the accuracy of the systems could increase.

It should be noted that although rating agencies claim that the experience and subjective judgment of their analysts are essential in determining the ratings, results seem indicate that a neural networks system based on a small set of input variables from publicly available financial and no financial information could provide accurate classifications of credit ratings. This result is in line with other researches where neural networks are successfully used to model different credit ratings [1,18,19].

From those remarks it seems possible to generalize the use of neural networks for the development of credit rating tools addressed to different types of products.

Furthermore, by means of neural networks and having access to expert's knowledge, it could be possible to create rating tools for other proposes such as the internal rating systems of banks. In this regard, and starting from Basel II proposals, it could worth a try to research on the suitability of banks using their own credit rating systems based on neural networks in determining the amount of capital they need to put aside for different types of loans.

Neural networks may also be used to predict future credit ratings, at it is possible to use predicted variables as inputs for trained neural networks systems. In this regard, credit rating forecasting may be really useful for issuers and investors, since it could be used by analysts as an informing and supporting tool for their decision making processes, in order to reduce the costs or risks of their financial operations.

However, it is still necessary further research in order to better understand the main variables that determine credit ratings provided by agencies, and to test different neural networks configurations and structures until better results are obtained. Furthermore, the same research should also be conducted using different credit ratings at corporations, sub national and supranational levels.

Acknowledgements. This work is supported by P1.1B2010-04 and P1.1B2010-13 research projects and the Master in Sustainability and CSR offered by Universitat Jaume I of Castellon, Spain.

References

1. Bennell, J.A., Crabbe, D., Thomas, S., ap Gwilym, O.: Modelling sovereign credit ratings: Neural networks versus ordered probit. Expert Systems with Applications 30, 415–425 (2006)
2. Taffler, R.: The assessment of company solvency and performance using a statistical model. Accounting and Business Research 13, 295–307 (1983)
3. Horton, J.J.: Statistical classification of municipal bonds. Journal of Bank Research 3(1), 29–40 (1970)
4. Trevino, L., Thomas, S.: Local versus foreign currency ratings: What determines sovereign transfer risk? Journal of Fixed Income 11(1), 65–76 (2001)
5. Bessis, J.: Risk management in banking, 2nd edn. Wiley, West Sussex (2002)
6. Trigueiros, D., Taffler, R.: Neural networks and empirical research in accounting. Accounting and Business Research 26, 347–355 (1996)
7. Vellido, A., Lisboa, P.J.G., Vaughan, J.: Neural networks in business: a survey of applications (1992-1998). Expert Systems with Applications 17, 51–70 (1999)
8. Malhotra, R., Malhotra, D.K.: Evaluating consumer loans using neural networks. Omega 31(2), 83–96 (2003)
9. Granger, C.W.J.: Developments in the nonlinear analysis of economic series. The Scandinavian Journal of Economics 93(2), 263–276 (1991)
10. Kim, J.W., Weistroffer, H.R., Redmond, R.T.: Expert systems for bond rating: A comparative analysis of statistical, rule-based and neural network systems. Expert Systems 10, 167–188 (1993)

11. Leshno, M., Spector, Y.: Neural network prediction analysis: The bankruptcy case. Neurocomputing 10, 125–147 (1996)
12. Ahn, B.S., Cho, S.S., Kim, C.Y.: The integrated methodology of rough set theory and artificial neural network for business failure prediction. Expert Systems with Applications 18(2), 65–74 (2000)
13. Baesens, B., Setiono, R., Mues, C., Vanthienen, J.: Using Neural Network Rule Extraction and Decision Tables for Credit-Risk Evaluation. Management Science 49(3), 312–329 (2003)
14. Chaveesuk, R., Srivaree-Ratana, C., Smith, A.E.: Alternative neural network approaches to corporate bond rating. Journal of Engineering Valuation and Cost Analysis 2(2), 117–131 (1999)
15. Kräussl, R.: Do credit rating agencies add to the dynamics of emerging market crises? Journal of Financial Stability 1, 355–385 (2005)
16. Cantor, R., Packer, F.: Differences in opinion and selection bias in the credit rating industry. Journal of Banking and Finance 21, 1395–1417 (1997)
17. Stolper, A.: Regulation of credit rating agencies. Journal of Banking and Finance 33, 1266–1273 (2009)
18. Falavigna, G.: Financial ratings with scarce information: A neural network approach. Expert Systems with Applications 39, 1784–1792 (2012)
19. Hájek, P.: Municipal credit rating modelling by neural networks. Decision Support Systems 51, 108–118 (2011)
20. Mora, N.: Sovereign credit ratings: guilty beyond reasonable doubt? Journal of Banking and Finance 30(7), 2041–2062 (2006)
21. Alsakka, R., ap Gwilym, O.: Leads and lags in sovereign credit ratings. Journal of Banking & Finance 34, 2614–2626 (2010)
22. Kim, S.-J., Wu, E.: Sovereign credit ratings, capital flows and financial sector development in emerging markets. Emerging Markets Review 9, 17–39 (2008)
23. Erb, C., Harvey, C.R., Viskanta, T.: Understanding Emerging Market Bonds. Emerging Markets Quarterly, 7–23 (Spring 2000)
24. Van Gestel, T., Baesens, B., Van Dijcke, P., Garcia, J., Suykens, J.A.K., Vanthienen, J.: A process model to develop an internal rating system: Sovereign credit ratings. Decision Support Systems 42, 1131–1151 (2006)
25. Cantor, R., Packer, F.: Sovereign Credit Ratings. Current Issues in Economics and Finance 1(3) (1995)
26. Basel Committee on Banking Supervision: International convergence of capital measurement and capital standards (2004)
27. Alsakka, R., ap Gwilym, O.: Heterogeneity of sovereign rating migrations in emerging countries. Emerging Markets Review 10, 151–165 (2009)
28. Cantor, R., Packer, F.: Determinants and inputs of sovereign credit ratings. FRBNY Economic Policy Review 2(2), 37–53 (1996)
29. Afonso, A.: Understanding the determinants of sovereign debt ratings: evident for the two leading agencies. Journal of Economics and Finance 27(1), 56–74 (2003)
30. Beers, D.T., Cavanaugh, M.: Sovereign credit ratings: A primer. Standard & Poor's Counterparty Ratings Guides, 1st Quarter (1999)
31. Truglia, V.J.: Moody's sovereign ratings: A ratings guide, Moody's Iinvestors Service Global Credit Research, special comment (March 1999)
32. World Economic Outlook Database: International Monetary Fund (September 2011)
33. AMECO database of the European Commission's Directorate General for Economic and Financial Affair (2011)

34. Hornik, K., Stinchcombe, M., White, H.: Multilayer feedforward networks are universal approximators. Neural Networks 2, 359–366 (1989)
35. Min, J.H., Lee, Y.-C.: Bankruptcy prediction using support vector machine with optimal choice of kernel function parameters. Expert Systems with Applications 28, 603–614 (2005)
36. Beale, M.H., Hagan, M.T., Demuth, H.B.: Neural Network Toolbox^{TM} User's Guide. The MathWorks, Inc., Natick M.A (2010)
37. Chauhan, N., Ravi, V., Chandra, D.K.: Differential evolution trained wavelet neural networks: Application to bankruptcy prediction in banks. Expert Systems with Applications 36, 7659–7665 (2009)

Measuring Errors with the OWA Operator

José M. Merigó

Department of Business Administration, University of Barcelona,
Av. Diagonal 690, 08034 Barcelona, Spain
jmerigo@ub.edu

Abstract. We study the use of the ordered weighted average in the mean error. We introduce the ordered weighted averaging error operator. It is an aggregation operator that provides a parameterized family of aggregation operators between the minimum and the maximum error. We study some of its properties and particular cases. We also develop several extensions and generalizations such as the ordered weighted averaging percentage error, the ordered weighted averaging square error and the induced ordered weighted averaging error operator.

Keywords: OWA operator, Error mean, Induced OWA operator, Aggregation operators.

1 Introduction

An important issue in statistics [2] and related areas when analyzing the information is to measure the errors between the available information and the forecasts. In the literature, we find a wide range of tools for doing so [2] such as the mean error, the mean squared error and the mean percentage error. Usually, when analyzing these errors, we use an aggregation operator to fuse them into a single representative one such as the arithmetic mean or the weighted average. An aggregation operator that is becoming more popular in the scientific community is the ordered weighted average (OWA) [7]. It provides a parameterized family of aggregation operators between the minimum and the maximum. Since its introduction, it has been studied by a lot of authors [1,10]. An interesting extension is the induced OWA (IOWA) operator [12] that uses order inducing variables in the aggregation process in order to deal with complex information. Furthermore, the OWA operator can be generalized by using generalized and quasi-arithmetic means forming the generalized OWA (GOWA) [12] and the quasi-arithmetic OWA (Quasi-OWA) operator [12]. Recently, Merigó and Gil-Lafuente [12] has suggested the induced generalized (OWA) operator that provides a further extension that uses induced and generalized aggregation operators in the same formulation. Since its introduction, these models have been studied by different authors by using fuzzy information [12], interval numbers [12] and continuous aggregations [12].

The aim of this paper is to analyze the use of the OWA operator in the calculation of errors. For doing so, we introduce the ordered weighted averaging error (OWAE) operator. It is an aggregation operator that provides a parameterized family of

K.J. Engemann, A.M. Gil Lafuente, and J.M. Merigó (Eds.): MS 2012, LNBIP 115, pp. 24–33, 2012.

aggregation operators between the minimum error and the maximum error. Its main advantage is that it can represent all the potential errors that may occur in a problem in a complete way considering all the potential situations and select the error that seems to be in closest accordance with the decision maker interests. We study some of its main properties and particular cases including the mean error, the maximum error and the average error (or mean error).

We further extend the OWAE operator by using percentage errors obtaining the OWA percentage error (OWAPE) operator. We also extend it by using the mean squared error forming the OWA squared error (OWASE) operator and by using generalized aggregation operators obtaining the generalized OWAE (GOWAE) operator. Next, we analyze the use of induced aggregation operators obtaining the induced OWAE (IOWAE) operator. We generalize this approach by using quasi-arithmetic means forming the induced quasi-arithmetic OWAE (Quasi-IOWAE) operator. The main advantage of these generalizations is that they provide a more general framework that represents the information obtained in the analysis of errors in a more complete way.

We study the applicability of the OWAE operator and its extensions and we see that it is very broad because all the previous studies that have been developed with the error mean can also be implemented with this approach. We focus on an application in the selection of the optimal experts for making a forecast for the next year. We analyze the forecasts given by several experts during the previous years in order to see who is giving the lowest errors.

This paper is organized as follows. In Section 2 we briefly describe some basic preliminaries. Section 3 introduces the OWAE operator and some other extensions. Section 4 studies the applicability of this new approach and Section 5 summarizes the main results of the paper.

2   Preliminaries

Some basic concepts useful for understanding this paper are the OWA operator and the mean error. The OWA operator can be defined as follows.

Definition 1. An OWA operator of dimension n is a mapping $OWA: R^n \to R$ that has an associated weighting vector W of dimension n with $w_j \in [0, 1]$ and $\sum_{j=1}^{n} w_j = 1$, such that:

$$OWA\,(a_1, ..., a_n) = \sum_{j=1}^{n} w_j b_j \,, \tag{1}$$

where b_j is the jth largest of the a_i.

The OWA operator can be extended by using order inducing variables that deal with complex reordering processes forming the induced OWA (IOWA) operator.

Definition 2. An IOWA operator of dimension n is a mapping $IOWA: R^n \times R^n \to R$ that has an associated weighting vector W of dimension n with $\sum_{j=1}^{n} w_j = 1$ and $w_j \in [0, 1]$, such that:

$$IOWA \left(\langle u_1, a_1 \rangle, \langle u_2, a_2 \rangle, \ldots, \langle u_n, a_n \rangle \right) = \sum_{j=1}^{n} w_j b_j , \tag{2}$$

where b_j is the a_i value of the IOWA pair $\langle u_i, a_i \rangle$ having the jth largest u_i, u_i is the order-inducing variable and a_i is the argument variable.

The OWA and the IOWA operator can be further generalized by using generalized and quasi-arithmetic means. By using generalized means we get the generalized OWA (GOWA) and the induced generalized (IGOWA) operator. The IGOWA operator can be defined as follows.

Definition 3. An IGOWA operator of dimension n is a mapping $IGOWA: R^n \times R^n \to R$ that has an associated weighting vector W of dimension n with $w_j \in [0, 1]$ and $\sum_{j=1}^{n} w_j = 1$, such that:

$$IGOWA \left(\langle u_1, a_1 \rangle, \langle u_2, a_2 \rangle \rangle, \ldots, \langle u_n, a_n \rangle \right) = \left(\sum_{j=1}^{n} w_j b_j^{\lambda} \right)^{1/\lambda} , \tag{3}$$

where b_j is the a_i value of the IGOWA pair $\langle u_i, a_i \rangle$ having the jth largest u_i, u_i is the order inducing variable, a_i is the argument variable and λ is a parameter such that $\lambda \in (-\infty, \infty) - 0$.

The mean error measures the average error of two set of arguments. It can be defined as follows.

Definition 4. A mean error of dimension n is a mapping $ME: R^n \times R^n \to R$ such that:

$$ME \left(\langle a_1, f_1 \rangle, \ldots, \langle a_n, f_n \rangle \right) = \left(\frac{1}{n} \sum_{i=1}^{n} e_i \right), \tag{4}$$

where e_i is the individual error of each $(a_i - f_i)$, a_i is the set of actual values and f_i the set of forecasts.

3 The OWA Error Operator

The OWAE operator is an aggregation operator that measures the errors of two set of arguments providing a parameterized family of aggregation operators from the minimum error to the maximum error. Its main advantage is that it permits to represent all the possible errors that may occur so the decision maker gets a complete picture of the available information. It can be defined as follows.

Definition 5. An OWAE operator of dimension n is a mapping OWAE: $R^n \times R^n \to R$ that has an associated weighting vector W, with $\sum_{j=1}^n w_j = 1$ and $w_j \in [0, 1]$ such that:

$$OWAE\ (\langle a_1, f_1 \rangle, \ldots, \langle a_n, f_n \rangle) = \sum_{j=1}^n w_j E_j , \tag{5}$$

where E_j represents the jth smallest of the individual errors $(a_i - f_i)$, a_i is the set of actual values and f_i the set of forecasts.

Note that the key feature of the OWAE operator is that we allow the errors to be either positive or negative. In the case that we give absolute values and all the differences are positive, then the OWAE operator becomes the OWA distance (OWAD) operator.

Note also that if $a_i = f_i$, for all i, then, the OWAE = 0. If the set of arguments of f is empty, then we get the OWA operator and if the set of arguments of a is empty, the negative OWA (–OWA). It is possible to distinguish between descending and ascending order by using $w_j = w^*_{n-j+1}$. The aggregation process can be characterized by analyzing the degree of orness and the entropy of dispersion [6-7] as follows. For the degree of orness we get:

$$\alpha(W) = \sum_{j=1}^n w_j \left(\frac{j-1}{n-1} \right). \tag{6}$$

And for the entropy of dispersion:

$$H(W) = - \sum_{j=1}^n w_j \ln(w_j). \tag{7}$$

By using a different manifestation in the weighting vector of the OWAE operator, we can form a wide range of particular cases. For example:

- The maximum error ($w_1 = 1$ and $w_j = 0$, for all $j \neq 1$).
- The minimum error ($w_n = 1$ and $w_j = 0$, for all $j \neq n$).
- The mean error ($w_j = 1/n$, for all i).
- The step-OWAE ($w_k = 1$ and $w_j = 0$, for all $j \neq k$).
- The olympic-OWAE ($w_1 = w_n = 0$, and $w_j = 1/(n-2)$ for all others).
- The general olympic-OWAE operator ($w_j = 0$ for $j = 1, 2, \ldots, k, n, n-1, \ldots,$ $n - k + 1$; and for all others $w_{j*} = 1/(n - 2k)$, where $k < n/2$).
- The S-OWAE ($w_1 = (1/n)(1 - (\alpha + \beta) + \alpha$, $w_n = (1/n)(1 - (\alpha + \beta) + \beta$, and $w_j = (1/n)(1 - (\alpha + \beta)$ for $j = 2$ to $n - 1$ where $\alpha, \beta \in [0, 1]$ and $\alpha + \beta \leq 1$).

The OWAE operator can also be studied in percentages by using the mean percentage error. Thus, we form the OWA percentage error (OWAPE) operator as follows:

$$OWAPE\left(\langle a_1, f_1 \rangle, \ldots, \langle a_n, f_n \rangle\right) = \sum_{j=1}^{n} w_j P_j, \tag{8}$$

where P_j represents the jth smallest of the individual percentage errors $\dfrac{a_i - f_i}{a_i} \times 100$.

A similar generalization can be constructed for the mean squared error. In this case, we get the OWA squared error (OWASE). It has similar properties than the OWAE with the difference that now we use the mean squared error instead of the mean error. It can be defined as follows.

Definition 6. An OWASE operator of dimension n is a mapping OWASE: $R^n \times R^n \rightarrow R$ that has an associated weighting vector W, with $\sum_{j=1}^{n} w_j = 1$ and $w_j \in [0, 1]$ such that:

$$OWASE\left(\langle a_1, f_1 \rangle, \ldots, \langle a_n, f_n \rangle\right) = \sqrt{\sum_{j=1}^{n} w_j S_j^2}, \tag{9}$$

where S_j represents the jth smallest of the individual errors $(a_i - f_i)$.

Furthermore, we can generalize the OWAE operator by using generalized aggregation operators [8] obtaining the generalized OWAE (GOWAE) operator. It is defined as follows.

Definition 7. A GOWAE operator of dimension n is a mapping GOWAE: $R^n \times R^n \rightarrow R$ that has an associated weighting vector W, with $\sum_{j=1}^{n} w_j = 1$ and $w_j \in [0, 1]$ such that:

$$GOWAE\left(\langle a_1, f_1 \rangle, \ldots, \langle a_n, f_n \rangle\right) = \left(\sum_{j=1}^{n} w_j E_j^{\lambda}\right)^{1/\lambda}, \tag{10}$$

where E_j represents the jth smallest of the individual errors $(a_i - f_i)$ and λ is a parameter such that $\lambda \in (-\infty, \infty) - \{0\}$.

As we can see, if $\lambda = 1$, we obtain the OWAE operator and if $\lambda = 2$, the OWASE operator. Some other interesting cases are the following:

- If $\lambda = 3$, we get the OWA cubic error (OWACE).

$$OWACE\left(\langle a_1,f_1\rangle, \ldots, \langle a_n,f_n\rangle\right) = \left(\sum_{j=1}^{n} w_j E_j^3\right)^{1/3}, \tag{11}$$

- If $\lambda = -1$, we form the OWA harmonic error (OWAHE).

$$OWAHE\left(\langle a_1,f_1\rangle, \ldots, \langle a_n,f_n\rangle\right) = \frac{1}{\sum_{j=1}^{n} \dfrac{w_j}{b_j}}, \tag{12}$$

- If $\lambda \to 0$, we obtain the ordered weighted geometric error (OWGE). Note that if an individual error is zero, we assume that its individual result is zero.

$$OWGE\left(\langle a_1,f_1\rangle, \ldots, \langle a_n,f_n\rangle\right) = \prod_{j=1}^{n} b_j^{w_j}, \tag{13}$$

Note that we can also distinguish between descending and ascending orders in the GOWAE operator and it accomplishes similar properties than the OWAE operator including monotonicity, idempotency, commutativity, reflexivity and the boundary condition.

A further interesting extension that can be developed in the OWAE operator is by using induced aggregation operators [3-5,9]. Thus, we get the induced OWAE (IOWAE) operator. It can be defined as follows.

Definition 8. An IOWAE operator of dimension n is a mapping IOWAE: $R^n \times R^n \times R^n \to R$ that has an associated weighting vector W such that $w_j \in [0, 1]$ and $W = \sum_{j=1}^{n} w_j = 1$, according to the following formula:

$$IOWAE\left(\langle u_1, a_1, f_1\rangle, \ldots, \langle u_n, a_n, f_n\rangle\right) = \sum_{j=1}^{n} w_j E_j, \tag{14}$$

where E_j is the individual error $(a_i - f_i)$ of the IOWAE triplet $\langle u_i, a_i, f_i\rangle$ having the jth smallest u_i, u_i is the order-inducing variable, a_i is the set of actual values and f_i the set of forecasts.

Note that the IOWAE operator accomplishes similar properties than the OWAE operator including the distinction between ascending and descending orders, reflexivity, commutativity and so on. Moreover, it is also possible to study similar extensions including the induced OWAPE (IOWAPE) operator, the induced OWASE (IOWASE) operator and a generalization with generalized and quasi-arithmetic means. By using quasi-arithmetic means we obtain the induced quasi-arithmetic OWAE (Quasi-IOWAE) operator. It can be formulated as follows:

$$Quasi\text{-}IOWAE\ (\langle u_1, a_1, f_1 \rangle, \ldots, \langle u_n, a_n, f_n \rangle) = g^{-1}\left(\sum_{j=1}^{n} w_j\, g(E_j) \right), \qquad (15)$$

where E_j is the individual error $(a_i - f_i)$ of the Quàsi-IOWAE triplet $\langle u_i, a_i, f_i \rangle$ having the jth smallest u_i, u_i is the order-inducing variable and g is a strictly continuous monotonic function.

- If $\lambda = 1$, we get the IOWAE operator.

- If $\lambda = 2$, we get the induced OWA squared error (IOWASE).

$$IOWASE\ (\langle u_1, a_1, f_1 \rangle, \ldots, \langle u_n, a_n, f_n \rangle) = \left(\sum_{j=1}^{n} w_j E_j^2 \right)^{1/2}, \qquad (16)$$

- If $\lambda \to 0$, we get the induced ordered weighted geometric error (IOWGE).

$$IOWGE\ (\langle u_1, a_1, f_1 \rangle, \ldots, \langle u_n, a_n, f_n \rangle) = \prod_{j=1}^{n} b_j^{w_j}, \qquad (17)$$

- If $\lambda = -1$, we get the induced OWA harmonic error (IOWAHE).

- If $\lambda = 3$, we get the induced OWA cubic error (IOWACE).

In a similar way, we could study many other particular cases by giving different representations to the strictly continuous monotonic function.

4   Illustrative Example

The OWAE operator and its extensions can be applied in a wide range of problems including statistics, economics and engineering. In this paper, we focus on an application regarding the selection of the best experts for investing in the stock market. We assume that an investor is looking for an expert in order to invest money in the stock market. In order to select an expert, the investor has the available results obtained by each expert in the stock market during the last years. He considers five potential experts and the results that each one provided each year. Note that each expert provided a price forecast regarding a general product in the future market. The analysis considers the final price found in the stock market each year. Then, if the price provided by the expert was higher, he lost money because at the end of the year the price is lower and he will have to sell the product at a lower price. However, if he predicted a lower price, he will sell the product at a higher price than expected and will earn a benefit. Thus, in this example the optimal choice is the one that provides the lowest results. The available information for each of the five experts is shown in Table 1.

Table 1. Historic results provided by each expert

	2005	2006	2007	2008	2009	2010	2011
E_1	76	79	73	75	78	81	77
E_2	72	76	77	80	82	84	81
E_3	67	70	73	69	76	78	80
E_4	74	73	78	80	82	77	81
E_5	71	74	77	79	81	78	77

Next, the investor analyzes the real results found at the end of the year at the stock market. The results are shown in Table 2.

Table 2. Real results at the end of the year

	2005	2006	2007	2008	2009	2010	2011
Result	76	77	74	76	79	80	77

With this information, the investor will calculate the error found each year by the forecast given by the experts. If the result is negative, it means that the expert made a good prediction and got a profit and if the result is positive he lost money. The results are shown in Table 3.

Table 3. Individual errors

	2005	2006	2007	2008	2009	2010	2011
E_1	0	2	−1	−1	−1	1	0
E_2	−4	−1	3	4	3	4	4
E_3	−9	−7	−1	−7	−3	−2	3
E_4	−2	−4	4	4	3	−3	4
E_5	−5	−3	3	3	2	−2	0

With this information, we can aggregate it in order to select the most appropriate expert. Since this information is given under uncertainty because next year we do not know who is going to provide the optimal result, we use different aggregation operators to analyze the information. The investor considers the minimum error for these seven years, the maximum error, the average error (AE), the weighted averaging error (WAE), the OWAE and the IOWAE operator. He assumes the following weighting vector for the WAE, OWAE and IOWAE: $W = (0.1, 0.1, 0.1, 0.1, 0.2, 0.2, 0.2)$. For the IOWAE operator, he also assumes the following order inducing variables: $U = (6, 3, 4, 5, 7, 9, 8)$. The results are shown in Table 4.

Table 4. Aggregated errors

	Min	Max	AE	WAE	OWAE	IOWAE
E_1	−1	2	0	0	−0.3	0
E_2	−4	4	1.85	2.4	1.1	1.9
E_3	−9	3	−3.71	−2.8	−4.9	−4.1
E_4	−4	4	0.85	1	−0.3	1
E_5	−5	3	−0.28	−0.2	−1.2	0.1

As we can see, each aggregation operator may provide different results. However, in this example it seems clear that in most of the cases expert 3 (E_3) is the one who makes the highest benefits in this financial problem since it has the lowest aggregated errors.

5   Conclusions

We have studied the use of the OWA operator in the analysis of errors between two set of arguments. We have presented the OWAE operator as an aggregation operator that represents the different potential errors of a problem in a more complete way. We have also extended the OWAE operator by using the percentage error forming the OWAPE operator, by using the mean squared error constructing the OWASE operator and by using the generalized mean obtaining the GOWAE operator. We have seen that these operators give a more general framework in the representation of errors. We have further extended this approach by using induced aggregation operators forming the IOWAE operator.

Finally, we have developed an illustrative example where we have seen the applicability of this new approach in an aggregation process. We have focused on a problem concerning the selection of the best experts in a forecasting process. In future research, we expect to develop further applications and extend this approach by using other aggregation operators that include the probability, the weighted average and the moving average.

Acknowledgments. Support from the project 099311 from the University of Barcelona is gratefully acknowledged.

References

1. Beliakov, G., Pradera, A., Calvo, T.: Aggregation functions: A guide for practitioners. Springer, Berlin (2007)
2. McClave, J.T., Sincich, T.: Statistics, 9th edn. Prentice Hall, Upper Saddle River (2003)

3. Fodor, J., Marichal, J.L., Roubens, M.: Characterization of the ordered weighted averaging operators. IEEE Trans. Fuzzy Syst. 3, 236–240 (1995)
4. Grabisch, M., Marichal, J.L., Mesiar, R., Pap, E.: Aggregation functions: Means. Inform. Sci. 181, 1–22 (2011)
5. Merigó, J.M.: A unified model between the weighted average and the induced OWA operator. Expert Syst. Applic. 38, 11560–11572 (2011)
6. Merigó, J.M., Casanovas, M.: The fuzzy generalized OWA operator and its application in strategic decision making. Cybernetics & Syst. 41, 359–370 (2010)
7. Merigó, J.M., Casanovas, M.: The uncertain induced quasi-arithmetic OWA operator. Int. J. Intelligent Syst. 26, 1–24 (2011)
8. Merigó, J.M., Casanovas, M.: Decision making with distance measures and induced aggregation operators. Computers & Industrial Engin. 60, 66–76 (2011)
9. Merigó, J.M., Casanovas, M.: Induced aggregation operators in the Euclidean distance and its application in financial decision making. Expert Syst. Applic. 38, 7603–7608 (2011)
10. Merigó, J.M., Gil-Lafuente, A.M.: The induced generalized OWA operator. Inform. Sciences 179, 729–741 (2009)
11. Merigó, J.M., Gil-Lafuente, A.M.: New decision-making techniques and their application in the selection of financial products. Inform. Sciences 180, 2085–2094 (2010)
12. Merigó, J.M., Gil-Lafuente, A.M.: Fuzzy induced generalized aggregation operators and its application in multi-person decision making. Expert Syst. Applic. 38, 9761–9772 (2011)
13. Merigó, J.M., Wei, G.W.: Probabilistic aggregation operators and their application in uncertain multi-person decision making. Techn. Econ. Develop. Economy 17, 335–351 (2011)
14. Wei, G.W.: FIOWHM operator and its application to group decision making. Expert Syst. Applic. 38, 2984–2989 (2011)
15. Xu, Z.S., Da, Q.L.: An overview of operators for aggregating information. Int. J. Intelligent Syst. 18, 953–969 (2003)
16. Xu, Z.S., Xia, M.: Induced generalized intuitionistic fuzzy operators. Knowledge-Based Syst. 24, 197–209 (2011)
17. Yager, R.R.: On ordered weighted averaging aggregation operators in multi-criteria decision making. IEEE Trans. Syst. Man Cybern. B 18, 183–190 (1988)
18. Yager, R.R.: Families of OWA operators. Fuzzy Sets Syst. 59, 125–148 (1993)
19. Yager, R.R.: Generalized OWA aggregation operators. Fuzzy Optim. Decision Making 3, 93–107 (2004)
20. Yager, R.R., Filev, D.P.: Induced ordered weighted averaging operators. IEEE Trans. Syst. Man Cybern. B 29, 141–150 (1999)
21. Yager, R.R., Kacprzyk, J.: The ordered weighted averaging operators: Theory and applications. Kluwer Academic Publishers, Norwell (1997)
22. Yager, R.R., Kacprzyk, J., Beliakov, G.: Recent developments on the ordered weighted averaging operators: Theory and practice. Springer, Heidelberg (2011)
23. Zhou, L.G., Chen, H.Y.: Continuous generalized OWA operator and its application to decision making. Fuzzy Sets Syst. 168, 18–34 (2011)

Solvency Capital Estimation and Risk Measures

Antoni Ferri[1,*], Montserrat Guillén[1], and Lluís Bermúdez[2]

[1] Departament d'Econometria, Estadística i Economia Espanyola, Riskcenter-IREA,
University of Barcelona, Spain
tonoferri@ub.edu
[2] Departament de Matemàtica Financera i Actuarial, Riskcenter-IREA,
University of Barcelona, Spain

Abstract. This paper examines why a financial entity's solvency capital estimation might be underestimated if the total amount required is obtained directly from a risk measurement. Using Monte Carlo simulation we show that, in some instances, a common risk measure such as Value-at-Risk is not subadditive when certain dependence structures are considered. Higher risk evaluations are obtained for independence between random variables than those obtained in the case of comonotonicity. The paper stresses, therefore, the relationship between dependence structures and capital estimation.

Keywords: Solvency II, Solvency Capital Requirement, Value-at-Risk, Tail Value-at-Risk, Monte Carlo, Copulas.

1 Introduction

Recent years have seen the developement of regulatory frameworks designed to guarantee the financial stability of banking and insurance entities around the world. Europe developed Basel II and, more recently, Basel III for its banking market, and parallel to these accords drew up the Solvency II directive in 2010 for its insurance market and the Swiss Solvency Test in Switzerland. The regulatory frameworks seek to stablish what might be considered a reasonably amount of capital (referred to as *Solvency Capital* in Basel II and III and as *Solvency Capital Requirement* in Solvency II and the Swiss Solvency Test) to put aside to ensure financial stability in the case of adverse fluctuations on losses. This quantity must reflect the entity's specific risk profile and, under the aforementioned frameworks, it can be arrived at by applying either the *Standard Model* proposed by the regulator or an *Internal Model* proposed by the entity itself. In the later case, a number of requirements must first be satisfied before the model can be used for the purposes of capital estimation. In the European frameworks is regulated by the calibration of a risk measurement given a confidence level over a given time horizon. In this paper we focus our attention on the European insurance market, and more specifically in non-life underwriting risk, in relation to the Solvency II and Swiss Solvency Test regulations. Under both frameworks, the

---

* Corresponding Author.

K.J. Engemann, A.M. Gil Lafuente, and J.M. Merigó (Eds.): MS 2012, LNBIP 115, pp. 34–43, 2012.

total capital estimation is obtained by aggregating individual capitals requirements arising from a company's various sources of risk, based on the correlation between them as defined by the *Standard Model*. As a means of aggregating risks, we proposed a simulation of a multivariate random variable where each marginal distribution function represents the claims of a given line of business. We simulate a sample of this multivariate random variable taking into account the correlation between lines of business and, then, we aggregate the results of each simulated claim by line of business in order to obtain the distribution of the total claims. Finally, we estimate the capital requirements by applying a risk measure over the total claims distribution.

By representing an example of a multivariate random variable simulation we show that under certain assumptions there are risk measures that fail to satisfy the subadditive property. This is tipically the case where there is a very heavy tailed or skewed distribution on the margin and/or in which a special dependence structure is assumed for its joint distribution. Such circumstances can lead to an underestimation of the solvency capital if we incorrectly assimilate a risk measurement to the capital requirements, i.e., the appropiate distribution is not fitted to the marginals or the joint behavior of the marginal distributions is unknown.

The paper also emphasizes typical misunderstanding in the meaning of risk measures and the relationship between these risk measures and the underlying dependence structures of the variables, which represents the sources of risk.

2 Misunderstanding in the Concept of Value at Risk as a Risk Measure

Risk measures are tipically employed to determine the amount of capital that should be aside to cover unexpected losses. In [1] was proposed a number of desirable properties that a risk measure should satisfy in order to be considered a *coherent* risk measure. One such property, that of *subadditivity*, captures the idea of diversification across random variables since in the words of the authors: "...*the merger [of two risks] does not create extra risk*". Suppose we have n random variables, $X_i, i = 1, \ldots, n$ and its sum, $S = \sum_{i=1}^{n} X_i$, then we say that risk measure ρ has the *subadditivity* property if and only if for all X_i

$$\rho(S) \leq \sum_{i=1}^{n} \rho(X_i). \tag{1}$$

Although several risk measures are available, here we focus on two that are on loss distribution, namely *Value-at-Risk* and *Tail Value-at-Risk*. These risk measures seek to describe how risky a portfolio is. Of the two, the most frequently adopted is *the Value-at-Risk* measure given that it is employed under Basel III and Solvency II as a tool for calibrating the solvency capital requirements. *Value-at-Risk* and its properties has been widely discussed (see [2]). The *Value-at-Risk* measure is simply a quantile of a distribution function. However, it is not a *coherent* risk measure since it does not satisfy condition (1) for all X_i, although it does in the case in which X_i are normally distributed and, more generally, in the case in which elliptical distributions are considered for X_i as is shown in [3] and [4]. Formally, given a confidence level $\alpha \in [0,1]$, *Value-at-Risk* is defined as the infimum value of the distribution of a random variable X such that the probability that a value x exceeds a certain threshold x is no greater than $1-\alpha$. Usually this probability is taken to be 0.05 or less.

$$VaR^{\alpha}(X) = infimum\{x/P(X > x) = 1-\alpha\} = F_X^{\leftarrow}(\alpha), \qquad (2)$$

where $F_X(x)$ denotes de distribution function of X and $F_X^{\leftarrow}(\alpha)$ the inverse distribution function of $F_X(x)$.

It is common to choose $1-\alpha$ based on a very large period. For example, under the Solvency II directive, this choice is made on the basis of an occurrence of an event every two hundred years, i.e., a probability of 0.5%. For this reason, this risk measure is known as a frequency measure. It describes the distribution up to the α-th percentile, but it provides no information as to how the distribution behaves at higher percentiles. Unlike the *Value-at-Risk* measure, the *Tail Value-at-Risk* describes the behavior of the tail of the distribution. This risk measure is defined, therefore, as the expected value of all percentiles higher than the *Value-at-Risk*. Formally,

$$TVaR^{\alpha}(X) = E[X \mid x \geq VaR^{\alpha}(X)], \qquad (3)$$

where $E[\cdot \mid \cdot]$ denotes the conditional expectation operator.

Thus, while two different distributions might have the same *Value-at-Risk* for a given a confidence level, their *Tail Value-at-Risk* may differ due to a different heaviness of the tail of the distribution. The fact of having to consider not just the choice of $1-\alpha$ but also the values of the distribution higher than the α-th percentile mean the *Tail Value-at-Risk* is known as a severity risk measure. Likewise, the *Tail Value-at-Risk* measure is a subadditive and *coherent* since it satisfies property (1), as is shown in [5].

Despite the mathematical principles underpinning the *Value-at-Risk* and *Tail Value-at-Risk* measures, a misunderstanding arises when seeking to apply them (specially, in the case of the former) to capital requirements. It is common to interpret *Value-at-Risk* as the value that will not be exceeded with a probability α. If, as is usual, the random variable is considerd as the loss of a portfolio, this definition is equivalent to the loss that will not be exceeded with a probability of α, or the maximum loss given α as pointed out by [2]. Yet, this interpretation is not stictly correct since maximum loss is not generally given by *Value-at-Risk* for a given confidence level as we shall see bellow.

Inequality (1) shows an upper bound for $\rho(S)$ which is $\sum_{i=1}^{n} \rho(X_i)$. A typical misunderstanding arises from this bound given that only in cases of diversifiable risks can such a bound ocurr. In instances of non-diversifiable risks this bound fails. As [6] showed, if we consider *Value-at-Risk* as a risk measure and a sequence of comonotonic random variables $X_i, i = 1, \ldots, n$ then

$VaR(S) = VaR(\sum_{i=1}^{n} X_i) = \sum_{i=1}^{n} VaR(X_i)$ which is known as the comonotonic bound.

But this is not the worst possible case since comonotonicity does not necessarily result in the worst loss a company might suffer. Dependence structures between random variables X_i might be found so that the *Value-at-Risk* of the sum of them exceeds the comonotonic bound. Translating this to the definition of *Value-at-Risk* we can conclude that it is possible to have greater losses than those arising from the comonotonic case given α. Then $\sum_{i=1}^{n} VaR(X_i)$ is not the maximum loss we could have given α.

Generally, *Value-at-Risk* fails to be a sub-additive risk measure in those cases where we have very heavy-tailed random variables, which is the case of those variables representing catastrophic or operational risks. It also fails in the case of skewed random variables and in some instances in which special dependence structures are imposed on the joint behavior of margins. In [5] it is shown that when Pareto random variables are considered, *Value-at-Risk* fails to be subadditive. Moreover, when the infinite mean case is considered, i.e., *tail-parameter* of Pareto distribution is equal to one for all random variables considered, *Value-at-Risk* fails to be sub-additive at all confidence levels, otherwise, there's a confidence level up to which *Value-at-Risk* is sub-additive and beyond which it is not.

In short, the estimation of capital requirements using a risk measure should be approached with caution since the resulting valuation might underestimate the capital needs depending on the joint distribution of the random variables representing the implicit risks within a company as well as on its individual statistical distributions. Therefore, risk measurements derived from non *coherent* risk measures could lead the company to financial and solvency instability.

3   A Non *sub-additivity* Example for Value a Risk

In this section we present an example in which we demonstrate that while the *Value-at-Risk* measure fails to be subadditive, the *Tail Value-at-Risk* measure satisfies the property of *subadditivity*. We use a historical non-life insurance market data set for Spain corresponding to the period 2000 to 2010. The data were obtained from public information published at the website of the Dirección General de Seguros (DGS)[1]. The data represents the yearly annual aggregate claims and three lines of business are considered (see Table 1 for their descriptive statistics[2]). A complete description of the risks included in each line of business can be found in [7].

---

[1] www.dgsfp.meh.es
[2] Minimum (Min.), Quartile (Qu.), Standard deviation (Sd), Maximum (Max.).

We assume that each line of business behaves statistically as a generalized Pareto random variable. The parameters resulting from fitting the data to the Pareto distribution are shown in Table 2.

Table 1. Descriptive statistics[*] of the yearly annual aggregate[**] claims by line of business

	Min.	1st Qu.	Median	Mean	Sd	3rd Qu.	Max.
Motor, third party Liability	4.286	4.904	5.182	5.085	0.332	5.318	5.439
Fire and other property damage	2.337	3.006	3.740	3.539	0.723	4.067	4.544
Third party liability	0.552	0.684	0.819	0.842	0.195	1.011	1.084

Source: Own source from DGS / [*]thousand millions Euros / [**] deflated

After fitting the data of each line of business to its corresponding distribution, we performed a multivariate Monte Carlo simulation and we compute the *Value-at-Risk* and the *Tail Value-at-Risk* for several confidence levels. For the joint behavior of risks we imposed several copulas, which in fact entails considering several different dependence structures. An introduction to multivariate models and dependence concepts and their properties can be found in [8] and [9].

Table 2. Generalized Pareto shape and scale[*] parameters

	Shape	Scale
Motor, third party liability	0.93	0.30
Fire and other property damage	0.95	0.23
Third party liability	0.75	0.19

[*]thousand millions Euros

First, we established two extreme cases of dependence: comonotonicity and independence. We used the upper Fréchet copula, which reflects the case of comonotonicity between margins, and then we used the independence copula, which reflects the case of independence between margins. Finally, we employed two further copulas, the Clayton copula, a very right- skewed distribution, and the Frank copula, a symmetric but very heavy tailed distribution, with two dependence parameters (θ) each one.

Table 3 shows the results of *Value-at-Risk* and *Tail Value-at-Risk* for several confidence levels. Under the comonotonic assumption, which leads to the comonotonic bound, *Value-at-Risk* underestimates the risk (i.e., fails to be subadditive) compared to the independence, the Clayton and Frank copula cases at all condidence level up to some point between 0.90 and 0.99, and beyond which it *Value-at-Risk* satisty the *subadditivity* property. The values in bold indicate where the *subadditivity* property fails to be satisfied compared with the case of comonotonicity.

Since, *Tail Value-at-Risk* is a subadditive measure of risk, all values under independence case, as well as those for the Clayton and Frank copula assumptions are, as expected, lower than those in the comonotonic case at all confidence level. In the *Tail Value-at-Risk* case, the more dependence assumption implies higher values of risk measurements at all confidence levels.

Table 3. VaR and TVaR[*] from a simulation[**] of a tridimensional multivariate distribution for several dependence structures and Pareto margins

c.l.	independence copula	Clayton copula		Frank copula		comonotonic copula
Value at Risk						
		$\theta = 1$	$\theta = 2$	$\theta = 1$	$\theta = 2$	
0.8	3.41	3.53	3.57	3.45	3.48	2.67
0.9	6.36	6.78	6.99	6.59	6.85	5.78
0.99	47.69	49.6	50.7	50.02	50.17	53.33
0.999	398.05	396.45	421.82	371.26	401.18	435.69

c.l.	independence copula	Clayton copula		Frank copula		comonotonic copula
Tail Value at Risk						
		$\theta = 1$	$\theta = 2$	$\theta = 1$	$\theta = 2$	
0.8	23.14	25.09	26.76	24.78	31.5	32.66
0.9	41.7	45.34	48.57	44.84	58.35	61.49
0.99	289.28	318.72	347.22	315.22	446.51	488.76
0.999	1868.51	2151.37	2419.51	2163.13	3409.88	3777.41

Source: Own source / [*]thousand millions Euros / c.l.: confidence level / [**]100.000 simulations

4 Issues from Risk Management

New regulatory frameworks in Europe establish the capital requirements that financial companies need to maintain in order to ensure acceptable levels of solvency. These requirements are calibrated by applying a risk measure based on the distribution of a random variable that represents the total losses a company would suffer.

Typically, in the field of risk management, managers are aware of the individual risks for which they have responsibility, but are unaware of the relationship between the sources of risk that they manage and those arising from other departments or other areas within the same company. This means that the joint behavior of the random variables representing the different sources of risk remains unknown. As such, the process of aggregating risk measurements to obtain an global capital requirement is complex and difficult to address.

As consequence of this, it's usual for risk managers to obtain the risks measurements of each variable that represent a single source of risk considered and after that add them up to obtain a single amount and assimilate it to the regulatory capital.

In so doing, if companies choose a non *coherent* risk measure (as defined by [1]) such as *Value-at-Risk*, (thereby adhering to Solvency II or Basel III criteria), they may overlook a number of crucial points. First, they may overlook the tail behavior of each random variable considerd individually, which means, they fail to incorporate the possibility of severe losses that could greatly undermine the company's solvency. Second, by summing up risk measurements, such as those resulting from the *Value-at-Risk*, a company implicitly assumes, perhaps inadvertenly, that its random variables are comonotonic, which is in conflict with the idea of their ignoring the joint behavior of variables. Even in those instance in which risk managers believe that the joint behavior is comonotonic, it is possible to suffer worse losses than those that are consequence of summing the *Value at Risk* measurements derived from comonotonic random variables, as shown in the example in the previous section (see Table 3). In this example, even when the random variables are independent, the *Value at Risk* of the sum of random variables is greater than it is in the case of comonotonicity at certain confidence levels.

This finding highlights not only the importance of choosing of a *coherent* risk measure, but also the need, first, to select the right fit and distribution function so as to reflect correctly the joint behavior of sources of risk and their dependence structures; and, second, to select the right fit and distribution functions so as to reflect the margin behavior of each random variable representing a single sources of risk.

In example presented in Table 3, we used Pareto distributions for margins. When considering this distribution, it is possible to find cases in which we might have infinite expected values or variance, which are extreme cases that cause non-*subadditivity* for the *Value-at-Risk* measure, but they are not the only cases in which this might occur. There are many risks, generally catastrophic risks, for which the Pareto distribution fits well, and for which the non-*subadditivity* property could fail when using non-*coherent* risk measures. Examples of the Pareto distribution being fitted to operational and catastrophic risks can be found in [10] and [11], respectively.

5   Conclusions

A number of lessons can be drawn by those involved in risk management from the preceding analysis. First, knowledge of the joint statistic behavior of risks is essential when considering a company's overall level of risk. This does not simply mean the need to estimate the correlation coefficients but also to consider underlying dependence structures. Second, although using a *coherent* risk measure such as the *Tail Value-at-Risk* leads to higher capital estimations than in the case when using the *Value-at-Risk* measure, it serves to ensure that capital requirements are not underestimated due to the property of subadditivity should the hypothesis regarding the joint behavior of sources of risk prove to be incorrect. However, errors in the model estimation could well lead to the real capital needs being underestimated, even though a *coherent* risk measure such as *Tail Value-at-Risk* has been used. Third, comonotonicity does not represent the worst possible scenario when estimating capital requirements using loss and risk measurements. This might appear counterintuitive, but managers of risks need to bear in mind that risks may well be superadditive as opposed to subadditive, and as such, no diversification effect occurs when several random variables are merged. Fourth, while it may appear obvious, differences in capital estimation derived from the application of different risk measures do not mean the portfolio has varying degrees of risk; rather what we see is simply different ways of measuring what might happen beyond a given threshold, i.e. the losses that would exceed the threshold in the case of the *Value-at-Risk* measure, and the severity of these losses beyond the threshold in the case of *Tail Value-at-Risk* measure.

Acknowledgments. This research is sponsored by the Spanish Ministry of Science ECO2010-21787.

References

1. Artzner, P., Delbaen, F., Eber, J.M., Heath, D.: Coherent Measures of Risk. Mathematical Finance 9(3), 203–228 (1999)
2. Jorion, P.: Value at Risk. The new Benchmark for measuring financial risks. McGraw Hill, NY (2007)
3. Fang, K.T., Kotz, S., Ng, K.W.: Symmetric Multivariate and Related distributions. Chapman and Hall, London (1990)
4. Embrechts, P., McNeil, A., Straumann, D.: Value at Risk and Beyond. Correlation and Dependency in Risk Management: properties and pitfalls. Cambridge University press, Cambridge (2002)
5. Embrechts, P., McNeil, A., Frey, R.: Quantitative Risk Management. Concepts, Techniques and Tools. Princeton University Press, Princeton (2005)
6. Embrechts, P., Höing, A., Juri, A.: Using copulae to bound the value at risk for functions of dependent risks. Finance Stochastic 7(2), 145–167 (2003)
7. CEIOPS, 5th Quantitative Impact Study - Technical Specifications, https://eiopa.europa.eu/consultations/qis/index.html
8. Nelsen, R.B.: An introduction to Copulas. Springer, USA (2006)
9. Joe, H.: Multivariate Models and Dependence Concepts. Chapman & Hall, UK (1997)
10. Guillén, M., Prieto, F., Sarabia, J.M.: Modeling losses and locating the tail with de Pareto Positive Stable distribution. Insurance: Mathematics and Economics 49(3), 454–461 (2011)
11. Sarabia, J.M., Prieto, F., Gómez-Déniz, E.: Análisis de riesgos con la distribución Pareto estable positiva. In: Cuadernos de la Fundación, vol. 136, pp. 191–205. Fundación Mapfre, Madrid (2009)

Appendix

A multivariate simulation of the random variable was conducted using the R-Project software, version 2.13.1 and the *copula* package implemented therein. Bellow we describe the simulation performance.

A random variable X has a Generalized Pareto distribution (GPD) if its distribution function is

$$
G_{X,\xi,\beta} = \begin{cases} 1 - \left(\dfrac{\xi \cdot x}{\beta}\right)^{\frac{-1}{\xi}} & \text{if} \quad \xi \neq 0 \\[2mm] 1 - \exp\left(\dfrac{-x}{\beta}\right) & \text{if} \quad \xi = 0 \end{cases}
$$

where ξ the shape parameter and $\beta > 0$ the scale parameter. The expexted value $E[X]$ is

$$E[X] = \frac{\beta}{1-\xi}$$

and the standard deviation is the scale parameter β.

The shape parameter can be estimated using maximum likelihood estimation or the method of moments. Using the moments estimation procedure, the shape parameter results in $E[X] = \frac{\beta}{1-\xi} \Rightarrow \xi = \frac{-\beta}{E[X]} + 1 = -CoVa[X] + 1$; being $CoVa[X]$ the coefficient of variation of X.

Having estimated the sample expected value and the standard deviation for each line of business considered we obtain the coefficient of variation and construct a 2x3 matrix which containing the coefficients of variation and standard deviation in each of the rows. The columns represent each one of the three line of business.

```
data<-read.table('CoV.csv', header=TRUE, sep=";")
data<-as.matrix(data)
```

The estimation of the shape and scale parameters becomes

```
shape1<- -data[1,1]+1 # the shape parameter for the first margin.
shape2<- -data[1,2]+1 # the shape parameter for the second margin.
shape3<- -data[1,3]+1 # the shape parameter for the third margin.
scale1<-data[2,1] # the scale parameter for the first margin.
scale2<-data[2,2] # the scale parameter for the second margin.
scale3<-data[2,3] # the scale parameter for the third margin.
```

The comonotonic bound are derived from the sum of the α-th quantiles of each marginal distribution. We compute the quantiles for each margin after simulating one thousand observations of the GPD for each margin given its corresponding parameters.

```
n=100000 # number of simulations.
l<-c(0.80,0.90,0.99,0.999) # vector of given confidence level.
```

The simulation for the first, second and third margin are

```
pareto1<-rGPD(n,shape1, beta=scale1) # for the first margin.
pareto2<-rGPD(n,shape2, beta=scale2) # for the second margin.
pareto3<-rGPD(n,shape3, beta=scale3) # for the third margin.
VaR1<-quantile(pareto1,l) # Value-at-Risk for the first margin.
VaR2<-quantile(pareto2,l) # Value-at-Risk for the second margin.
VaR3<-quantile(pareto3,l) # Value-at-Risk for the third margin.
VaR_com<-(VaR1+VaR2+VaR3) # comonotonic bound vector.
```

When considering independence between the random variables, we obtain *the Value-at-Risk* and the *Tail Value-at-Risk* by simulating the three-dimensional Gaussian copula with Pareto margins given shape and scale parameters and dependence parameters (linear correlations) for the copula equal to zero.

```
param1<-list(shape1,scale1)
param2<-list(shape2,scale2)
param3<-list(shape3,scale3)
corr<-c(0,0,0) # vector of linear correlations for the Gaussian copula.
copulagaussiana<-mvdc(normalCopula(corr,dim=3,
dispstr="un"),c(rep("GPD",3)),list(param1,param2,
param3))
gaussian.sample<-rmvdc(copulagaussiana,1000000)
# simulation of the copula.
gauss.sample.aggrega<-rowSums(gaussian.sample)# distribution of
total losses.
VaR_ind<-quantile(gauss.sample.aggrega,1)
TVaR_ind<-rep(0,4)
for(i in 1:4){
ES<-mean(gauss.sample.aggrega[gauss.sample.
aggrega>VaR_ind[i]])
TVaR_ind[i]<-ES}
VaR_ind # Value-at-Risk vector under independence assumption.
TVaR_ind # Tail Value-at-Risk vector under independence assumption.
```

The *Tail Value-at-Risk* under the comonotonicity assumption between random variables is obtained by simply setting a new correlation vector for the Gaussian copula, i.e, corr<-c (0.9999,0.9999,0.9999) as the comonotonic copula can be obtained when the dependence parameters of the Gaussian copula tend to one.

The *Value-at-Risk* and *Tail Value-at-Risk* for the Clayton and Frank copulas are obtained in a similar way but changing the kind of copula used and the corresponding dependence parameter. We show the Clayton copula case for a dependence parameter $\theta = 1$.

```
theta=1 # dependence parameter for the Clayton copula.
Copulaclayton<-mvdc(claytonCopula(theta,dim=3),
c(rep("GPD",3)),list(param1,param2,param3))
clayton.sample<-rmvdc(copulaclayton,1000000)
clayton.sample.aggrega<-rowSums(clayton.sample)
VaR_cl<-quantile(clayton.sample.aggrega,1)
TVaR_cl<-rep(0,4)
for(i in 1:4){
ES<-mean(clayton.sample.aggrega[clayton.sample.
aggrega>VaR_cl[i]])
TVaR_cl[i]<-ES}
VaR_cl # Value-at-Risk vector under Clayton copula assumption.
TVaR_cl #Tail Value-at-Risk vector under Clayton copula assumption.
```

Decision Making of Customer Loyalty Programs to Maintain Consumer Constancy

Aras Keropyan and Anna M. Gil-Lafuente

Department of Economics and Business Organization,
University of Barcelona, Barcelona, Spain

Abstract. Companies realized the importance of well-managing their relationships with their customers. Customer relationship management (CRM) allows companies to manage their marketing strategies and deliver specific services to clients with different values. Customers with higher values may deserve better service while less service should be delivered to customers with lower value for the company. In this article our objective to provide companies a model that facilitates to decide what kind of customer loyalty programs they should apply to what kind of clients. In order to do that we present a fuzzy based Hungarian method that allow assigning different loyalty programs to customers with different values for the companies.

Keywords: Marketing Strategy, Customer Loyalty, Customer Relationship Management, CRM, Marketing.

1 Introduction

Decision making is one of the most important activities for managers. Over the years, researchers have discussed the influence of the ability of managers on organizational outcomes. Some authors have argued that managers have a remarkable impact on organizational performance. S. Robins (1999) describes in his book the manager's impact as the essence of the manager's job and a critical element of organizational life. Meanwhile Rowe (1994) suggests that decision making is synonymous with managing. Different kinds of computer-based information systems have been developed to support decision making and decision support systems, group support systems and executive information systems. In order to be more competitive organizations in though market conditions, it is widely agreed that managers must make good decisions which affect their organizations significantly.

We believe that by using fuzzy logic methodology, we can propose good examples of decision making in marketing and present a useful application. In this study we are going to discuss different customer loyalty programs and try to propose the most adequate models to maintain the consumer's constancy. We are going to leave topics for further research so that our model can be approved afterwards by applying data to different techniques of fuzzy logic.

K.J. Engemann, A.M. Gil Lafuente, and J.M. Merigó (Eds.): MS 2012, LNBIP 115, pp. 44–53, 2012.
© Springer-Verlag Berlin Heidelberg 2012

2   Customer Relationship Management

Companies realized the importance of well-managing their relationships with their customers. Customer relationship management (CRM) allows companies to manage their marketing strategies and deliver specific services to clients with different values. Customers with higher values may deserve better service while less service should be delivered to customers with lower value for the company.

Customer Relationship Management (CRM) is a term that has been first proposed by Varadarajan (1986). According to Varadarajan, "CRM is a joint-transaction model, which is an alignment of product and donation through the partnership between company and non-profit organization, and ultimately gains interests for both parties." Varadarajan and Menon (1988) supposed that CRM helps companies to achieve many of their substantial benefits, such as helping them to improve performance, strengthen corporate and brand image and expand the target market.

According to Peppers and Rogers (1999), CRM is the management of relationships between companies and individual customers with the aid of (customer) databases and interactive and mass customization technologies.

Swift (2001) defines CRM as "an enterprise approach to understanding and influencing customer behavior through meaningful communications in order to improve customer acquisition, customer retention, customer loyalty, and customer profitability."

According to Kincaid (2003) CRM is "the strategic use of information, processes, technology, and people to manage the customer's relationship with the company across the whole customer life cycle." Ko et. al (2004) defines CRM as an integration of customer management strategy of firms that allows them to manage customers efficiently by supplying customized goods and services and maximizing the lifetime value of customers.

As Kincaid (2003) and other authors emphasize in their studies, CRM approach firstly started to take place in company's agenda when they had discovered that all the customers didn't have the same value and profitability. Then, companies realized that it would be more effective to deliver distinct services and develop specific strategies for customers who have different importance for the company instead of treating them all equally. The basic goal of customer relationship management is to achieve a competitive advantage in customer management and as a result increase profit levels of the customers.

Today companies use CRM to obtain some useful information of current and prospective customers. Through this relevant information they can improve the service that they offer to their customers, give more attention to better customers, may abandon unprofitable customers and attract some new good potential clients.

CRM is known as the new basis of marketing strategies which companies should well-manage and maintain its sustainability. On the other hand, as Peppers and Rogers (1999) indicate, it is widely known as a software-based approach which allows companies to supply information about customers and manage this information to develop and improve their marketing processes. Among the most important goals of CRM are to offer better services to customers, increase profitability and use customer contact information to manage marketing processes more efficiently.

By using customer information contained in databases, companies can invest in the customers that are potentially valuable for the company, but also minimize their investments in non-valuable customers. Figures on the turnover of each customer or customer profitability are often used as segmentation variables to distinguish between valuable and non-valuable customers (Donkers, Verhoef and Jong, 2007).

Ravald and Grönroos (1996), point in their previous studies that marketing is facing a new concept, which is named relationship marketing and they underline the evolution from the activity of attracting customers to activities which concern of having customers and taking care of them in marketing. According to Ravald and Grönroos (1996), the main basis of relationship marketing are relations, maintenance of relations between the company and the customers, suppliers, market intermediaries, the public, etc. These authors define the value concept as an important element of relationship marketing which enables companies to provide superior services to its customers.

Customer satisfaction is one of the key objectives of marketing. To increase the satisfaction level of customers and maintain it, customer behavior should be well-known by the firm. Customer behavior is related to customer loyalty issue. With that reason, Jackson (1985) divided industrial buyers into two major categories: lost-for-good and always-a-share. The lost-for-good category assumes that a customer is either totally committed to the company or totally lost and committed to some other company. The second category, always-a-share includes the buyers which can easily switch their vendor to another one. We can say that customers who are in the lost-for-good category are satisfied, loyal and will remain in the company. But the customers who are in always-a-share category would easily go to another company and would not retain its products for long terms. So customer behavior should be well-known in order to measure their retention correctly.

3   Customer Loyalty

Loyalty marketing is an approach to marketing, based on strategic management, in which a company focuses on growing and retaining existing customers through incentives. Branding, product marketing and loyalty marketing all form part of the customer proposition – the subjective assessment by the customer of whether to purchase a brand or not based on the integrated combination of the value they receive from each of these marketing disciplines. Evans, Stuart (2007). It is very important to keep customers satisfied for companies since in the majority of sectors there is a very high competition and customers are inclined to move easily from one to another company when there is a better service or price in the competitor. The effort and money should be inverted to existing clients in many cases as it is done to capture new clients in order to have loyal customers as they are important income generators for the company.

4   Fuzzy Logic

Zadeh (1965) has published first fuzzy set theory. Zimmermann (1991) explained fuzzy set theory as a strict mathematical framework in which vague conceptual phenomena was precisely and rigorously studied. The theory can also be thought as a modeling language which suited well for situations that were containing fuzzy relations, criteria and phenomena. Afterwards, Rowe (1994) have proved the portfolio matrix and 3Cs model which were enabling companies to analyze their strategic business units and projects, and providing strategic directions in an efficient way. This hasn't worked very well. Certain values in the decisions making aren't always correct. Because there are always vague processes and it is difficult to estimate decision making processes with an exact numerical value. Pap and Bosnjak (2000) defined the main problem of using the classical portfolio matrix as the precise determination of the numerical value for the criteria. As a result, it would be useful to use the linguistic assessments which have been introduced by Zadeh (1965) and Bellman (1970) instead of numerical indicators.

Fuzzy Number and Linguistic Variable

Dubois and Prade (1970) have defined the fuzzy numbers. They have described its meaning and features. A fuzzy number $\tilde{N}$ is a fuzzy set which membership function is $\mu_{\tilde{N}}(y): R \to [0,1]$. A triangular fuzzy number $\tilde{N} = (a,b,c)$ can conform to different set of a, b, c characteristics. If we explain those characteristics in management terms, a value is the optimistic estimate, when everything goes great. The value b is the most likely estimate, which implies to the situation not very good either very bad. The c value is a pessimistic estimate, when everything goes badly.

Zadeh and Bellmann (1970) defines a linguistic variable as a variable whose values are not numbers but words or phrases in a natural or synthetic language. In a problem when we are working on linguistic variables we can present their means. At that moment, we can rate and weight the various conditions by using the fuzzy numbers and linguistic variables. Linguistic variables represent the relative importance and appropriateness of each ranking method that simultaneously considers the metric distance and fuzzy mean value is proposed. The distance from the ideal solution and the fuzzy mean value are usual criteria for ranking fuzzy numbers.

Moon, Lee and Lim defines fuzzy numbers as if Y is a collection of objects represented of generated of y's, then a fuzzy set $\tilde{N}$ in Y is a set of ordered pairs:

$$\tilde{N} = \{(x, \mu_{\tilde{N}}(y)) \mid y \in Y\}$$

$\mu_{\tilde{N}}(y)$ is the membership function or grade of membership of y in $\tilde{N}$ that maps Y to the membership space N (when N contains only the two points 0 and 1, $\tilde{N}$ is no fuzzy and $\mu_{\tilde{N}}(y)$ is identical to the characteristic function of a no fuzzy set). The range of the membership function is a subset of the nonnegative real numbers whose supreme is finite. Elements with a membership of zero degrees are normally not listed. The authors characterize a linguistic variable by a quintuple $(y, F(y), A, B, \tilde{N})$ in which y is the name of the variable; $F(y)$ denotes the term of y set; for example the set of names of linguistic values of y, with each value being a fuzzy variable denoted generically by Y and ranging over a universe of discourse A that is associated with the base variable a; B is a syntactic rule for generation of the name, Y, of values of y; and $\tilde{N}$ is a semantic rule for associating with each Y its meaning $\tilde{N}(y)$ which is a fuzzy subset of A.

When it comes to take objective decisions in management we know the hardness to evaluate them by binary definite numbers 0 and 1s. Therefore in this study we use transform linguistic expressions which can be transformed to numerical values easier. We propose following semantics for the set of three terms to point different loyalty programs on different segments of customers:

A) HIGH = (High-High, High-Medium, High-Low)
B) MEDIUM= (Medium-High, Medium-Medium, Medium-Low)
C) LOW = (Low-High, Low-Medium, Low-Low)

Each of these three semantics also includes three other semantics which enables us to evaluate the decisions in wider intervals. This approach facilitates us to value easier the relationships between loyalty programs and customers when it is hard to link them in an objective way.

In this study we represent every linguistic semantic by following numeric values:

A) HIGH = (0.9, 0.8, 0.7)
B) MEDIUM= (0.6, 0.5, 0.4)
C) LOW = (0.3, 0.2, 0.1)

We use triangular fuzzy numbers and therefore we present the following semantics in Figure 1:

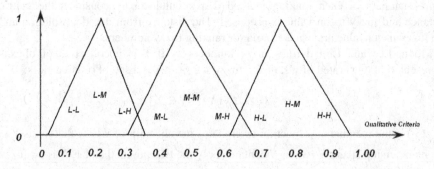

Fig. 1. The membership functions for fuzzy numbers according to the qualitative criteria

5   The Hungarian Method for the Assignment Problem

In this study the objective is to assign the most adequate loyalty program to each segment of clients. The Hungarian Method is going to be used for assigning the most appropriate campaign action that is going to be carried to a specific group of clients.

The Theorem

The Hungarian Method: The Hungarian method is an algorithm which finds an optimal assignment for a given loyalty programs matrix. In order to find the proper assignment it is essential for us to know the Hungarian method. This method is dependent upon two vital theorems, stated as below (Kuhn, 2005).

Theorem 1: If a constant is added (or subtracted) to every element of any row (or column) of the loyalty programs matrix [cij] in an assignment problem then an assignment which minimizes the total programs for the new matrix.

Theorem 2: If all $c_{ij} \geq 0$ and there exists a solution $x_{ij} = X_{ij}$ such that

$$\sum_i c_{ij} \text{ and } x_{ij} = 0.$$

then this solution is an optimal solution, i.e., minimizes z. And the matrix of that solution:

$$C = \begin{pmatrix} c_{11} \, c_{12} \, c_{13} \, \cdots \, c_{1n} \\ c_{21} \, c_{22} \, c_{23} \, \cdots \, c_{2n} \\ \cdot\cdot \quad \cdot\cdot \quad \cdot\cdot \, \cdots \, \cdots \\ c_{31} \, c_{32} \, c_{33} \, \cdots \, c_{3n} \end{pmatrix}$$

The Hungarian Algorithm

The Hungarian algorithm is an algorithm for solving a matching problem or more generally an assignment linear programming problem. The Hungarian Algorithm is actually a special case of the Primal-Dual Algorithm. It takes a bipartite graph and produces a maximal matching.

According to Hungarian Algorithm we can handle our customer loyalty decision making problem easier. In order to do that all the variables should be reduced to the matrices and the following steps should be followed as given by Kuhn.

For customer loyalty we assume different customer loyalty programs. Let's consider that there are n "customers" and n "customer loyalty programs".

A) If necessary, the problem should be converted from a maximum assignment into a minimum assignment. This operation is done by assigning C = maximum value in the assignment matrix. From here, each cij should be replaced with $C - c_{ij}$.

B) After the replacement from each row the row min, and from each column the row column min has to be subtracted.

After steps A and B we suppose that we are using k lines.

If k < n, m has to be let as minimum uncovered number. From every uncovered number m has to be subtracted and then it has to be added to every number covered with two lines.

If k = n, then;

C) Starting with the top row, work your way downwards as you make assignments.

An assignment can be (uniquely) made when there is exactly one zero in a row. Once the assignment is made, that row has to deleted and columned from the matrix.

The operation between row assignments and column assignments have to be repeated until a unique assignment remains. If still there is no unique assignment either with rows or columns, one arbitrary cell with a zero in it should be selected.

The matrix interpretation of the technique is presented below, so that the case can be explained easier.

Given n customer and customer programs, and an n×n matrix containing the loyalty action of assigning each program to a customer, the objective is to find the most adequate loyalty program for each segment of customer minimizing the assignments. First the problem is written in the form of a matrix as given below:

To start, the row operations on the matrix should be performed. To do this, the lowest of all p_i (i = 1, 2, .. , 4) is taken and is subtracted from each element in that row. This will lead to at least one zero in that row (We get multiple zeros when there are two equal elements which also happen to be the lowest in that row). This procedure is repeated for all rows. We now have a matrix with at least one zero per row. Now we try to assign programs to customer segments. This is illustrated below:

$$Q = \begin{bmatrix} 0 & p_2 & 0 & p_4 \\ q_1 & q_2 & 0 & q_4 \\ r_1 & 0 & r_3 & r_4 \\ 0 & s_2 & s_3 & s_4 \end{bmatrix}$$

The zeros that are indicated as 0 are the assigned programs.

Sometimes it may turn out that the matrix at this stage cannot be used for assigning, as is the case in for the matrix below:

$$Q = \begin{bmatrix} 0 & p_2 & p_3 & p_4 \\ q_1 & q_2 & 0 & q_4 \\ r_1 & 0 & r_3 & r_4 \\ 0 & s_2 & s_3 & s_4 \end{bmatrix}$$

In cases as the matrix above it is not possible to make any assignments. Both a and c can't be assigned to the same program. To overcome this, we repeat the above

procedure for all columns (i.e. the minimum element in each column is subtracted from all the elements in that column) and then check if an assignment is possible.
In most situations this will give the result, but if it is still not possible to assign then the procedure described below must be followed.

After here all rows having no assignments should be indicated (row 1) and all columns having zeros in that row(s) should be marked (column 1). Then all rows having assignments in the given column (row 3) have to be marked until a closed loop is obtained and drawn lines through all marked columns and unmarked rows.

$$Q = \begin{bmatrix} 0 & p_2 & p_3 & p_4 \\ q_1 & q_2 & 0 & q_4 \\ r_1 & 0 & r_3 & r_4 \\ 0 & s_2 & s_3 & s_4 \end{bmatrix}$$

From the elements that are left, the lowest value should be found. After that, the lowest value should be subtracted from every unmarked element and added to every element covered by two lines. The all steps should be repeated until an assignment is possible; this is when the minimum number of lines used to cover all the 0's is equal to the max (number of programs, number of assignments), assuming dummy variables are used to fill in when the number of programs is greater than the number of assignments.

6   Application

In this study, our objective is to find out the most adequate customer loyalty programs to maintain the consumer fidelity.

To be able to the present the most adequate loyalty program to the right customer, four different actions are going to be taken into account for a specific sector:

P1) Discounts
P2) Loyalty Programs and cards
P3) Priority
P4) Free coupons and checks

which would be offered to n number of clients C_i where i = 1,2, 3, ..., n.

$$C_i = \begin{bmatrix} C_{p1} & C_{p2} & C_{p3} & C_{p4} \end{bmatrix}$$

From here it is aimed to provide recommendations to the company regarding the most adequate loyalty programs that it should carry to each segment of its clients. In order to be able to that we are going to reduce all these variables to matrices for each customer and we are going to try to assign the most appropriate program to each client using the fuzzy based Hungarian algorithm. We are benefiting from linguistic variables and fuzzy numbers because when it comes to take objective decisions we it is not easy to represent the links with definite numbers 0 and 1s. Therefore in this study we use transform linguistic expressions which can be transformed to numerical values easier.

7   Conclusion and Further Research

In this study, we are proposing a fuzzy based model which can be applied for a decision making process in marketing. We present 4 different customer loyalty programs. In further studies, those programs and customer from different segments are going to be correlated based on the Hungarian algorithm model. Fuzzy numbers are going to be used when those correlations are calculated as it is difficult to correlate the objective outlines by binary definite numbers 0 and 1s. For that reason when it comes to apply the model to a real life problem, the linguistic expressions $\{High, Medium, Low\}$ are going to be utilized and subsequently transformed to numerical values as it is given following: H = High = (H-H, H-M, H-L) = (0.9, 0.8, 0.7); M = Medium = (M-H, M-M, M-L) = (0.6, 0.5, 0.4); L = LOW = (L-H, L-M, L-L) = (0.3, 0.2, 0.1). In future studies we are going to analyze those programs for different sectors and for different customer profiles; taking into account gender, employment type, age ranges and social level factors.

References

1. Moon, C., Lee, J., Lim, S.: A performance appraisal and promotion ranking system based on fuzzy logic: An implementation case in military organizations. Applied Soft Computing 10, 512–519 (2009)
2. Martinsons, M.G., Davison, R.: Strategic Decision Making and Support Systems: Comparing American, Chinese and Japanese Management. Decision Support Systems 43, 284–300 (2007)
3. Robbins, S.: Management, vol. 6. Prentice Hall, Englewood Cliffs (1999)
4. Zimmermann, H.J.: Fuzzy Set Theory and Its Application. Kluwer Academic Publishing, Boston (1991)
5. Zadeh, L.A.: Fuzzy sets. Information and Control 8, 338–353 (1991)
6. Pap, E., Bosnjak, Z., Bosnjak, S.: Application of fuzzy sets with different t-norms in the interpretation of portfolio matrices in strategic management. Fuzzy Sets and Systems 114, 123–131 (2000)
7. Bellman, R.E., Zadeh, L.A.: Decision-making in a fuzzy environment. Management Science 17, 141–164 (1970)
8. Schwenk, C.R.: Strategic Decision Making. Journal of Management 21, 471–493 (1995)
9. Leonard, N.H., Scholl, R.W., Kowalski, K.B.: Information processing style and decision making. Journal of Organizational Behavior 20, 407–420 (1999)

10. Papadakis, V.M., Lioukas, S., Chambers, D.: Strategic Decision-Making Processes: The Role of Management and Context. Strategic Management Journal 19, 115–147 (1998)
11. Kuhn, H.W.: The Hungarian Method for the assignment problem. Naval Research Logistic Quarterly 2, 83–97 (1955)
12. Lam, S., Shankar, V., Erramilli, M., Murthy, B.: Customer Value, Satisfaction, Loyalty, and Switching Costs: An Illustration From a Business-to-Business Service Context. Journal of the Academy of Marketing Science 32(3), 293–311 (2004)
13. Kim, Y., Lee, J.: Relationship between corporate image and customer loyalty in mobile communications service markets. African Journal of Business Management 4(18), 4035–4041 (2010)
14. Waarden, L.M.: The effects of loyalty programs on customer lifetime duration and share of wallet. Journal of Retailing 83, 223–236 (2007)
15. Karjaluoto, H., Leppäniemi, M.: Factors influencing consumers' willingness to accept mobile advertising: a conceptual model. International Journal of Mobile Communications 3(3), 197–213 (2005)
16. Merigó, J.M., Gil Lafuente, A.M.: The Generalized Adequacy Coefficient and its Application in Strategic Decision Making. Fuzzy Economic Review 8, 17–36 (2008)
17. Merigó, J.M., Gil Lafuente, A.M., Gil-Aluja, J.: Decision Making with the Induced Generalized Adequacy Coefficient. Applied and Computational Mathematics 10(2), 321–339 (2011)

User and Usage Constraints in Information Systems Development – Application to University: A Case Study of Lagos State University, Lagos, Nigeria

Moses A. Akanbi[1,*] and Amos A. David[2]

[1] Department of Mathematics, Lagos State University, P.M.B. 0001 LASU Post Office
Lagos, Nigeria
akanbima@gmail.com

[2] Laboratoire Lorraine de Recherché en Informatique et ses Applications (LORIA)
Campus Scientifique, BP 239 54506 Vandoeuvre - Lès - Nancy, France
Amos.David@loria.fr

Abstract. An application's conceptual model is the "mental map" that the designers employ to present information logically. A hallmark of usable design is when both the designer's conceptual model and the end users' conceptual models are in alignment. In this paper, we present the User and Usage Constraints in Information Systems Development as applicable to University. A special case study of Lagos State University Information Systems development is considered.

Keywords: Information Systems, Economic Intelligence, Entity Relation, Course Unit system, Data modeling, conceptual model.

1   Introduction

A database is an integrated collection of data, usually stored on secondary storage devices such as disks or tapes. It is possible to maintain this data either as a collection of operating systems files, or stored in a Database Management Systems (DBMS). A DBMS is a set of software programs that controls the organization, storage and retrieval of data in a database.

Its advantages include data independence and efficient access, reduced application development time, data integrity and security, data administration; and concurrent access and crash recovery. Organizations depend on their ability to acquire and use information to support planning and decision making. It has been estimated that the world's total store of knowledge is doubling every four years (CSIRO).

Thus, organizations need to know how to find the right information, analyze it correctly and draw meaningful conclusions.

---

[*] The corresponding author is the pioneer and current Head of University Data Processing Unit, Vice Chancellor's Office, Lagos State University, Lagos, Nigeria where this Information System has been deployed since October 2005.

K.J. Engemann, A.M. Gil Lafuente, and J.M. Merigó (Eds.): MS 2012, LNBIP 115, pp. 54–61, 2012.
© Springer-Verlag Berlin Heidelberg 2012

2   Economic Intelligence Systems

Economic Intelligence deals with "the process of collection, processing and diffusion of information that has as an objective, the reduction of uncertainty in the making of all strategic decisions" [3].

An Information System (IS) that consists of strategic information and that also permits the automation of the organization to better satisfy the objective of the management is called Strategic Information System (SIS) [4]. Economic Intelligence Systems (EIS) is the combination of SIS and user modeling domains [1].

The main objective here is to help the user or decision maker in his decision process. Four main stages identified in the architecture of an EIS proposed by SITE [4], are selection, mapping, analysis and interpretation. In this process three major actors were identified. The major actors are decision maker, information watcher and end user.

3   Information System Development in Lagos State University

The University runs Course Unit System [5, 6]. This is an operational system in which the entire package of courses required by a Student for a particular degree, is bundled into a number of modules each consisting of a prescribed number of units and status.

Usually one module is to be offered in one Semester. During the semester or at the end, each of these courses is assessed, either by written examination, presentation of seminar, project defense, teaching practice and industrial training assessment, and so on. There are many significant documents that result from the assessment, such as Course Mark Sheet, Broad Sheet, Profile, Academic Transcript and so on.

In the recent past, candidates seeking admissions for further study home or abroad, or employments into multinational organizations, have one sad story or the other to tell due to late release or non-release of such essential and fundamental documents. In order to abate the trend, the University decided to set up a Central Data Processing Unit powered by Information Systems.

However, the development is not without its "pros and cons". This work presents the user and usage constraints in the development of the Information System.

3.1   Applications Specification

The application is designed with the aim of retrieving students' examination scores from Course Lecturers for onward processing by the central Data Processing Unit (DPU) of the University. The University's highest academic board is the Senate [5, 6]. The University comprises of Schools and Faculties. A Department belongs to one and only one Faculty or School.

The department runs academic or professional programs. The programs at times may be in different Disciplines. A student registers for only one program at a time, but he may pursue another program after the completion of the previous one. The departments through the faculties recommend students to senate for the award of certificates, diplomas and degrees of their programs. Diploma can be either of

Sub-Degree or Post-Graduate category. Degree can be one of Bachelor, Master, or Doctor of Philosophy.

The Department determines the graduation requirements of each category of program. The graduation requirement of each program is specific and unique. The graduation requirement consists of the following:

- Minimum Cumulative Number of Units required passing (MCNURP) at the end of the program
- List of compulsory courses. These are courses that must be taken and passed. These courses are specific
- List of elective courses. These are courses that can be taken to meet up the MCNURP. It may or may not be specific.
- Minimum Cumulative Number of Units required passing (MCNURP) at the end of each session. This is specific to number of session spent so far but general to all degree programs
- Minimum Cumulative Grade Point Average at the end of every session. This is general to all degree programs.
- Minimum and Maximum Number of Units required registering for each semester and session. This is also general to all degree programs.

A course and other related entities can be described as follows:

- It is floated / based in only one department.
- It may involve lectures, practical, project write up, defense, seminar, teaching practice or industrial training
- It is taught by one or more Lecturers
- It is coordinated by one Lecturer
- A Lecturer belong to only one department
- The Lecturer(s) may be from the Department and / or other Departments
- It can be registered for by students of the Department and / or other Departments
- It has unique credit units called Number of units.
- It has status of compulsory or elective depending on the student's program
- It is graded – examined, assessed or defended – and recorded in Mark Sheet.
- There is only one Mark Sheet for a course in a session
- A student registers once for a course in a session. However, it is also possible to register for the same course as many times as possible till he passes the course, if the course, for example has the status of compulsory.
- A Lecturer in the department heads the department, from a start date to an end date
- A Lecturer in the faculty is appointed as the dean of the faculty, from a start date to an end date

3.2   Entity Relation Model

An application's conceptual model is the "mental map" that the designers employ to present information logically. A hallmark of usable design is when both the designer's conceptual model and the end-users' conceptual models are in alignment.

An abstract model (or conceptual model) is a theoretical construct that represents something, with a set of variables and a set of logical and quantitative relationships between them. Models in this sense are constructed to enable reasoning within an idealized logical framework about these processes and are an important component of scientific theories.

A logical model provides key semantic and logical information around data.

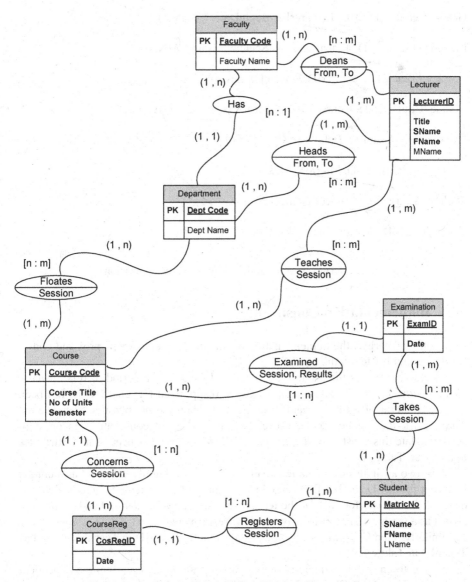

Fig. 1. Entity Relation Model of LASU Mark Sheet Application

Faculty (FacultyCode, FacultyName, DeanID)

Deans (DeanID, FacultyCode, LecturerID, BeginDate, EndDate)

Department (DeptCode, DeptName, FacultyCode)

Floats (FloatCode, DeptCode, CourseCode, Session)

Heads (HeadID, DeptID, LecturerID, BeginDate, EndDate)

Lecturer (LecturerID, TeachID, Title, SName, FName, MName)

Teaches (TeachID, LecturerID, CourseID, Session)

Course (CourseCode, CourseTitle, No of Units, Semester, ExamID)

Examination (ExamID, Date)

CourseReg (CosRegID, MatricNo, Date)

Student (MatricNo, SName, FName, LName)

Takes (TakenID, Session, MatricNo, ExamID)

Fig. 2. Relational Model of LASU Mark Sheet Application

4   Implementation Constraints

As mentioned earlier, the implementation of this model has its own "pros and cons", which are highlighted in this section.

A lecturer oversees the affairs of a faculty as **Dean,** for a period of time, usually called *tenure*, but a particular lecturer may be reappointed for tenure. If he has been assigned a unique identifier, DeanID the first time, then it is not possible to update the database for the second tenure, especially when it is not consecutive. In order to accommodate this constraint, *BeginDate* and *Enddate* are introduced as new attributes to the Dean Table.

Thus, two attributes of the table namely DeanID and BeginDate are set as a unique identifier for any Dean. If a Dean should serve more than one tenure, there is always a different BeginDate for each of the tenure. In a similar vein, *HeadID* and *BeginDate* are set as a unique identifier for any Head of Department.

Similarly, FloatCode was set as key in Floats table, TeachID in Teaches, as well as TakenID in Takes table.

In other to ensure data integrity, new entities were used to regroup the dependent attributes. There are Course Codes in different Faculties and Departments with different Course Titles and Number of Units.

Whenever a Student takes Courses more than the Maximum Number of Units, the penalty is not stated.

4.1   Constraints Associated With Matriculation Number

Every student is usually assigned a unique identifier called *Matriculation Number*. This Matriculation Number is assigned after the University Matriculation ceremony. The identifier is the means by which every Mark is released/submitted after the University semester examinations.

However, there are many problems arising from this identifier viz;

Delay/Non issuance of Matriculation Number at the beginning of the first semester examination. This makes some students to write examination without the identifier and as such the results of such examinations become untraceable or missing. At times, some students copy the wrong Matriculation from the notice board, while some others cannot remember their Matriculation during the first few examinations. During examinations there are cases where two students write the same Maticulation numbers. By so doing it create integrity problems in the database.

Some other issues associated with Matriculation number is the change in Matriculation number pattern. Since Matirculation number is also the means by which the discipline of the student is also recognised, it became necessary that the discipline of a student is identified uniquely by his Matriculation number. The Matriculation pattern in the University in the pre-computerisation era is as shown in A below.

The constraint with this pattern is that there are many departments that award degrees in many disciplines. For instance, Department of Curriculum studies awards Bachelor degree of Science Education (B. Sc. Ed.) in Biology, Chemistry, Mathematics and Physics, such that the faculty and departmental code in the Matriculation numbers are the same for students in these disciplines, yet they are to satisfy different graduation requirements.

Thus, we changed the Departmental code to one digit and introduce unit code as the second digit to the departmental code.

Table 1. Samples of Faculty, Departmental and Unit Codes during the pre-computerization and computerization era

S/N	PRE-COMPUTERISATION ERA CODES			COMPUTERISATION ERA CODES			
	Faculty	Dept.	Combined	Faculty	Dept.	Unit	Combined
1	01	01	0101	01	1	1	0111
2	01	01	0101	01	1	2	0112
3	01	01	0101	01	1	3	0113
4	01	01	0101	01	1	4	0114

4.2   Non-existing / Inconsistent Matriculation Number Pattern

In the course of processing the 2007/2008 Harmattan Semester results, it was observed that some students' scores were submitted with inconsistent / non-existing Matriculation Numbers.

By inconsistent/non-existing Matriculation Numbers, it means that the number does not conform to the University's new system of Matriculation Numbering (shown below) and in most cases such Matriculation Numbers do not exist in the University's database.

The examples of correct Matriculation Numbers in Botany Department at different sessions are given below.

Table 2. Samples of correct Matriculation Numbers in Botany Department at different academic sessions

A. BEFORE 2005 / 2006 SESSION			
Year	**Faculty**	**Department**	**Student**
04	05	02	001

B. DURING 2005 / 2006 SESSION				
Year	**Faculty**	**Department**	**Unit**	**Student**
05	05	2	0	001

C. AFTER 2005 / 2006 SESSION				
Year	**Faculty**	**Department**	**Unit**	**Student**
07	05	2	1	001

Some specific cases are:

1. Use of old pattern A for students of 2005/2006 or or later academic sessions
2. Use of new pattern B or C retrospectively, that is, for students who had matriculated before 2005/2006 academic session

Since Matriculation Number is the only primary identifier for students, it is imperative that there is a convincing understanding of this concept by **ALL** involved.

In addition, by regulation, once a student is matriculated in a University, he cannot matriculate again or be given another Matriculation number, even if he chooses to study another course. The implication of this is that such student's second discipline cannot be uniquely identified with the former Matriculation number.

5   Conclusion

This paper discussed the constraints encountered in the development of an Information System for processing students' results at Lagos State University. The significance of our development is the alignment of the designer's conceptual model and the end-users' conceptual models.

The Information System presented here has reduced "bottle necks" and time wasted in processing students' results drastically. In the nearest future, other emerging constraints which are being resolved shall be discussed in details.

References

1. Afolabi, B., Thiery, O.: Considering users' behaviors in improving the responses of an information base. In: 1st International Conference on Multidisciplinary Information Sciences and Technologies, Merida, Spain (October 2006)
2. Afolabi, B.: La conception et l'adaptation de la structure d'un système d'intelligence économique par l'observation des comportements de l'utilisateur. Ph.D. Nancy 2 University, France (2007)
3. Revelli, C.: Intelligence stratégique sur Internet, Paris, Dunod (1998)
4. SITE. Modeling and Development of Economic Intelligence Systems. Activity Report 2006 (2007), http://www.inria/recherche/equipes/site.en.html
5. Lagos State University Calendar 1995/1997 academic session. Lagos State University Press, Lagos (1995)
6. Lagos State University External System Undergraduate Prospectus 2008-2012. Press De Oak, Ogba (2008)

Online Business Registration and Congolese Entrepreneurs in Cape Town

Alain M. Momo[1] and Wilfred I. Ukpere[2]

[1] Cape Peninsula University of Technology, Faculty of Business, Cape Town-RSA
momomike161@gmail.com
[2] University of Johannesburg, Department of Industrial Psychology & People Management
Faculty of Management, Johannesburg, RSA
wilfredukpere.ajbm@gmail.com, pastorwilfred@yahoo.co.uk,
wiukpere@uj.ac.za

Abstract. It has been conceived that migration, which has impacted human history is God ordained. However, rationalism, capitalism, liberalisation and technological innovation are also central driving forces of globalization and global migration. Regardless of their entrepreneurial engagement, the presence of immigrants, including those from Congo-Brazzaville in South Africa, is often associated with an increasing unemployment rate; therefore, Congolese immigrants have become easy targets of blame for everything that has gone wrong in Cape Town. Therefore, the current authors believe that technological innovation, namely e-commerce could enable Congo-Brazzaville immigrant entrepreneurs, who are often accused of taking South African jobs, to plan for returning home. The main objective of this study was to analyse the benefits of online business registration for Congo-Brazzaville immigrant entrepreneurs in Cape Town. In order to obtain a clearer idea, the study focused on the junction of two strands, namely online company registration from both inductive and deductive perspectives. This paper utilised both qualitative and quantitative research methods. For research purposes, a self-administered questionnaire was utilised. The research target population included Congo-Brazzaville immigrant entrepreneurs in Cape Town, which is where the study was conducted. The purposeful sampling method was utilised with a sample size N = 116. The samples were sought from the general population of Congolese immigrant entrepreneurs who are also members of the Congolese Association of Cape Town.

Keywords: Center for business and administrative procedures (CFBAP) Congo-Brazzaville, e-commerce, online business registration, Cape Town.

1 Introduction

Various authors have enumerated key drivers of global immigration, but have often ignored the supreme mandate of God. [36] argues that Zoroastrianism and Buddhism were of the first religions to provide a slight hint of globalism during the fifth and sixth centuries and, later on, rationalistic global consciousness was also reflected in

K.J. Engemann, A.M. Gil Lafuente, and J.M. Merigó (Eds.): MS 2012, LNBIP 115, pp. 62–77, 2012.
© Springer-Verlag Berlin Heidelberg 2012

the building of the pyramids in Egypt. Furthermore, he states that the Jews were the first to provide a transparent expression of a global community that united their diaspora [36] [34]. God has ordained migration [36] and migrations have positively altered human history [9]. [36] has weighted two strands of migration by citing [32] position, which states that rationalism, technological innovation and capitalism are major forces in the process of globalisation. Conversely, the anti-globalists' cited in [16] position is that capitalism, liberalisation and Third World debt are the central driving forces of globalization, which have also spurred global immigration. Regardless of the positive aspects of immigration and entrepreneurial prospects and initiatives of immigrants, the presence of Congolese immigrants in South Africa is often perceived with resentment amongst the indigenous, as they regard them as job takers and a source of increasing unemployment. Hence, Congo-Brazzaville immigrants have become easy targets of blame for everything that has gone wrong in Cape Town [21]. Therefore, the current authors believe that technological innovation, namely e-commerce could enable Congo-Brazzaville immigrant entrepreneurs who are often accused of taking South African jobs to plan to resettle at home.

The Internet has created a 'borderless' virtual business platform where suppliers, customers, competitors and network partners can freely interact without experiencing pre-defined channels on the value chain. Members of the same business network or of different networks can by-pass traditional interaction patterns and form virtual value chains in most developed and some developing countries [12]. Although, traditional business registration in Congo-Brazzaville has been widely touted as a key element for Congolese SMMEs' support, empirical evidence suggests that both Congolese entrepreneurial activities and poverty alleviation have a functional relationship with online business registration, which has been claimed by Congo-Brazzaville immigrant entrepreneurs. Early findings of this paper reveal that 84.5% of Congolese immigrant entrepreneurs believe that online business registration in Congo-Brazzaville will improve supply chain management; 82.8% believe that online company registration will increase businesses' profits; and 94 % of them completed higher education. Furthermore, 94.8 % of Congolese immigrant entrepreneurs answered *yes* to expand their business in Brazzaville, and 85.3% of them believe that the potential Centre for Business and Administrative Procedures' (CFBAP) e-commerce website will attract them to expand business activities in Brazzaville. It is, therefore, in the interest of CFBAP to implement an e-commerce website in order to attract Congolese immigrant entrepreneurs' businesses in Congo-Brazzaville. If sufficient attention is not paid to the necessity of implementing an e-commerce website at CFBAP, Congo-Brazzaville immigrant entrepreneurs will not secure a source of income back home. Therefore, the high cost of living in Congo-Brazzaville will create hindrances for Congolese immigrants to return home, which may compromise the country's human resources.

2   Conceptual Background

Despite being attracted to South Africa, Congo-Brazzaville immigrant entrepreneurs have now realised the need to invest back home in Congo-Brazzaville owing to certain reasons, namely xenophobia, a lack of access to resources, political uncertainty in South Africa, relative political stability in Congo-Brazzaville and a high crime rate, including armed robbery, rape, child abuse, prostitution, pick-pocketing, shop-lifting, car hijacking and hijacking of cash-in-transit vehicles, which were the most frequently cited reasons for professional South African citizens emigrating to other countries [31].

According to [35], currently, the South African Correctional Services have complained of congested jails around the country owing to rising levels of crime within the country. Considering the saying that *an idle mind is a devil's workshop* [35], less fortunate South Africans who are victims of delayed political promises vent their anger on Black foreigners by looting their belongings, which is a practice commonly known as xenophobia. However, instead of promoting foreign Black SMMEs' initiatives, the South African government, has not only placed South Africans first, but has also adopted a double- edged immigration policy, which is based on the 1998 legislation of the Refugee Act and the 2002 Immigration Act, which appeared ineffective for Black African migrants, including Congo-Brazzaville immigrant entrepreneurs. The inefficiency of the aforementioned policies, which have failed most Black African immigrants, resonates Charles Darwin's wise words: *"if the misery of our poor be caused not by the laws of nature, but by our institutions, great is our sin"* [35].

A report, which was released by the Paris-based International Federation for Human Rights stated that South Africa's immigration policy is still like that of the previous governments' in terms of the harassment of so-called illegal immigrants [31]. Furthermore, the report criticised the policy, which has criminalised immigration and fuelled xenophobia. In addition, South Africa's immigration policy remains focused on security concerns and population control. This approach is based on the premise that considerable numbers of economic migrants intend to enter into South Africa and should be classified either under the Immigration Act (IA) of 2002 or the Refugee Act (RA) of 1998. The former is for people who come to work, start a business, or study, while the latter is for people who have fled to South Africa because of political circumstances in their own countries [31].

Early findings of this paper reveal that a majority of Black foreigners classified under both the IA and the RA who sought a better life in South Africa had difficulty obtaining legal status even though one - third of them possessed education or skills that are much needed in South Africa. Therefore, the South African double-edged immigration policy, which has made it impossible for a majority of Black foreigners to fully access available resources in South Africa, has also forced Congolese immigrant entrepreneurs to invest back home in Brazzaville. At a leadership level of the country, the power tussle during the election that brought President Jacob Zuma to power, which also coincided with the xenophobic attack, caused a sharp increase in the number of affluent Congo-Brazzaville immigrants looking to go back home for good, either before the 2010 Soccer World Cup or soon after. Finally, the last 2009 presidential election in Congo-Brazzaville, which re-elected president Sassou to power, did not raise any concern, violence or opposition protest. This tendency attracted companies such as Warid Telecom, MTN South Africa and several others to invest in Brazzaville [10].

3   Research Hypotheses

In developed countries dynamic arguments for the existence of SMMEs have been stressed in terms of their innovative capacities and, potentially, that they merge to create larger firms. In contrast, Congolese immigrant entrepreneurs' SMMEs are increasingly taking the role of primary vehicles for the creation of employment and income generation through self-employment and, therefore, are among tools, which enable poverty alleviation in South Africa. In addition, Congolese SMMEs in Cape Town operate in the retail and services sectors. However, owing to their characteristics, which include size of the capital investment, number of employees, turnover, management style, market share and various hindrances of immigration policies, Congolese immigrant entrepreneurs agree that business expansion in Brazzaville, which creates domestic linkages with other SMMEs in Congo-Brazzaville in terms of diversification strategies, becomes imperative for business survival [37]. From the above, the current authors note that business expansion through e-commerce technologies will enable Congolese immigrants' SMMEs to save on supply chain management costs. Therefore, the rapidly accelerating rate of technological innovation has forced Congolese immigrant entrepreneurs to demand a business-to-business (B2B) e-commerce website at CFBAP, which should enable online business service. B2B e-commerce, which is fully automated in South Africa, is still not applicable in Congo-Brazzaville. In South Africa, businesses rely on computer technology in almost every area of the corporate life cycle [26]. Reliance on high-speed digital computers is so complete that South African commerce would come to a sudden stop if computers were removed from business cycles [7]. As a result, technologies were introduced at the South African Department of Trade and Industry (DTI), enabling business people to register their businesses online. According to [38], e-commerce can reduce transaction-level costs, improve time-scale and reduce errors. The authors further argue that e-commerce will redesign CFBAP's interface and that of its partners. Sources of delay in offline business registration processes at CFBAP will be corrected, while redundancy, and unnecessary delays will be improved [38]. Congolese immigrants' perceived online business registration benefits include areas of finance, management, marketing, and logistics and being a part of the Congo-Brazzaville economy. Finance has been identified as the most important factor, which determines the survival, growth and expansion of Congolese immigrants' businesses in Cape Town. The first role of e-commerce for Congolese immigrant entrepreneurs in finance is to increase the speed of financial operations between Congo-Brazzaville and South Africa [37] [17]. Based on the above, the current authors have hypothesised that:

H1: Increased speed of financial operations is positively related to Congolese immigrant entrepreneurs' business expansion in Brazzaville.

H2: Online company registration will contribute to job creation, poverty alleviation and the country's economic growth.

Online VAT declaration is another benefit of e-commerce for Congolese immigrant entrepreneurs [6]. Hence, the authors assume that:

H3: CFBAP's e-commerce website will enable fast online VAT payment.

In addition, Congolese immigrants' businesses will increase visibility via a CFBAP e-commerce website by using various technologies, namely website page landing, Google Pay per Click (PPC) and banner [8], [1] and [20]. Due to the fact that a majority of Congolese immigrant entrepreneurs' SMMEs face challenges related to poor resources at different levels, including finance and management, to overcome these managerial challenges with the aid of the Internet and its use for commercial purposes, the current authors hypothesised that:

H4: CFBAP's e-commerce website will improve Congolese immigrant entrepreneurs' business marketing functions.

Furthermore, managerial benefits of online business registration at the CFBAP's website will strengthen business relationships amongst Congo-Brazzaville SMMEs [37]. SMMEs in Congo-Brazzaville also suffer from management constraints that lower their resilience to risk and prevent them from growing and attaining economies of scale. E-commerce adoption at CFBAP will enable Congolese immigrant entrepreneurs to overcome management constraints related to customs declaration and the bureaucratic process of business environment-related permit applications. Moreover, financial and accounting records within some Congolese SMMEs are rarely in place, and where they are available, their accuracy is usually doubtful. In instances where bank loans are provided, they are mostly granted to the most organised and profitable SMMEs. Considering that in Congo-Brazzaville SMMEs' bank loans are of a short duration owing to the inflation rate and SMMEs' general characteristics, it may be difficult for borrowers to secure collateral and realise high returns to finance repayments. Hence, it is proposed that:

H5: online business registration at CFBAP will improve Congolese SMMEs' administration.

Furthermore, some SMMEs in Congo-Brazzaville employ less than five people, mostly family members who are not legally registered with the *Onemo*: the Congo-Brazzaville Labour Department. Not surprisingly, such SMMEs in Congo-Brazzaville apply simple and relatively backward technology in production and, therefore, the quality of their products is likely to be poor. Such SMMEs suffer from limited market access and face fierce competition from many local produce [37]. Hence, the authors propose that:

H6: B2B e-commerce adoption at CFBAP will generate a culture of Life Long Learning (LLL) amongst employees.

Although, shipping costs can increase the cost of many products that Congolese immigrants may purchase via e-commerce and add substantially to the final price, distribution costs will be significantly reduced for some Congolese immigrants' products and services, namely financial services, business registration, software, and travel, which are important segments of brick-and-mortar commerce in Congo-Brazzaville [24], [30]. Indeed, e-commerce will reduce the internal costs of many transactions of Congolese entrepreneurs and change the cost structure that dictates Congolese immigrants' businesses relationships with other businesses [28], [24]. In addition, e-commerce will have impact intermediaries [28] who help producers sell to distributors such as wholesalers and retailers [24]. Therefore:

H7: B2B e-commerce adoption at CFBAP will improve Congolese SMMEs' supply chain management.

However, customer care improvement will be another benefit of B2B e-commerce adoption for Congolese immigrants' businesses in Congo-Brazzaville. In today's Knowledge Based Economy (KBE), which is dominated by sophisticated products and services, after-sales services is a major cost for many Congolese SMMEs. Traditionally, Congolese SMMEs place service personnel in the field to visit clients and obtain feedback about products and services' performance [24]. Based on the above, the current authors propose that:

H8: online business registration in Congo-Brazzaville will enable Congolese SMMEs to improve customer relationship management.

Furthermore, a survey conducted by [32] concerning the role of e-commerce in the economies of various countries indicates that, compared to "traditional" commerce, e-commerce raises some expectations at a national level, including increased productivity, reduced costs for producers and consumers and increased accessibility [2]. According to [32], e-commerce adoption in Congo-Brazzaville will have various positive impacts on the economy, which include increased marketplace economic interactivity, an increased catalytic ICT role, increased economic openness of technology, while it will also alter time importance in the country's economy for sustainable development [24]. Traditionally, 80 per cent of businesses in Congo-Brazzaville have been established in the cities of Brazzaville and Pointe-Noire, which are close and located in the south of the country [10]. However, adoption of e-commerce will replace traditional distribution channels, create new products and market development, connect all Congolese cities to the rest of the world, globalise the Congo-Brazzaville economy, and replace *brain drain* with *brain gain*, in respect of higher-skilled workers. In addition, there will be a shift in the role of Congo-Brazzaville consumers, who will be increasingly implicated as partners in product design and creation, which will enable consumers to conduct transactions around the clock. Hence, the current authors propose that:

H9: e-commerce adoption at CFBAP is positively related to the country's economic growth.

However, Machiavelli cited in [38] asserts that there are challenges related to e-commerce adoption in Congo-Brazzaville owing to its newness and states that "*there is nothing more difficult to plan, more doubtful of success, nor more dangerous to manage than the creation of a new system.*" From the above and owing to the country's MDG policy, Congolese immigrant entrepreneurs should contain e-commerce related challenges, namely culture [14], IT infrastructure [11], [5], security and privacy [18], [5], [11], illiteracy, organisation [3], fraud [4], [25], [15] and public policy [13] based on the fact that CFBAP policymakers would not invest 100 percent in an e-commerce website that might be used by too few people. Hence, the authors propose that:

H10: e-commerce users' readiness will positively influence potential e-commerce providers' intentions to offer a B2B website.

However, having assessed the above-mentioned hypotheses, the current authors aimed to determine the relationships between the constructs, which would validate the hypotheses.

Table 1. Summary of hypotheses testing

Hypothesis	Relationship	Correl.	Direction	Results
1	Increased speed of financial operations → Congolese immigrant entrepreneurs' business expansion in Brazzaville.	1	+	Accepted
2	Online company registration → Job creation, poverty alleviation and the country's economic growth.	1	+	Accepted
3	Potential CFBAP's e-commerce website → online VAT payment.	1	+	Accepted
4	Potential CFBAP's e-commerce website Congolese → immigrant entrepreneurs' business marketing function.	1	+	Accepted
5	Online business registration at CFBAP Congolese → SMMEs' administration.	1	+	Accepted

Table 1. *(continued)*

6	B2B e-commerce adoption at CFBAP generate ⟶ employees' LLL culture.	1	+	Accepted
7	B2B e-commerce adoption at CFBAP Congolese ⟶ SMMEs' supply chain management.	1	+	Accepted
8	Online business registration in Congo-Brazzaville ⟶ improve customer relationship management.	1	+	Accepted
9	E-commerce adoption at CFBAP ⟶ The country's economic prosperity.	1	+	Accepted
10	E-commerce users' readiness ⟶ Positively influences potential e-commerce providers' intention to offer a B2B website.	1	+	Accepted

4   Research Methodology

This research utilised both qualitative and quantitative research methods. For research purposes, a self-administered questionnaire was distributed to the research population. The designed questionnaire was divided into two sections, namely a demographic information section (1); and content-based questions in section (2). Demographic information comprised of variables such as gender, age, education and dependents, while content based questions focused on the benefits of online business registration in Brazzaville for Congo-Brazzaville immigrant entrepreneurs. The research target population included Congo-Brazzaville immigrant entrepreneurs who live in Cape Town, where the study was conducted. The purposeful sampling method was utilised, with a sample size N = 116 comprising Congo-Brazzaville immigrant entrepreneurs who were sought from the general population amongst Congolese immigrant entrepreneurs who are also members of the Congolese Association of Cape Town.

The survey questions were translated from English into French and French into English by a sworn translator to aid respondents' understanding, as French is their first language. A pilot study was conducted among five Congolese immigrants, namely two workers and three businesspersons in Cape Town before distributing the questionnaire to larger population. For respondents' convenience, the researchers were involved in the distribution of 116 questionnaires, and they were given a minimum of two weeks to complete the questionnaires before collection. Ethical considerations were also taken into account when collecting and analysing the data. The Congolese Association of Cape Town's members who participated in the survey were guaranteed anonymity, whilst confidentiality of information was also guaranteed. The main objective of this study was to analyse the benefits of online business registration for Congo-Brazzaville immigrant entrepreneurs in Cape Town. Therefore, the study focused on the junction of two strands, namely both inductive and deductive reasoning for online company registration purposes. [19] states that inductive reasoning goes from the specific to the general, whereas deductive reasoning goes from the general to the specific. Careful scrutiny of inductive reasoning, which included various empirical factors such as respondents' gender, education level, business experience, business expansion, potential attraction to a CFBAP e-commerce website, profits, risks and e-commerce legislation in Congo-Brazzaville enabled the researcher to explain what the study entails. Furthermore, deductive reasoning, which included e-commerce and business expansion, Congolese immigrant entrepreneurs' education level e-commerce adoption and business expansion as well as e-commerce legislation in Congo-Brazzaville, equipped the researcher to address how the study would be conducted. [19] furthermore explains that inductive and deductive reasoning methods are best utilised when the researcher tries to describe an empirical problem based on "what" questions are addressed, "how" the solution to the problem will be implemented. In addition, the benefits of online company registration for Congolese immigrant entrepreneurs, which is an empirical problem, was described according to "what", and resolved according to "how". From an interpretive point-of- view, the above reasoning enabled the researchers to make sense of statistics inferences based on relevant data, which was collected [23]. Furthermore, the originality of this research study is that it examines how Black African immigrant entrepreneurs can utilise e-commerce to plan to return home, which creates an opportunity for further constructive debate.

5   Analysis and Results

In order to address the research problem that was identified, the researchers proceeded with e-commerce inductive and deductive reasoning. The researchers also made an earnest effort to discover the reasons that led to Congolese immigrant entrepreneurs, to demand for online business registration service at CFBAP, whilst being established in Cape Town.

Inductive Reasoning Statistical Analysis

This inductive reasoning was based on the quantitative data that was collected, which enabled a more objective evaluation of research variables [22]. In addition, various components depicted in Figure 1, namely business experience, respondents' intention to expand businesses in Brazzaville, potential attraction to a CFBAP e-commerce website, profits, supply chain management, e-commerce legislation in Congo-Brazzaville and respondents' Internet related risks, were further analysed.

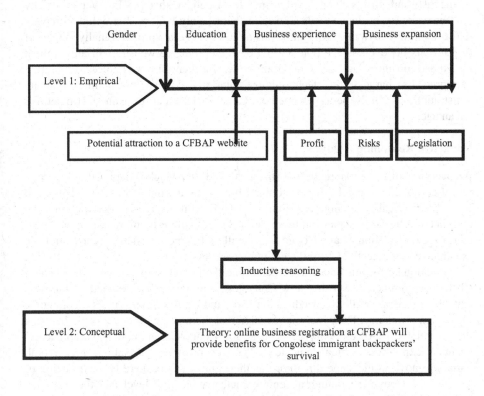

Fig. 1. Online business registration inductive reasoning; Source: [23]

Based on the above online company registration theory building concept, it is clear that the increased utilisation of e-commerce technology is currently modifying the way people transact, which was formerly done by face to face contact within and between nations [27]. From the 116 respondents, 67 % are male and 49 % are female. This shows that male respondents were dominant in this sample. In addition, 46.6 % completed college and 47.4 % completed university, which totals 94 % of respondents (46.6 plus 47.4) who have higher education qualifications. Regarding business experience, 13.8 % of respondents have been operating their businesses for less than

2 years, and 45.7% of respondents have operated their businesses for between 3 and 5 years. However, 35.3 % of respondents spent between 6 and 8 years in business and, finally, 5.2% of respondents have been in business for 12 years or more. Moreover, 94.8 % of respondents answered *yes* to expand their business in Brazzaville; whereas 92% of respondents believe that Internet legislation should be adopted in Congo-Brazzaville in order to build trust among potential users and to guarantee online traffic on the potential CFBAP e-commerce website. Hence, most computer specialists, whether for fun or profit, invest in committing Internet crime. Regarding online risk and profits, 95.6 % of respondents believe that the Internet is a risky environment, for business while 82.8% of them firmly believe that online company registration in Brazzaville will enable businesses to generate profits. Finally, 85.3% of respondents believe that the potential CFBAP e-commerce website will attract them to expand business activities in Brazzaville. Furthermore, Congolese immigrant entrepreneurs harness the technological platform known as online company registration, which is being claimed at CFBAP, in order to gain a competitive advantage.

Deductive Reasoning Analysis

An overview of deductive reasoning statistical analysis, which is depicted in Figure 2, was used for this research. The reasoning proposed that from level 1 (conceptual) to level 2 (empirical), e-commerce adoption at CFBAP in Congo-Brazzaville could be correlated to business expansion (correlation 1), Congolese immigrant entrepreneurs' education (correlation 2) and Congo-Brazzaville's Internet legislation (correlation 3), as perceived by potential CFBAP's e-commerce users.

The rationale for this data was to determine the relationship between the potential CFBAP e-commerce website and Congolese immigrant entrepreneurs' business expansion in Brazzaville, their education level and Congo-Brazzaville's e-commerce legislation efficiency. Data analysis revealed that there is a correlation of +1 between the potential CFBAP e-commerce website and Congolese immigrant entrepreneurs' business expansion, which indicates a perfect positive correlation, hence both variables move in the same direction together. Furthermore, there is a correlation of +1 between Congolese immigrant entrepreneurs' education level and their internet skills, which indicates a perfect positive correlation, which means that both variables, again, move in the same direction together. Hence, data analysis reflected that there are opposite correlations of +1 and -1 between Congolese immigrant entrepreneurs' online business expansion in Brazzaville and Congo-Brazzaville e-commerce's legislation reliability, which indicates imperfect causation, which means that both variables do not move in the same direction together. The result of the analysis is discussed in the next section.

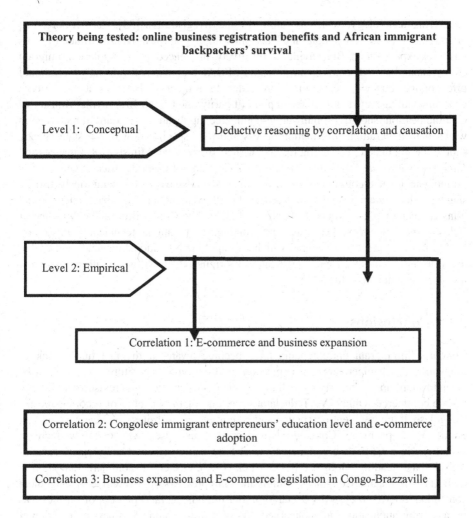

Fig. 2. Online business registration deductive reasoning. Source: [23]

6   Discussion of Results

The business connotation of online company registration and supply chain management imply that the CFBAP e-commerce website will enable Congolese immigrant entrepreneurs to attract new customers, select suppliers, reduce operating costs, and declare both VAT and customs electronically; which, in turn, is positively related to businesses' profit. However, Congolese immigrant entrepreneurs require legal insurance so that electronic transactions, which are conducted on the CFBAP e-commerce website, are safe, whilst collective data is kept private. However, experienced, Congolese immigrant entrepreneurs are aware that a lack of innovation can hamper business expansion in Brazzaville unless steps are taken at the right time to shore up the underlying technology, which is online business registration [33]. Furthermore, this interpretation leads to formulation of the following hypotheses: overall attitude towards business expansion in Brazzaville is positively influenced by Congolese immigrant entrepreneurs' attitude towards the potential CFBAP e-commerce website, which should offer online business registration. In order to project CFBAP as a tech savvy governmental agency and to enable a point of parity, the Congo-Brazzaville Ministry of Communication should promulgate e-commerce application standard directories in which data integration, information access and data specification in the state agency should be regulated. Furthermore, the Congo-Brazzaville Ministry of Finance and *Banque des Etats de l'Afrique Centrale* (BEAC): Bank of Central African States should promulgate legal e-commerce texts, which guide businesses to unanimously use IT standards that are applicable to the sectors of card payment, and inter-bank international transfer, including e-customs declarations. Finally, the Congo-Brazzaville Ministry of Industry and Trade should establish and promulgate the national technical regulation on EDI to issue a certificate of origin, which will apply to all national agencies and enable them to issue an electronic certificate of origin (eCoSys) when all e-commerce transactions will be conducted [33].

7   Conclusions

Congolese immigrant entrepreneurs have become aware of the fact that a lack of innovation can hamper business expansion in Brazzaville. Therefore, steps should be taken to shore up the underlying technology such as online business registration through B2B e-commerce. Online VAT declaration is one of the benefits of e-commerce for Congolese immigrant entrepreneurs and, e-commerce websites is a catalyst of fast online VAT payment. Congolese immigrants' businesses will certainly increase visibility via a CFBAP e-commerce website by using various technologies, such as website page landing, Google Pay Per Click (PPC), etcetera. E-commerce adoption at CFBAP will enable Congolese immigrant entrepreneurs to overcome management constraints related to customs declaration and the bureaucratic process of business, as far as permit applications are concerned. In other words, online business registration at CFBAP will improve Congolese SMMEs' administration. Since there will be training on how to utilise the new online technologies, it will generate a culture of Life Long Learning (LLL) amongst employees at CFBAP. More than that, B2B e-commerce will significantly reduce distribution cost of some Congolese immigrants' products and services, such as finance, business registration, software and travel, which are important elements of commerce. Indeed, e-commerce will reduce the internal costs of various transactions of Congolese entrepreneurs and change the cost structure that dictates Congolese immigrants' businesses relationships with other businesses. Hence, B2B e-commerce adoption at CFBAP will improve Congolese SMMEs' supply chain management. In addition, customer care improvement will be another benefit of B2B e-commerce adoption by Congolese immigrants' businesses in Congo-Brazzaville. Traditionally, Congolese SMMEs place service personnel in the field to visit clients and obtain feedback about products and services' performance. However, in today's

Knowledge Based Economy (KBE), which is dominated by sophisticated products and services, after-sales services can be provided online at a reduced cost of operation. Therefore, online business operation by small businesses will certainly increase the speed of financial and marketing operations, which will positively enhance Congolese immigrant entrepreneurs' business expansion to Congo-Brazzaville. Business expansion is indeed potent for job creation, poverty alleviation and economic growth of any nation including Congo-Brazzaville.

References

1. Ahmed, H.: Benefits of Banner Ads (2010),
   http://ezinearticles.com/?Benefits-of-Banner-Ads&id=589677
   (accessed on December 20, 2010)
2. Bharati, P., Chaudhury, A.: Current status of technology adoption: Micro, small and medium manufacturing firms in Boston. Communications of ACM 49(10), 88–93 (2006)
3. Bester, A.: Information systems performance in the construction sector: the role of the chief executive officer. Cape Town: Cape Peninsula University of Technology. Unpublished MTech thesis (2006)
4. Byers, S., Rubin, A.D., Kormann, D.: Defending against an Internet-based attack on physical World. ACM Transaction on the Internet Technology (TOIT) 4(3), 239–254 (2004)
5. Cap Gemini: Online Availability of Public Services: How Is Europe Progressing? (2006),
   http://europa.eu.int/information_society/eeurope/i2010/docs/
   benchmarking/online_availability_2006.pdf (accessed on March 18, 2010)
6. Clear Books. Online VAT return (2010),
   http://www.clearbooks.co.uk/partners/case-study/
   online-vat-return/ (accessed on December 20, 2010)
7. E-commerce Juice. Landing pages can increase your conversions exponentially overnight (2010),
   http://www.ecommercejuice.com/2010/06/
   landing-pages-can-increase-your-conversions-exponentially-
   overnight/ (accessed on October 20, 2010)
8. E-commerce Juice. Using Google Pay Per Click (PPC) to get immediate e-commerce sales (2010),
   http://www.ecommercejuice.com/2010/06/
   using-google-pay-per-click-ppc-to-get-immediate-ecommerce-
   sales (accessed on October 20, 2010)
9. Esperanza USA. A Biblical Basis on Immigration Reform, http://www.esperanza.us (accessed on May 15, 2011); Forum des Jeunes Entreprises du Congo (2010), Les PME de Brazzaville et de Pointe-Noire (2011), http://www.fjec.org (accessed on March 21, 2010)
10. Forum des Jenunes Entreprises du Congo. Les PME de Brazzaville et de Pointe-Noire (2010), http://www.fjec.org (accessed on March 21, 2010)
11. Furnell, S., Tsagani, V., Phippen, A.: Security beliefs and barriers for novice Internet users. Computers & Security 27(7-8), 235 (2008)
12. Graham, M.: Disintermediation, altered chains and altered geographies: the internet in the Thai silk industry. The Electronic Journal on Information Systems in Developing Countries 45(5), 1–25 (2011)

13. Importance of computers from 1946 to date.. (2010),
    http://en.allexperts.com/q/Computer-Science-3197/
    importance-computers-1946-date.html (accessed on December 04, 2010)
14. Javalgi, R.G.: The export of e-services in the age of technology transformation: challenges and implications for international service providers. Journal of Services Marketing 18(7), 560–573 (2004)
15. Kuchinskas, S.: Fraud Chewing E-commerce Profits (2005),
    http://www.ecommerce-guide.com/news/
    research/article.php/3563526 (accessed on July 04, 2010)
16. Kirkbride, P. (ed.): Globalization: the external pressures. John Wiley, Chichester (2001)
17. Lipsky, J.: International financial markets: stability and transparency in 21st century. International monetary fund (2007),
    http://www.imf.org/external/np/speeches/2007/062007.html
    (accessed on December 20, 2010)
18. Liebermann, Y., Stashevsky, S.: Perceived risks as barriers to Internet and e-commerce usage. Qualitative Market Research: An International 5(4), 291–300 (2002)
19. Lith, I.: Inductive reasoning vs. Deductive reasoning (2011),
    http://everything2.com/title/
    Inductive+reasoning+vs.+Deductive+reasoning
    (accessed on February 19, 2011)
20. Maksimovic, Z.: What Is Return On Investment (ROI)? (2010),
    http://www.buzzle.com/articles/
    what-is-return-on-investment-roi.html (accessed on December 20, 2010)
21. Mahony, R.: Standing with the eleven million (2010),
    http://www.justiceforimmigrants.org (accessed on March 16, 2011)
22. McClelland, J.: Reaping the benefits of computers: factors encouraging and discouraging computer use. JOE Technical Support (2010) ISSN: 1077-5315,
    http://www.joe.org/joe/1986summer/a3.php (accessed on January 22, 2011)
23. New York Education: What is research design? (2011),
    http://www.nyu.edu/classes/bkg/methods/005847ch1.pdf
    (accessed on January 31, 2011)
24. OECD : Central Europe B2C E-commerce report 2010 (2010),
    http://www.slideshare.net/ReportLinker/
    central-europe-b2c-ecommerce-report-2010 (accessed on January 21, 2011)
25. Pathak, J.: Guest editorial: Risk management, internal controls and organisation vulnerabilities. Managerial Auditing Journal 20(6), 569–577 (2004)
26. Pillai, P.: Uses of computer (2010), http://www.buzzle.com/articles/
    uses-of-computer.html (accessed on December 12, 2010)
27. Remenyi, D., Sherwood-Smith, M.: Maximize Information Systems value by continues Participative Evaluation. Logistics Information Management 12(2), 14–31 (1999)
28. Seddon, P.B.: A Re-specification and extension of the Delone and McLean Model of IS Success. Information Systems Research 8(3), 240–253 (1997)
29. Stair Jr., R.M.: Computers in today's World. Irwin, Homewood (1986)
30. Stock, J.R., Lambert, D.M.: Strategic logistics management, 4th edn. McGraw-Hill, New York (2001)
31. South African Survey: Demographics. South African Institute of Race Relations: Johannesburg (2008)

32. Šumak, B., Polančič, G., Heričko, M.: Towards an e-service knowledge system for improving the quality and adoption of e-services. In: 22nd Bled e-Conference, Bled, Slovenia, June 14-17 (2009)
33. Sidhu, D.: ECommerce Business - Importance of Supporting Expansion With Technology (2011), http://www.ezinearticles.org (accessed on February 07, 2011)
34. Scholte, J.: Globalisation: a critical introduction. Macmillan, Basingstoke (2000)
35. Ukpere, W.: Distinctiveness of globalisation and its implications for labour markets: an analysis of economic history from 1990-2007. The Indian Economic Journal 56(4), 1–20 (2009)
36. Ukpere, W.I.: Rationalism, technological innovations and the supreme mandate in the process of globalisation. African Journal of Business Management 4(4), 467–474 (2010)
37. United Nations Conference on Trade and Development: Improving the competitiveness of SMEs in developing countries: The role of finance to enhance enterprise development. United Nations, New York (2001)
38. Vogt, J.J., Pienaar, W.J., De Wit, P.W.C.: Business logistics management: Theory and practice. Oxford University Press, Cape Town (2003)

The Algorithm of Optimal Polynomial Extrapolation
of Random Processes

Igor P. Atamanyuk[1], Volodymyr Y. Kondratenko[2], Oleksiy V. Kozlov[3],
and Yuriy P. Kondratenko[4]

[1] Mykolaiv State Agrarian University, Commune of Paris str. 9,
54010 Mykolaiv, Ukraine
atamanyuk_igor@mail.ru
[2] Department of Mathematical and Statistical Sciences
University of Colorado Denver, Denver, CO 80217-3364, USA
volodymyr.kondratenko@email.ucdenver.edu
[3] National University of Shipbuilding, Geroiv Stalingrada ave. 9,
54025 Mykolaiv Ukraine
kozlov_ov@ukr.net
[4] Petro Mohyla Black Sea State University, 68-th Desantnykiv str. 10,
54003 Mykolaiv, Ukraine
y_kondrat2002@yahoo.com

Abstract. This work deals with the modelling and prediction of the realizations of random processes in corresponding future time moments. The extrapolation algorithm of nonlinear random process for arbitrary quantity of known significances and random relations used for forecasting has been received on the basis of mathematical instrument of canonical decomposition. The received optimal solutions of the nonlinear extrapolation problem, as well as the canonical decomposition, that was use as a base for optimal solution, does not set any substantional restrictions on the class of investigated random process (liniarity, Markov processes propety, stationarity, monotonicity etc.). Theoretical results, block-diagrams for calculation procedures and the analysis of applied applications, especially for the prediction of economic indexes and parameters of technical devices, are under discussions.

Keywords: random process, canonical decomposition, extrapolation algorithm.

1   Introduction

A solution of problems concerning modelling and prediction of realizations of random processes in corresponding future time moments is an actual direction of modern scientific researches, as most of the physical, technical, economic and other real processes have a stochastic character. There are a large number of different methods of extrapolation of random processes taking into account different real assumptions. Presently the forecast theory, taking into account an exceptional meaningfulness of

K.J. Engemann, A.M. Gil Lafuente, and J.M. Merigó (Eds.): MS 2012, LNBIP 115, pp. 78–87, 2012.

the problem, is constantly complemented by new algorithms that extend the class of the investigated random processes and conditions for problem solutions.

2   Problem Statement

Let a random process $X(t)$ in the fixed set of points $t_i, i = \overline{1,I}$ be fully defined by means of the digitized moment functions:

$$M\left[X^v(i)\right], M\left[X^v(i)X^\mu(j)\right], \quad t_i, t_j = \overline{1,I}; \quad v, \mu = \overline{1,N}.$$

For the known values $x^\mu(j)$, $t_j = \overline{1,k}$, $\mu = \overline{1,N}$ of the investigated realization $x(t)$ of the random process $X(t)$ it is necessary to forecast the values of this realization in future moments of time t_i, $i = \overline{k+1,I}$.

In [1] a universal solution of the problem of extrapolation of a realization of the random process has been received in the following recurrent form

$$m_x^{(\mu)}(i) = \begin{cases} M\left[X(i)\right], \ \mu=0, \ i=\overline{1,I} \\ m_x^{(\mu-1)}(i) + \left[x(\mu) - m_x^{(\mu-1)}(\mu)\right]\varphi_\mu(i), \ \mu=\overline{1,k}, \ i=\overline{\mu+1,I} \end{cases} \tag{1}$$

or in a vivid form

$$m_x^{(k)}(i) = M\left[X(i)\right] + \sum_{j=1}^{k}\left(x(\mu) - M\left[x(\mu)\right]\right)f_\mu^{(k)}(i), \ i=\overline{k+1,I}; \tag{2}$$

$$f_\mu^{(k)}(i) = \begin{cases} f_\mu^{(k-1)}(i) - f_\mu^{(k-1)}(k)\varphi_k(i), \ \mu \le k-1 \\ \varphi_k(i), \ \mu = k \end{cases} \tag{3}$$

where $\varphi_\mu(i)$, $\mu = \overline{1,k}$ - are coordinate functions of a canonical expansion [1,2] of the random process $X(t)$, based on the points t_i, $i = \overline{1,I}$:

$$X(i) = \sum_{v=1}^{i} V_v \varphi_v(i), \ i=\overline{1,I}. \tag{4}$$

The parameters of a canonical expansion (4) are defined by the following recurrent relations:

$$V_i = X(i) - \sum_{v=1}^{i-1} V_v \varphi_v(i), \ i = \overline{1,I}; \tag{5}$$

$$D_V(i) = M\left[V_V^2\right] = M\left[X^2(i)\right] - M^2\left[X(i)\right] - \sum_{v=1}^{i-1} D_V(v)\varphi_v^2(i), \quad i = \overline{1,I}; \tag{6}$$

$$\varphi_v(i) = \frac{1}{D_V(v)}\left[M\left[X(v)X(i)\right] - M\left[X(v)\right]M\left[X(i)\right] - \sum_{j=1}^{v-1} D_V(j)\varphi_j(v)\varphi_j(i)\right],$$
$$v = \overline{1,I}, \quad i = \overline{v,I}. \tag{7}$$

Expressions (1), (2), within the framework of the linear approximation, determine a posterior mathematical expected value of the random process $X(t)$ under the condition that $X(\mu) = x(\mu)$, $\mu = \overline{1,k}$, which in other words give the undisplaced estimation $m_x^{(k)}(i)$, $i = \overline{k+1,I}$ of future values $x(i)$, $i = \overline{k+1,I}$ of the extrapolated realization, and provide a minimum of the mean-square error of the extrapolation $E_x^{(k)}(i)$, which is equal to the dispersion $D_x^{(k)}(i)$ of a posteriori random process $X^{(k)}(i)$, where in particular:

$$E_x^{(k)}(i) = M\left[\left|m_x^{(k)}(i) - X(i)\right|^2\right], \quad i = \overline{k+1,I}; \tag{8}$$

$$E_x^{(k)}(i) = D_x^{(k)}(i) = \sum_{v=k+1}^{i} D_V(v)\varphi_v^2(i), \quad i = \overline{k+1,I}; \tag{9}$$

$$X^{(k)}(i) = X(i / x(j), j = \overline{1,k}) = m_x^{(k)}(i) + \sum_{v=k+1}^{i} V_v\varphi_v(i), \quad i = \overline{1,I}. \tag{10}$$

Thus in (1) and (2) probabilistic connections of higher-orders $M[X^v(i)X^\mu(j)]$, $v + \mu \geq 3$ of the random process $X(t)$ are not used and as result this limits exactness of the extrapolation. The removal of the indicated defect is possible using the forecast algorithm on the base of the canonical expansion [5,6] in which the information about the investigated process is fully considered in a discrete set of points t_i, $i = \overline{1,I}$:

$$X(i) = M\left[X(i)\right] + \sum_{v=1}^{i}\sum_{\lambda=1}^{N} W_v^{(\lambda)}\beta_{1v}^{(\lambda)}(i), \quad i = \overline{1,I}. \tag{11}$$

The elements of the canonical expansion (11) are defined by the following recurrent relations:

$$W_v^{(\lambda)} = X^\lambda(v) - M\left[X^\lambda(v)\right] - \sum_{\mu=1}^{v-1}\sum_{j=1}^{N} W_\mu^{(j)}\beta_{\lambda\mu}^{(j)}(v) - \sum_{j=1}^{\lambda-1} W_v^{(j)}\beta_{\lambda v}^{(j)}(v), \quad v = \overline{1,I}; \tag{12}$$

$$D_\lambda(v) = M\left[\left\{W_v^{(\lambda)}\right\}^2\right] = M\left[X^{2\lambda}(v)\right] - M^2\left[X^\lambda(v)\right] -$$

$$-\sum_{\mu=1}^{v-1}\sum_{j=1}^{N} D_j(\mu)\left\{\beta_{\lambda\mu}^{(j)}(v)\right\}^2 - \sum_{j=1}^{\lambda-1} D_j(v)\left\{\beta_{\lambda v}^{(j)}(v)\right\}^2, \; v=\overline{1,I}; \tag{13}$$

$$\beta_{hv}^{(\lambda)}(i) = \frac{M\left[W_v^{(\lambda)}\left(X^h(i) - M\left[X^h(i)\right]\right)\right]}{M\left[\left\{W_v^{(\lambda)}\right\}^2\right]} = \frac{1}{D_\lambda(v)}\left(M\left[X^\lambda(v)X^h(i)\right] -\right.$$

$$-M\left[X^\lambda(v)\right]M\left[X^h(i)\right] - \sum_{\mu=1}^{v-1}\sum_{j=1}^{N} D_j(\mu)\beta_{\lambda\mu}^{(j)}(v)\beta_{h\mu}^{(j)}(i) - \tag{14}$$

$$-\sum_{j=1}^{\lambda-1} D_j(v)\beta_{\lambda v}^{(j)}(v)\beta_{hv}^{(j)}(i), \; \lambda=\overline{1,h}, \; v=\overline{1,i}, \; h=\overline{1,N}.$$

In the canonical expansion (11) the random process $X(t)$ in the investigated row of points is presented by means of N massives $\left\{W^{(\lambda)}\right\}$, $\lambda=\overline{1,N}$ of the uncorrelated centred random coefficients $W_i^{(\lambda)}$, $i=\overline{1,I}$. These coefficients $W_i^{(\lambda)}$ contain information about the values $X^\lambda(i)$, $\lambda=\overline{1,N}$, $i=\overline{1,I}$, and coordinate functions $\beta_{hv}^{(\lambda)}(i)$, $\lambda,h=\overline{1,N}$; $v,i=\overline{1,I}$ describe probabilistic connections of the order $\lambda+h$ between the sections of the random process in discrete moments of time t_v and t_i, $v,i=\overline{1,I}$.

Block-diagram of the procedure for calculating the parameters of the canonical decomposition is shown in Fig. 1.

We suppose that the value $x(1)$ of the process $X(t)$ at the point t_1 is known, as a result of measuring. Consequently, the values are known:

$$w_1^{(\lambda)} = x^\lambda(1) - M\left[X^\lambda(1)\right] - \sum_{j=1}^{\lambda-1} w_1^{(j)}\beta_{1v}^{(j)}(1), \; v=\overline{1,I} \tag{15}$$

for the set of coefficients $W_1^{(\lambda)}$, $\lambda=\overline{1,N}$.

The substitution of the value $w_1^{(1)}$ in the equation (11) allows us to get a polynominal canonical expansion of a posteriori random process $X^{(1,1)}(i) = X(i/x_1(1))$:

$$X^{(1,1)}(i) = X(i/x(1)) = M\left[X(i)\right] + \left(x(1) - M\left[X(1)\right]\right)\beta_{11}^{(1)}(i) +$$

$$+\sum_{\lambda=2}^{N} W_1^{(\lambda)}\beta_{11}^{(\lambda)}(i) + \sum_{v=2}^{i}\sum_{\lambda=1}^{N} W_v^{(\lambda)}\beta_{1v}^{(\lambda)}(i), \; i=\overline{1,I}. \tag{16}$$

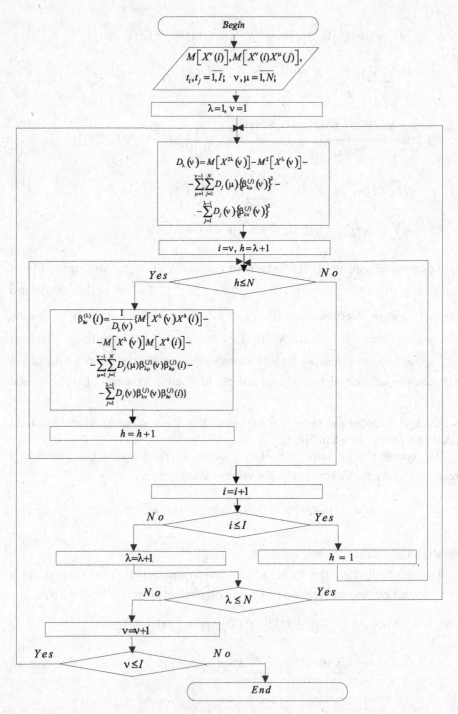

Fig. 1. Block-diagram of the procedure for calculating the parameters of the canonical decomposition

Applying expected value operation to (16) gives the optimal (by the minimum mean-square error of extrapolation criterion) estimation of future values of the random process $X(t)$ under the condition that for the determination of this estimation only value $x(1)$ is used:

$$m_x^{(1,1)}(1,i) = M\left[X(i/x(1))\right] = M\left[X(i)\right] + \left(x(1) - M\left[X(1)\right]\right)\beta_{11}^{(1)}(i),\ i = \overline{1,I}. \qquad (17)$$

Considering that coordinate functions $\beta_{hv}^{(\lambda)}(i)$, $\lambda, h = \overline{1,N}$; $v, i = \overline{1,I}$ are determined by the minimum mean-square error of the approximation in the intervals between arbitrary values $X^{\lambda}(v)$ and $X^h(i)$, expression (17) can be generalized in case of the prediction of higher order parameters $x^h(i)$, $h = \overline{1,N}$, $i = \overline{2,I}$:

$$m_x^{(1,1)}(h,i) = M\left[X^h(i/x(1))\right] = M\left[X^h(i)\right] + \left(x(1) - M\left[X(1)\right]\right)\beta_{h1}^{(1)}(i),\ i = \overline{1,I}. \qquad (18)$$

The usage of $w_1^{(2)}$ in (16) gives a canonical expansion of a posteriori process $\{X^{(1,2)}\} = X(i/x(1), x(1)^2)$:

$$X^{(1,2)}(i) = X(i/x(1), x(1)^2) = M\left[X(i)\right] + \left(x(1) - M\left[X(1)\right]\right)\beta_{11}^{(1)}(i) + \qquad (19)$$

$$+\left(x^2(1) - \left(x(1) - M\left[X(1)\right]\right)\beta_{21}^{(1)}(1)\right)\beta_{11}^{(2)}(1) + \sum_{\lambda=3}^{N} W_1^{(\lambda)}\beta_{11}^{(\lambda)}(i) + \sum_{v=2}^{i}\sum_{\lambda=1}^{N} W_v^{(\lambda)}\beta_{1v}^{(\lambda)}(i),\ i = \overline{1,I}.$$

Applying the operation of evaluation of the mathematical expection, that uses expression (18), to the equation (19), we receive a model of extrapolation of the investigated realization of the random process on two values $x_1(1), x_1(1)^2$:

$$m_x^{(1,2)}(h,i) = M\left[X^h(i/x(1), x(1)^2)\right] = m_x^{(1,1)}(h,i) + \left[x^2(1) - m_x^{(1,1)}(2,i)\right]\beta_{11}^{(2)}(1),\ i = \overline{1,I}. \qquad (20)$$

This generalization of the approach allows to pattern the algorithm of prediction for the arbitrary number of the known values $x^{\mu}(j)$, $t_j = \overline{1,k}$, $\mu = \overline{1,N}$:

$$m_x^{(\mu,l)}(h,i) = \begin{cases} M\left[X^h(i)\right], \mu = 0 \\ m_x^{(\mu,l-1)}(h,i) + \left(x^l(\mu) - m_x^{(\mu,l-1)}(l,\mu)\right)\varphi_{h\mu}^{(l)}(i),\ l \neq 1 \\ m_x^{(\mu-1,N)}(h,i) + \left(x^l(\mu) - m_x^{(\mu-1,N)}(l,\mu)\right)\varphi_{h\mu}^{(l)}(i),\ l = 1 \end{cases} \qquad (21)$$

The parameter $m_x^{(\mu,l)}(h,i) = M\left[X^h(i)/x^v(j),\ j = \overline{1,\mu-1};\ v = \overline{1,N};\ x^v(\mu), v = \overline{1,l}\right]$ for $h = 1$, $l = N$, $\mu = k$, calculated using (21), is the undisplaced optimal estimation $m_x^{(k,N)}(1,i)$ of the future value $x(i), i = \overline{k+1,I}$, where, to determine this estimation, values $x^v(j)$, $v = \overline{1,N}$, $j = \overline{1,k}$ are used, i.e. the results of measuring of the random process $X(t)$ in points t_j, $j = \overline{1,k}$ are known.

Block-diagram of the calculation procedure of the future values of the random process by the algorithm (21) is shown in Fig. 2.

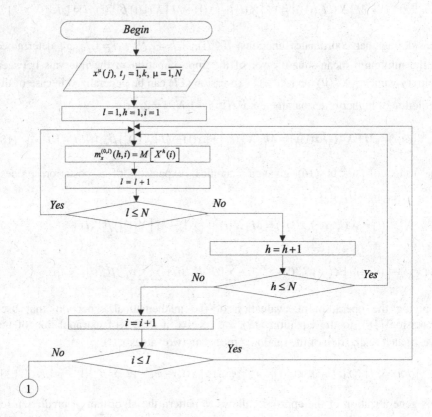

Fig. 2. Block-diagram of the calculation procedure of the future values of the random process by algorithm (21)

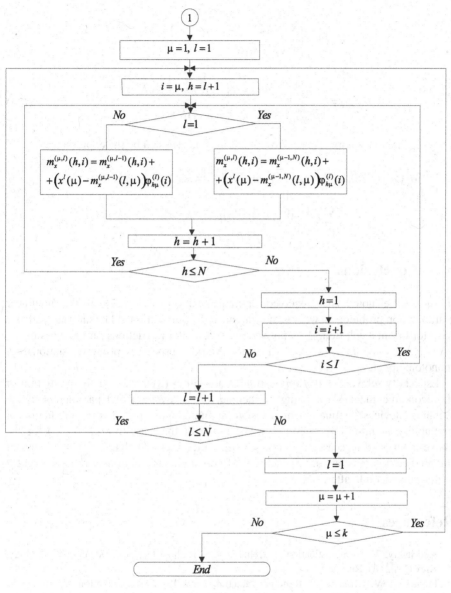

Fig. 2. (*continued*)

Expression (21) for the estimation of $m_x^{(k,N)}(1,i)$ can be transformed to the next simple form:

$$m_x^{(k,N)}(1,i) = M\left[X(i)\right] + \sum_{j=1}^{k}\sum_{v=1}^{N}\left(x^v(j) - M\left[x^v(j)\right]\right)S_{((j-1)N+v)}^{(kN)}((i-1)N+1), \quad (22)$$

$$\text{where } S_{\lambda}^{(\alpha)}(\xi) = \begin{cases} S_{\lambda}^{(\alpha-1)}(\xi) - S_{\lambda}^{(\alpha-1)}(\alpha)\gamma_k(i), & \lambda \le \alpha\text{-}1 \\ \gamma_\alpha(\xi), & \lambda = \alpha \end{cases}; \tag{23}$$

$$\gamma_\alpha(\xi) = \begin{cases} \beta_{1,[\alpha/N]+1}^{(\text{mod}_{N-1}(\alpha))}([\alpha/N]+1), & \text{for } \xi \le kN \\ \beta_{1,[\alpha/N]+1}^{(\text{mod}_{N-1}(\alpha))}(i), & \text{if } \xi = (i-1)N+1 \end{cases}. \tag{24}$$

Thus, the mean –square error of extrapolation is determined by the expression:

$$M\left[\left\{X(i/x^{\nu}(j), \nu=\overline{1,N}, j=\overline{1,k})-m_x^{(k,N)}(1,i)\right\}^2\right] = M\left[X^2(i)\right]-M^2\left[X(i)\right]-$$
$$-\sum_{j=1}^{k}\sum_{\nu=1}^{N}M\left[(W_j^{(\nu)})^2\right]\left\{\beta_{1j}^{(\nu)}(i)\right\}^2, \ i=\overline{k+1,I}. \tag{25}$$

3   Conclusions

It should be noted that received optimal solution (21), (22) of the nonlinear extrapolation problem, as well as the canonical decomposition (11), that was used as a base for optimal solution, does not set any substantial restrictions on the class of the investigated random processes (linearity, Markov processes property, stationarity, monotonicity etc.).

Especially relevant is the application of algorithm (21), (22) for the prediction of the economic indexes (planning of income, gross revenue) and parameters of the technical devices (estimation of the state at future moments of time, prediction of probability of no-failure operation). The usage of the algorithm (21), (22) for the above-mentioned forecast problems is based on the fact that changes of the values of the predicted parameters are realizations of non-stationary, nonlinear random process with a considerable aftereffect.

References

1. Kudritskiy, V.: Prognostication of Reliability of Radio Electronic Devices. Technology, Kiev (1982) (in Russian)
2. Pugachev, V.: Theory of Random Functions and Its Implementention. Physmathgis, Moscow (1962) (in Russian)
3. Kudritskiy, V., Atamanyuk, I., Ivashchenko, E.: Optimal Linear Extrapolation of the Realization of a Random Process with the Filtration of Errors of Correlated Measurings. J. Cybernetics and System Analysis 1, 99–107 (1995) (in Russian)
4. Kudritskiy, V., Atamanyuk, I.: Optimal Linear Filter-Extrapolator for Prognostication of Technical State of a Control Object. J. Technical Diagnosis and Undestroyed Control 1, 23–30 (1995) (in Russian)

5. Atamanyuk, I.: The Polinomial Canonical Curriculum of a Scalar Random Process of Changing Parameters of Radioelectronical Devices. J. Visnyk ZHITI 13, 99–110 (2000) (in Russian)
6. Atamanyuk, I., Kondratenko, Y.: Algorithm of Extrapolation of Nonlinear Casual Process on the Base of Its Canonical Decomposition. In: First International Workshop Critical Infrastructure Safety and Security, vol. 2, pp. 308–314. N.E.Zhukovsky National Aerospace University, Kharkiv (2011)

An Analytic Framework for Assessing Automobile Sales Loss from Recalls: A Toyota Case Study

Heechang Shin, Robert Richardson, and Oredola Soluade

Hagan School of Business, Iona College, New Rochelle, New York
715 North Avenue, New Rochelle, NY 10801-1890

{hshin,rrichardson,osoluade}@iona.edu

Abstract. Automobile companies recall vehicles whenever they are perceived to have defects that affect safety. Between September 2007 and February 2011, there were a total of seventy five recall incidents of various models of Toyota. The volume of recalls ranged from 7,000 vehicles to about 4 million. Based on the size of the recall, the impact on sales is measured in order to understand the customer's reaction under these circumstances. The well-publicized recalls in November 2009 and January 2010 are the focus of this study. The conclusion from the model indicates that major recalls have a negative impact on the market share with three-month lag, while minor recalls showed a positive impact on the market share after four months. Future studies are planned.

Keywords: Recall, Toyota, Sales.

1 Introduction

An automobile safety recall is based on defects identified in a motor vehicle. Such recalls are either conducted by the manufacturer or ordered by the National Highway Traffic Safety Administration (NHTSA). In either case, the manufacturer files a public report describing the safety-related defect, the vehicle population affected by the recall, the major events that resulted in the recall determination, a description of the remedy, and a schedule for the recall. NHTSA monitors each safety recall to ensure the manufacturers provide owners safe, free, and effective remedies in accordance with the Safety Act and Federal regulations [12].

The manufacturer estimates the direct cost of recall as a business expense. To measure the full impact of a recall, management must assess the major cost components of the repair – such as: (1) notification cost (including the development of a complete list of current owners from corporate files or state DMV records, the composing and mailing of the letters to each owner), (2) the cost of training technicians to make repairs, (3) labor cost and other expenses associated with the time required to resolve the issue, (4) cost of the parts - including manufacturing cost and shipping cost, (5) the management cost associated with scheduling and supervising staff, and (6) sales loss - the decline in sales as a result of the problem. These components of recall cost are calculated separately and then combined, because they

K.J. Engemann, A.M. Gil Lafuente, and J.M. Merigó (Eds.): MS 2012, LNBIP 115, pp. 88–97, 2012.
© Springer-Verlag Berlin Heidelberg 2012

are independent and vary widely from one company to another [11]. The notification, training, labor, parts, and management expenses are identifiable out-of-pocket costs that are derived directly from historical financial records available to management. In contrast, the lost sales must be estimated, since they represent an opportunity cost not derived from financial statements. A recall has a negative impact on sales, since there are fewer customers inclined to purchase an automobile from a manufacturer that is producing a defective product. The decline in sales is measured by assessing the company's market share or number of cars sold.

A recent literature review [8] identified several studies on product recalls and sales. Early studies [20, 7] found that only the most severe recalls influenced customer demand for a new car. Severe problems are defined as those that involve fires or loss of control of the vehicle. Reilly and Hoffer [14] showed that the competition's sales increased after a competitor had a recall.

Recently, Toyota recalled 5.77 million vehicles for floor mat issues and 4.45 million vehicles for the sticky-pedal fix during year 2009 and 2010 [15] after reports that several vehicles experienced unintended acceleration. These recalls resulted from a car accident on August 28, 2009 which killed four people including the driver, an officer with the San Diego California Highway Patrol [1]. Before the accident, his wife called 911 stating that she and three others were in a car with an accelerator pedal stuck and traveling at more than 100 miles per hour. This accident led to intense media coverage. As a California Highway Patrol Officer, he was an experienced driver with special training. This casts doubt on whether or not driver error was the main cause of the recalls in Toyota's previous sudden acceleration cases.

This study focuses on Toyota recalls over the last five years and its economic impact on the performance of Toyota. The following research questions are addressed: (1) How does Toyota's market shares change over time? (2) Do market share trajectories differ prior to, and after the accident on August 28, 2009? (3) What is the effect of recalls on the company's market share? An analytic framework for estimating the critical opportunity cost associated with the loss in sales encountered as the result of a recall is presented. A model is developed to evaluate the sales loss derived for different-size recalls. Finally, a discussion of the application of the model, its limitations, and opportunities for further research is presented.

2   Toyota Recalls

On August 28, 2009, a two-car collision killed four people riding in a Lexus dealer-provided loaner car in San Diego, California [1]. NHTSA released a safety investigation report on October 25, 2009 finding that the accident vehicle was wrongly-fitted with all-weather rubber floor mats meant for the RX 400hSUV, and that these mats were not secured by either of the two retaining clips [4]. The report stated that the accelerator pedal's hinge did not allow for relieving obstructions. NHTSA investigators recovered the accident vehicle's accelerator pedal, which was still "bonded" to the SUV floor mat [4]. The report concluded that the pedal entrapment caused the unintended acceleration resulting in the accident. The accident prompted a massive recall on November 2, 2009, to correct a possible incursion of an incorrect or out-of-place front driver's floor mat into the foot pedal well.

However, there were still some crashes that were shown not to have been caused by floor mat incursion, but instead, by a possible mechanical sticking of the accelerator pedal causing unintended acceleration. The second recall began on January 21, 2010 to address this sticking accelerator pedal issue. Toyota recalled 2.3 million vehicles on that day [10].

In fact, this sudden acceleration has roots in the 2002 redesigned Toyota Camry sedan, which featured a new type of gas pedal. Instead of physically connecting to the engine with a mechanical cable, the new pedal used electronic sensors to send signals to a computer controlling the engine in order to increase fuel efficiency. That same technology migrated to other cars - including Toyota's luxury Lexus ES sedan [10]. By early 2004, NHTSA was getting complaints that the Camry and ES sometimes sped up without the driver hitting the gas, and an investigation was launched [10]. In 2005 and 2006, Safety Research & Strategies, a consumer-safety research firm, reported that the NHTSA received hundreds of reports of unintended acceleration involving Toyotas. On two occasions, Toyota filed responses arguing that no defect or trends could be found in the complaints [10]. On September 26, 2007, Toyota recalled Camrys and ES350s from 2007 and 2008 model years to get new floor mats which Toyota blamed for the acceleration problem. Reports continued indicating that it may not have resolved the issue: one major case was 2008's fatal crash in Michigan, where the driver had removed her floor mats days before the accident [10].

In addition to recalls related to the floor mat and acceleration pedals, the following recalls continued. On February 8, Toyota announced a recall of a total of 133,000 Prius models that received a software update for the brakes and its ABS system [17]. On February 12, 2010 Toyota recalled approximately 8,000 Tacoma trucks for potential front drive shaft issues, and the recall involved inspecting a drive shaft component which if cracked would be replaced [19]. On October 21, 2010 Toyota announced a recall of 750,000 vehicles in the U.S. for addressing brake fluid leakage from the master cylinder [18]. On November 9, 2011 Toyota announced a recall of 447,000 vehicles in the U.S for a steering problem caused by the misalignment of the inner and outer rings of the crankshaft pulley, which could cause a noise or the Check Engine light to illuminate; if this problem was not corrected, the power steering belt could fall off the pulley, which would cause a sudden loss of power assist [3].

3   Model Development

3.1   Dependent Variable: Product Sales

The sample data from Automotive News Data Center [13] includes Toyota's monthly sales and market share in the United States between January 2005 and February 2011. Consistent with prior studies of auto recalls [5, 7, 14, 16], *month* was used as the unit of analysis because a longer period makes it difficult to detect the effect of a recall event, since there are many other intervening events that could affect sales.

Market share was used as the dependent variable in this analysis because market shares, unlike sales unit, are not affected by temporal fluctuations such as economic cycles. The technique of multiplicative classical decomposition was incorporated in the model to remove the effect of seasonality on the market share. This was accomplished by extracting the seasonal (monthly) factors from the data, and then deseasonalizing the market share. This makes for a smoother fluctuation of the data, and minimizes the errors associated with the statistical projections of the model.

3.2   Independent Variable: Product Recall

Product recall information was obtained from the NHTSA reports provided by the U.S. Department of Transportation. Toyota experienced total of 75 recalls over the period of study, for an average of 254,554 cars potentially affected.[1] Figure 1 presents the total number of potentially affected cars by month between January 2005 and February 2011. The number of cars in a recall indicates the number of autos originally sold to consumers. Over time, this number is reduced by those destroyed in accidents. The total number of affected cars appear to fluctuate, with no clear linear trend over the period studied. Since market responses to a product recall may be different depending on the number of affected vehicles in the recall, the analysis separated the recalls into *major* events with more than 500,000 cars per recall, and *minor* recalls with less than 500,000 cars per recall.

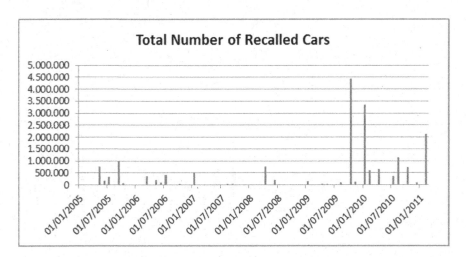

Fig. 1. Total Number of Recalled Cars for Toyota

3.3   Independent Variable: Accident Factor

As stated in the Introduction, on August 28, 2009, a collision killed four people riding in a dealer-provided loaner car in San Diego [1]. Since the accident, people's

---

[1] NHTSA database does not distinguish between the number of recalls for Lexus, Scion and Toyota. This figure represents the total number of affected cars for all three divisions.

perception of Toyota might have changed. An analysis of the sales determines whether this incident caused an immediate shift in market share, and how the trajectories of market share change prior to and after the incident.

3.4   Independent Variable: Media Coverage

Figure 2 presents the media coverage for Toyota recalls over the duration of the study. It is evident that media coverage surged after August 2009 and peaked by February 2010. The media coverage data was collected from the LexisNexis Database [9]. An interesting observation is that there were several instances where the number of recalled cars is over 500,000 before September 2009, but these recalls were not heavily covered by major media as observed in Figure 2.

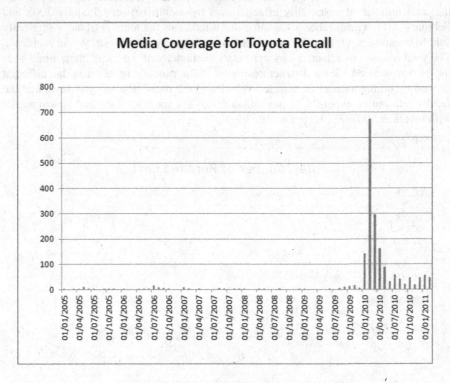

Fig. 2. Media Coverage for Toyota Recall

3.5   Model Specification: Development of the Mathematical Model

The underlying assumption, on which this model is developed, is that market share is a function of product recall, accident factor, and media Coverage. However, because there is usually a delay between when a recall takes place and when its effect on market share is manifested, a lag was introduced into the model to account for any delay.

The model of Toyota's market share uses the following function:

$$Y_i = \pi_0 + \pi_1 TIME_i + \pi_2 POSTTIME_i + \pi_3 ACCIDENT_i$$
$$+ \pi_4 MEDIA\_COVERAGE_i + \pi_5 TIME_i \times RECALL_i \qquad (1)$$
$$+ \pi_6 POSTTIME_i \times RECALL_i + \epsilon_i$$

The model tests how market shares changes over time, i.e., what is the market share trajectory during the time of the study? The study focuses on testing to see if the market share trajectory prior to the accident of the off-duty patrol car officer, (i.e., prior to September 2009) differs from that of the post-accident (i.e., post September 2009) trajectory. For the analysis, two predictors $TIME_i$ and $POSTTIME_i$ are used: $TIME_i$ is the specific moment in the duration of study, and $POSTTIME_i$ clocks "TIME" since the incident of the crash on August 2009 in the same cadence as $TIME_i$, and $POSTTIME_i$ captures the "extra slope" of the post-accident event. $ACCIDENT_i$ refers to a dummy variable indicating 0 if $TIME_i$ is prior to September 2009, and 1 otherwise; in order to test if the trajectory of market share includes a discontinuity in elevation, not slope. For the pre-accident period (i.e., $ACCIDENT_i = 0$) Equation (1) becomes:

$$Y_i = \pi_0 + \pi_1 TIME_i + \pi_2 POSTTIME_i + \pi_4 MEDIA\_COVERAGE_i$$
$$+ \pi_5 TIME_i \times RECALL_i + \pi_6 POSTTIME_i \times RECALL_i + \epsilon_i \qquad (2)$$

and, for the post-accident period (i.e. $ACCIDENT_i = 1$), Equation (1) becomes

$$Y_i = (\pi_0 + \pi_3) + \pi_1 TIME_i + \pi_2 POSTTIME_i + \pi_4 MEDIA\_COVERAGE_i$$
$$+ \pi_5 TIME_i \times RECALL_i + POSTTIME_i \times RECALL_i + \epsilon_i \qquad (3)$$

Therefore, if π_3 is statistically significant, Toyota's market share trajectory would include a discontinuity in elevation, not slope. $MEDIA\_COVERAGE_i$ refers to the number of media coverage regarding the product recall. $RECALL_i$ refers to a vector of dummy variables indicating where Toyota experienced (*i*) a major product recall (1) or not (0) and (*ii*) a minor product recall (1) or not (0) during months $(i - 1)$ (i.e., one month lag to the market share) and $(i - 6)$ (i.e., six month lags to the market share), because there could be an effect of product recalls over a longer period. ϵ_i is the error term.

3.6   Results

Table 1 presents the results from the ordinary least squares estimates of market share model shown in Equation (1). Model 5 (i.e., Equation (4)) in Table 1 produces the best model fit. The effect of TIME and POSTTIME is significant for all models, indicating that there was a positive linear relationship between TIME and the market share prior to September 2009, but this market share trajectory changed to a negative linear relationship post September 2009. However, ACCIDENT shows non-significant result indicating that there was no elevation in market share post September 2009. Also, media coverage on product recalls do suffer market penalty (i.e., decreased market share) about -0.009% for each news event on a product recall. When Toyota issued a major product recall, this recall did suffer market penalty (i.e., decreased market share) of Toyota after three months from this recall, since the information on the recall requires time to reach potential buyers. However, minor recalls produced no damage. Rather, it increased the market share after four months of the minor recalls about .1% as Model 5 indicates. This effect may be due to such recalls being perceived by consumers as a signal of an automaker's diligence in attending to quality control issues, as they found and fixed a potential (but minor) problem. Thus, it might be the case that such minor recalls help Toyota maintain public confidence in their product [16].

Table 1. Estimates of Market Share Change

	Model	Coefficients		t	Sig.
		B	Std. Error		
1	(Constant)	11.564	.265	43.669	.000
	Time	.036	.006	5.823	.000
2	(Constant)	10.851	.228	47.655	.000
	Time	.069	.007	10.349	.000
	POSTTime	-.210	.030	-7.078	.000
3	(Constant)	10.811	.221	48.918	.000
	Time	.072	.007	10.945	.000
	POSTTime	-.197	.029	-6.706	.000
	MEDIA_COVERAGE	-.009	.004	-2.405	.019
4	(Constant)	10.816	.216	50.016	.000
	Time	.073	.006	11.273	.000
	POSTTime	-.176	.030	-5.806	.000
	MEDIA_COVERAGE	-.009	.004	-2.323	.023
	MAJOR_LAG3 * TIME	-.012	.006	-2.030	.046
5	(Constant)	10.800	.209	51.597	.000
	Time	.068	.007	10.296	.000
	POSTTime	-.171	.029	-5.798	.000
	MEDIA_COVERAGE	-.007	.004	-2.025	.047
	MAJOR_LAG3 * TIME	-.015	.006	-2.523	.014
	MINOR_LAG4 * TIME	.011	.005	2.393	.019

Model 5 is defined as

$$Y_i = \pi_0 + \pi_1 TIME_i + \pi_2 POSTTIME_i + \pi_4 MEDIA\_COVERAGE_i$$
$$+\pi_{5,1} TIME_i \times MAJOR\_RECALL_{i-3} + \pi_{5,2} TIME_i \times MINOR\_RECALL_{i-4} \quad (4)$$
$$+ \epsilon_i$$

The parameters, illustrated in Figure 3, are as follows:

- $\pi_0 = 10.8$: initial market share (on January 2005) is estimated as 10.8%.
- $\pi_1 = 0.068$: rate of change in market share for pre-accident before September 2009

- $\pi_2 = -.171$: slope differential pre-post accident on and after September 2009

- $\pi_4 = -.007$: rate of change in market share for media coverage in product recall of Toyota.

- $\pi_{5,i-3} = -.015$: rate of change in market share by major recall status in the previous 3 months.

- $\pi_{5,i-4} = .011$: rate of change in market share by minor recall status in the previous 4 months.

This model suggests that the initial market share on January 2005 of Toyota is $\pi_0 = 10.8\%$, and it continues to increase until September 2009 with the monthly rate of change in market share $\pi_1 = 0.068\%$. However, on and after September 2009, the rate of change in market share became $\pi_1 + \pi_2 = 0.068 - .171 = -.103$. In other words, the market share of Toyota is increased about 0.068% per each month since January 2005, but after September 2009, it is decreased about .103% monthly. Also, for each media mention of a recall, the market share is dropped about $\pi_4 = -.007\%$. $\pi_{5,i-3}$ shows that the rate of change in market share of a major recall results in $\pi_1 + \pi_{5,i-3} = 0.068 - .015 = 0.053$, but the rate of change in market share of a minor recall is $\pi_1 + \pi_{5,i-4} = 0.068 + .011 = 0.079$.

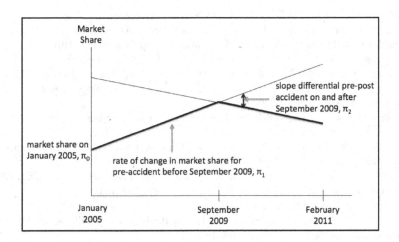

Fig. 3. Market Share Trajectories for Toyota

4 Conclusion

Recalling a vehicle can be costly for auto makers, both financially and in terms of reputational damage. In this paper, an analytic framework is presented for estimating the critical opportunity cost associated with the loss in sales encountered as the result of a recall. The study focuses on Toyota recalls over the last five years. The model suggested that the market share of Toyota was increasing about 0.068% per each month since January 2005, but after September 2009, it decreased about .103% monthly. Also, the influence of media coverage was investigated. The model suggested that for each media coverage of a Toyota recall, market share is dropped about.007%. Finally, we modeled and tested different market share lags of product recalls. The model suggests that the major recalls have negative impact on the market share with three month lag, while minor recalls showed positive impact on the market share after four months.

In future studies, the impact of Toyota recalls during 2009 and 2010 will be analyzed in greater detail. Toyota recalled 5.77 million vehicles for floor mat issues, and 4.45 million vehicles for the sticky-pedal fix during this period. An analysis of the economic impact of Toyota recalls on their competitors will be estimated. For example, what companies took advantage of Toyota's loss in sales during 2009 and 2010. Also, an examination of public image of Toyota and/or Japanese manufacturers resulting from the accident will be evaluated. In general, Japanese and German automakers possess higher reputations than the automakers from other countries [16]. Will the general public still considers Japanese cars as reliable compared to others after Toyota's massive recalls during 2009 and 2010?

References

1. 10News.com. 4 Killed In Fiery Santee Crash Believed Identified,
   http://www.10news.com/news/20609225/detail.html
2. The Associated Press, Toyota recalls 550,000 cars for steering issue (2010)
3. (November 9, 2011), http://cf.newsday.com/classifieds/cars/
   toyota-recalls-550-000-cars-for-steering-issue-1.3307313
4. Bensinger, K., Vartabedian, R.: New details in crash that prompted Toyota recall. Los Angeles Times (October 25, 2009),
   http://www.latimes.com/news/nationworld/nation/
   la-na-toyota-crash25-2009oct25,0,2288195.story
5. Borenstein, S., Zimmerman, M.: Market incentives for safe commercial airline operation. American Economic Review 78, 913–935 (1988)
6. Chernoff, A.: Toyota's big recall halts sales, production of 8 models. CNN Money (2010),
   http://money.cnn.com/2010/01/26/news/companies/
   toyota_recall/index.html
7. Crafton, S.M., Hoffer, G.E., Reilly, R.J.: Testing the impact of recalls on the demand for automobiles. Economic Inquiry, 19694–19703 (1981)
8. Etayankara, M., Bapuji, H.: Product recalls: a review of literature. In: Proceedings of the Annual Conference of ASAC, Strategic Division, vol. 30, pp. 44–59 (2009)
9. LexisNexis (2011), Database site: http://www.lexisnexis.com
10. Linebaugh, K., Searcey, D., Shirouzu, N.: Secretive Culture Led Toyota Astray. Wall Street Journal (February 10, 2010)
11. McDonald, K.M.: Do Auto Recalls Benefit the Public? Regulation, 12–17 (Summer 2008)
12. NHTSA: National Highway Traffic Safety Administration,
    http://www.odi.nhtsa.dot.gov/ivoq/
13. Raetz, M.: Automotive News Data Center, Automotive News, Detroit, MI (2011),
    http://www.autonews.com/section/DATACENTER

14. Reilly, R.J., Hoffer, G.E.: Will retarding the information flow on automotive recalls affect consumer demand? Economic Inquiry 21, 444–447 (1983)
15. Rechtin, M.: Toyota image surges on NASA study. Automotive News (January 17, 2012)
16. Rhee, M., Haunschild, P.R.: The liability off good reputation: a study of product recalls in the U.S. Automobile Industry. Organization Science 17(1), 101–117 (2006)
17. Saefong, M.: Toyota announces recall of 400,000 vehicles worldwide. Market Watch (February 9, 2010), http://www.marketwatch.com/story/toyota-plans-sai-lexus-hybrid-output-halt-report-2010-02-08
18. Tabuchi, H.: 1.5 Million Toyotas Recalled for Brake and Fuel Pump Problems (October 21, 2010), http://www.nytimes.com/2010/10/22/business/global/22toyota.html
19. Toyota Press. Toyota Announces Voluntary Recall on 8,000 2010 Model Year Tacoma 4WD Trucks to Inspect the Front Drive Shaft/Toyota (February 12, 2010), http://pressroom.toyota.com
20. Wynne, J.A., Hoffer, G.E.: Auto recalls: do they affect market share? Applied Economics 8, 157–163 (1976)

A Method for Uncertain Sales Forecast
by Using Triangular Fuzzy Numbers

Salvador Linares-Mustarós[1], José M. Merigó[2], and Joan Carles Ferrer-Comalat[1]

[1] Department of Business Administration, University of Girona, Campus de Montilivi s/n,
71017 Girona, Spain
{salvador.linares,joancarles.ferrer}@udg.edu
[2] Department of Business Administration, University of Barcelona, Av. Diagonal 690,
08034 Barcelona, Spain
jmerigo@ub.edu

Abstract. This work provides a tool to assess, from sales forecasts obtained by experts, the degree to which the sales forecast of a company is a specific value. It proposes different possibilities of triangular fuzzy number assignment whose vertices are obtained by aggregation functions that act on experts' forecast sales. The method offers too the option that the entrepreneurs or business owners remove or mitigate extreme values based in his personal opinion, thus enabling them to provide knowledge of the company not known to experts. With the possibility of allowing the entrepreneurs or business owners to incorporate or not extreme values, it opens the way to allow them to finally decide the characteristic function of the sales forecast, making the prediction an absolutely personal estimate.

Keywords: OWA operator, triangular fuzzy number, sales forecast.

1 Introduction

A basic problem confronting entrepreneurs who want to create a new company, or business owner who want to expand the market for their products, is determining the sales forecast for the first few years. A correct estimate is essential to identifying, among other things, the initial assets or the initial cash forecast.

A common approach to the problem is to assume that the firm will mimic the sales trend of a similar company in the sector. Company data can be obtained from the annual accounts of the registers, or from the databases created from them, such as the SABI database, and an estimate of future sales for the following year can be determined using statistical techniques such as a straight line extrapolation or another regression function.

Many factors can alter the forecast thus obtained and actual results are often very far from the estimated reserve. Among others, they include the location of new business, as for example more or less facilities for parking can cause an increase or a decrease in the amount of sales of the company, the sales of its competitors, the number of workers, the value of the new product, or the appearance or disappearance of substitutes.

K.J. Engemann, A.M. Gil Lafuente, and J.M. Merigó (Eds.): MS 2012, LNBIP 115, pp. 98–113, 2012.

It is therefore desirable that business owners or entrepreneurs make predictions about different companies, taking into account the various differentials of each in order to obtain a greater range of possibilities. The key advantage in getting them to analyze the results grouped by factors is, undoubtedly, the enormous increase of knowledge they get about the current market situation and its possible evolution.

A natural extension of the above method is to statistically infer the average of the forecast data used as the estimates obtained. In most cases, a proper study can give us an idea of future values adjusted in the event that there is no change in the current context. It seems sensible to think that predicting the future of a company looking at data of similar companies is not the best option in a situation of likely changes in the industry or in a situation with a sample of selected values, that may not be a hundred percent reliable.

In these cases a technique known as Delphi [1, 2, 3], which uses presupposed forecasts from experts, is employed to obtain an idea of the range of possible future prospects.

The main objective of this study is to develop a technique to determine, from the values predicted by a group of experts and based on the expectation of certain future events held by entrepreneurs or business owners, the possibility that a company reaches a certain value in its sales forecast. The entrepreneurs' or business owners' incorporation of subjectivity about the future has to allow them to interact with data using knowledge of their company not shown or considered by experts as well as to introduce into a system solution the intuitive trust in experts.

To achieve this objective, in the second section we review the main theoretical foundations and basic theory of fuzzy subsets, paying particular attention to a type of fuzzy subset known as triangular fuzzy number, which will be used to determine the degree of possibility of a sales forecast. In the third section we define the OWA aggregation functions that are likely to be incorporated into the sales forecast. The development of certain properties to be discussed in this section will ensure that the triangular fuzzy numbers to be proposed in the following section are defined correctly. Finally, in the fourth section we show how the tendency of business owner or entrepreneurs to expect that the future will bring favorable or unfavorable results can modify the estimated sales forecasts by experts, and we choose among different triangular fuzzy numbers, whose ends are determined by OWA operators, to obtain any characteristic functions for triangular fuzzy number "sales forecasting".

2   Fundamentals of Fuzzy Subsets

The intuitive definition of set, common in dictionaries, encyclopedias and mathematics books, is:

"A collection into a whole and certain other objects of our perception or our thought, called the elements of the set."

We identify a set expression by extension, listing each and every one of its elements, or by intension, indicating a property that serves to determine all the elements of the set, which is called the predicate.

If every member of set A is also a member of a referential set E, then A is said to be a subset of E, written A $\subseteq$ E (also pronounced A is contained in E). We can determine too by extension or by intension a subset of a set of reference.

From the construction of a subset we can show that every subset A of E has an associated reference function from the set reference on the set {0, 1} defined by:

$$\mu_A : E \rightarrow \{0, 1\}$$

(1)

$$x \rightarrow \mu_A(x) = \begin{cases} 1 & \text{if x is an element of set A} \\ 0 & \text{if x is not an element of set A} \end{cases}$$

and known as the characteristic function of the subset A in the reference set E.

Consider an example in order to understand the nomenclature:

Given the following characteristic function:

$$\mu_A : R \rightarrow \{0, 1\}$$

(2)

$$x \rightarrow \mu_A(x) = \begin{cases} 0 & si \ x < 100 \\ 1 & si \ 100 \leq x \leq 500 \\ 0 & si \ x > 500 \end{cases}$$

A can only be the closed interval [100,500] of the set of real numbers.

It is convenient to discuss the development of work when the whole universe is the set of real numbers. This is usually represented by a straight line and all of the subset A is represented by a series of dots or colored intervals as illustrated by Figure 1.

E   A

Fig. 1. Representation of the closed interval [100,500] of the set of real numbers

The theory of fuzzy subsets was proposed by Zadeh [4]. Due to its strong mathematical formalization and its wide applicability in problems of engineering or social science, it has become a theory with great potential. The main idea of the theory is allowing the image of the characteristic function to take values in the interval [0,1]. Based on this representation, this work incorporates the sense of confidence of an entrepreneur or business owner into the possible values of a sales forecast proposed by experts.

Let us define the subsets we work with throughout this paper.

We define a fuzzy number as a fuzzy subset of the actual reference (E = R), such that the membership function satisfies the following conditions:

1)   There is at least one value x such that $\mu_{\tilde{A}}(x) = 1$ (normality condition)

2) $\mu_{\tilde{A}}$ is increased until the minimum value x of R such that $\mu_{\tilde{A}}(x) = 1$ and decreasing from the maximum x value of R such that $\mu_{\tilde{A}}(x) = 1$ (convexity condition). This condition is expressed formally by:

$$\mu_{\tilde{A}}(s) \geq \min\left(\mu_{\tilde{A}}(x_1), \mu_{\tilde{A}}(x_2)\right) \forall x_1, x_2 \in \Re \text{ and } \forall s \in [x_1, x_2] \tag{3}$$

One can see that the condition itself requires that if two values x_1, x_2 such that $\mu_{\tilde{A}}(x_1) = 1$ and $\mu_{\tilde{A}}(x_2) = 1$, the image of the characteristic function for any value between them also is the value 1

3) $\mu_{\tilde{A}}$ is a continuous or semicontinuous at each point.

A common way of representing a fuzzy number is to draw its characteristic function in a Cartesian coordinate system. Figure 2 shows an example of a graphical representation of a fuzzy number whose shape resembles a triangle. This type of fuzzy number, which can be represented by four rays, two of which are horizontal at 0, is called a triangular fuzzy number. It is usually used to represent magnitudes of very uncertain economic reality, because it leads to a very quick way to get an idea of the range of possible values to be obtained, as well as a maximum value of trust.

By geometric reasoning we can show that any triangular fuzzy number is a characteristic function of the following type:

$$\mu_{\tilde{A}}(x) = \begin{cases} 0 \text{ si } x < a \\ \dfrac{x-a}{b-a} \text{ si } a \leq x \leq b \\ \dfrac{c-x}{c-b} \text{ si } b \leq x \leq c \\ 0 \text{ si } x > c \end{cases} \tag{4}$$

The values of a, b and c, $a \leq b \leq c$, characterizing the triangular fuzzy number, allows to write the triangular fuzzy subset as:

$$\tilde{A} = (a, b, c) \tag{5}$$

The treatment of uncertainty with fuzzy triangular numbers is a common technique today.

Its application in Spain in operational forecasting originated in 1986 with the joint publication by Kaufmann and Gil Aluja [5].

The ideas presented by the authors to model reality with fuzzy triangular numbers still have enormous potential to generate new models of financial or accounting problems concerning the estimates of subjective uncertainty data [6, 7, 8, 9, 10, 11].

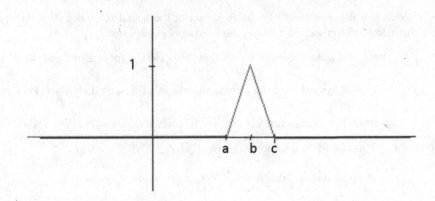

Fig. 2. We see that the representation allows us to quickly get both an idea of the range of possible values and identify our maximum level of presumption. In this sense, the vision offered by this system of representation of fuzzy numbers is still clarifying for teachers Arnold Kaufmann, Jaime Gil Aluja and Antonio Terceño [12]. In their view, a fuzzy number is the combination of two concepts: the confidence interval, linked to a concept of uncertainty, and the presumption level, linked to a concept of subjectivity.

3   OWA Operators and Aggregation Functions

When a problem requires several steps to focus decision making, it is common to use the arithmetic mean ($\overline{x}$) or the weighted arithmetic mean (WA). However, other interesting possibilities can be taken from the parameterization of the results sorted from minimum to maximum. They are known as ordered weighted average operators (OWA) [13] and multiple examples of applications exist [14, 15, 16, 17, 18, 19, 20].

An OWA operator is defined as a function OWA: $R^n \rightarrow R$, where n is the amount of data we want to add, such that W has an associated vector of length n, denoted by

$W = [w_1, w_2, ..., w_n]$ where w_j belongs to the interval [0,1] $\forall j$ and $\sum_{j=1}^{n} w_j = 1$

of the expression:

$$OWA(a_1, a_2, ..., a_n) = \sum_{j=1}^{n} w_j b_j \qquad (6)$$

and b_j is the j-th smallest value in the finite sequences a_i.

Example 1. Let W = [0.5, 0.3, 0.1, 0.1] that we can interpret as a time ordered series, the lowest value of this has an importance or weight of 50%, the second lowest 30% and the two largest 10% .
    Then

OWA(0.6, 1, 0.4, 0.2) = (0.5)(0.2) + (0.3)(0.4) + (0.1)(0.6) + (0.1)(1) = 0.33
OWA (1, 0.6, 0.4, 0.2) = (0.5)(0.2) + (0.3)(0.4) + (0.1)(0.6) + (0.1)(1) = 0.33
OWA (0.2, 0.4, 1, 0.6) = (0.5)(0.2) + (0.3)(0.4) + (0.1)(0.6) + (0.1)(1) = 0.33

One can argue that, as a result of changes in the series, the OWA does not depend on the function of permutations in the arguments, as shown in the example. This causes the items that add a_i to not be associated with weight w_i, but weight w_j is associated with the position j in the ordered series.

We note three special cases according to the weighting function [13]:

1. OWA^* if we take $W = (1,0,...,0)$ and consequently $OWA^* = Max(a_i)$.
2. OWA_* if we take $W = (0,0,...,1)$ and consequently $OWA_* = Min(a_i)$.
3. OWA_{ave} if we take $W = \left(\dfrac{1}{n},\dfrac{1}{n},...,\dfrac{1}{n}\right)$ and consequently $OWA_{ave} = \overline{x}(a_i)$.

OWA operators are a particular type of function used to calculate the mean or average that meets characteristic properties that define a family of functions known as aggregate functions.

We say that a function $f : [0,\infty)^n \rightarrow [0,\infty)$ is an aggregate function if it satisfies the following three conditions:

i) $f(0,0,...,0) = 0$

ii) $\displaystyle\lim_{x_1 \rightarrow \infty, x_2 \rightarrow \infty,...,x_n \rightarrow \infty} f(x_1,x_2,...,x_n) = \infty$

iii) $x_i \leq y_i$ for all $i \in \{1,2,...,n\}$ implies
$$f(x_1,x_2,...,x_n) \leq f(y_1,y_2,...,y_n)$$

We can see that we restrict ourselves to the interval $[0,\infty)$ because the sales can not be negative.

It is trivial to verify that the function OWA meets all three conditions and are therefore aggregation functions.

The conditions allow us to ensure consistency in the construction of triangular fuzzy numbers to forecast sales.

Finally, to close this section, we express another property necessary to ensure the perfect construction of such triangular fuzzy numbers:

$\forall$ OWA: $[0,\infty)^n \rightarrow [0,\infty)$ and $\forall x \in [0,\infty)^n$ is true:

$$OWA^*(x) \leq OWA(x) \leq OWA_*(x) \qquad (7)$$

4   Proposal for the Creation of a Triangular Fuzzy Number Concerning Sales Forecast from the Opinions of an Expert Group

Imagine a series of sales forecasts obtained by a group of experts and denoted by the set of positive real numbers or zeros $a_1, a_2,..., a_n$. Figure 3 shows an example of a forecast made by the set of positive real numbers or zeros.

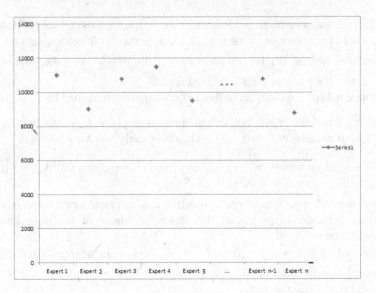

Fig. 3. Example of sales forecast by a group of n experts

A first and obvious option provided by a triangular fuzzy number for the sales forecast using the number given by experts is obtained from the following three values of aggregation: $OWA_*(a_1, a_2,..., a_n), OWA_{ave}(a_1, a_2,..., a_n), OWA^*(a_1, a_2,..., a_n)$:

$$\widetilde{P} = (OWA_*(a_1, a_2,..., a_n), OWA_{ave}(a_1, a_2,..., a_n), OWA^*(a_1, a_2,..., a_n)) \qquad (8)$$

whose characteristic function can be written as follows:

$$\mu_{\widetilde{P}}(x) = \begin{cases} 0 & si\ x < b_1 \\[2mm] \dfrac{x - b_1}{\dfrac{1}{n}\sum\limits_{j=1}^{n} b_i - b_1} & si\ b_1 \leq x \leq \dfrac{1}{n}\sum\limits_{j=1}^{n} b_i \\[4mm] \dfrac{b_n - x}{b_n - \dfrac{1}{n}\sum\limits_{j=1}^{n} b_i} & si\ \dfrac{1}{n}\sum\limits_{j=1}^{n} b_i \leq x \leq b_n \\[4mm] 0 & si\ x > b_n \end{cases} \qquad (9)$$

Note than in equation (9) we represented equation (8) as

$$\widetilde{P} = (b_1, \frac{1}{n}\sum_{j=1}^{n} b_i, b_n) \qquad (10)$$

It seems sensible to use the first option when the business owner or entrepreneur considered all the possible outcomes predicted by experts to be valid.

Now, imagine an unrestricted forecast, like the example in Figure 4.

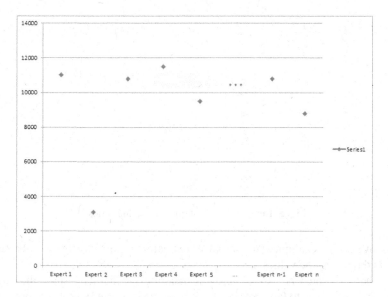

Fig. 4. Extreme values far from most predictions

In principle, one might think that the more optimistic end value is a value that could be discarded because it is far from the other values and indicates an extreme view that is difficult to achieve and not shared by the vast majority of the group. If the entrepreneur or business owner does not consider such provision valid, either because they think it can not reach that sales figure, or because they rely on the opinion of the expert, the fuzzy number of sales forecasting should not directly incorporate the value. In the event that all values are not considered valid, various different proposals are presented.

The first proposal we make is the use of quartiles to remove these extremely remote values.

From the series, we can calculate the lower quartile, denoted q1, that corresponds to the value that leaves below it 25% of the data of the series.

Analogously we calculate the upper quartile, denoted q3, that corresponds to the value that leaves above it 25% of the data of the series.

Because values of the quartiles need not correspond with any of the values of a series, or with representatives by segments such and as you can see in Figure 5.

In order to eliminate extreme values that from now on will be called outlier values in the first and second proposal, we can choose a classic statistical method of removing the values of the quartiles located at a distance greater than 1.5 times the distance between the two quartiles referred to as the interquartile range (IQR).

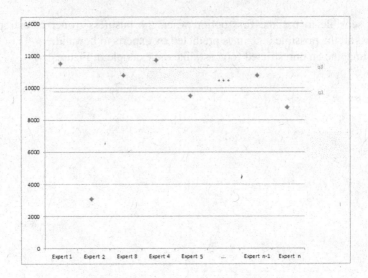

Fig. 5. Representation of quartiles in the series

Thus by way of example, we can have a situation similar to figure 6 that would eliminate three extreme values.

Note that one must be very careful in handling the outliers, since they do not necessarily correspond to large errors, but may be possible but improbable opinions of an expert.

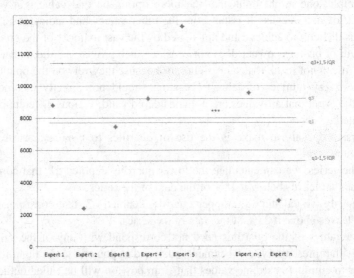

Fig. 6. Example of forecast values with three extremely remote values

With these outlier values of the series eliminated, say the first s and last t, we can choose to create the triangular fuzzy number:

$$\widetilde{P}_{IQRx} =(OWA_*(b_{s+1}, b_{s+2},..., b_{n-t-1},b_{n-t}), OWA_{ave}(b_{s+1}, b_{s+2},..., b_{n-t-1},b_{n-t}),$$

$$OWA^*(b_{s+1}, b_{s+2},..., b_{n-t-1},b_{n-t})) \tag{11}$$

whose characteristic function can be written in the following manner:

$$\mu_{\widetilde{P}_{IQRx}}(x) = \begin{cases} 0 & si\ x < b_{s+1} \\[2ex] \dfrac{x - b_{s+1}}{\dfrac{1}{n-s-t}\sum\limits_{i=s+1}^{n-t} b_i - b_{s+1}} & si\ b_{s+1} \le x \le \dfrac{1}{n-s-t}\sum\limits_{i=s+1}^{n-t} b_i \\[3ex] \dfrac{b_{n-t} - x}{b_{n-t} - \dfrac{1}{n-s-t}\sum\limits_{i=s+1}^{n-t} b_i} & si\ \dfrac{1}{n-s-t}\sum\limits_{i=s+1}^{n-t} b_i \le x \le b_{n-t} \\[3ex] 0 & si\ x > b_{n-t} \end{cases} \tag{12}$$

The second proposal presented here is not to lose the position information from expert group estimates. To do this, instead of eliminating outlier values, we will move the outliers values to the extremes of the range of values not considered outliers. The second option is more correct if you completely trust the opinion of the experts who have made extreme forecasts. Figure 7 shows an example showing that we have not completely lost the position information from the predictions, just as in the previous case.

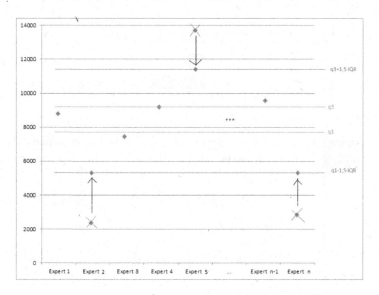

Fig. 7. The extreme values are moved to the upper and lower bounds

Suppose as before that first s and last t values of the ordered set -the outliers- are extreme values. Modifying these values of the series, we obtain a new set of cardinal set as the first cardinal set but in which the first s values have the value "$q_1 - 1,5 \cdot IQR$" and the last t "$q_3 + 1,5 \cdot IQR$". From the new series we propose the creation of triangular fuzzy number:

$$\tilde{P}_{IQR_{\uparrow}^{\downarrow}} = (OWA_{n*}, OWA_{n\ ave}, OWA_n^{\ *}) \tag{13}$$

whose characteristic function depends on the number of extreme values modified.

Thus, we distinguish four cases:

i) If we have not modified any value, since the extreme values are located within the zone of no modification, the characteristic function has the form:

$$\mu_{\tilde{P}_{IQR_{\uparrow}^{\downarrow}}}(x) = \begin{cases} 0 & si\ x < b_1 \\ \dfrac{x - b_1}{\left(\dfrac{1}{n}\sum\limits_{j=1}^{n} b_j\right) - b_1} & si\ b_1 \leq x \leq \dfrac{1}{n}\sum\limits_{j=1}^{n} b_j \\ \dfrac{b_n - x}{b_n - \left(\dfrac{1}{n}\sum\limits_{j=1}^{n} b_j\right)} & si\ \dfrac{1}{n}\sum\limits_{j=1}^{n} b_j \leq x \leq b_n \\ 0 & si\ x > b_n \end{cases} \tag{14}$$

ii) Suppose we have only increased the $s > 0$ earlier predictions. Then the characteristic function can be calculated using the following expression:

$$\mu_{\tilde{P}_{IQR_{\uparrow}^{\downarrow}}}(x) = \begin{cases} 0 & si\ x < q_1 - 1,5 \cdot RI \\ \dfrac{x - (q_1 - 1,5 \cdot RI)}{\dfrac{1}{n}\left(s \cdot (q_1 - 1,5 \cdot RI) + \sum\limits_{j=s+1}^{n} b_j\right) - (q_1 - 1,5 \cdot RI)} & si\ q_1 - 1,5 \cdot RI \leq x \leq \dfrac{1}{n}\left(s \cdot (q_1 - 1,5 \cdot RI) + \sum\limits_{j=s+1}^{n} b_j\right) \\ \dfrac{b_n - x}{b_n - \dfrac{1}{n}\left(s \cdot (q_1 - 1,5 \cdot RI) + \sum\limits_{j=s+1}^{n} b_j\right)} & si\ \dfrac{1}{n}\left(s \cdot (q_1 - 1,5 \cdot RI) + \sum\limits_{j=s+1}^{n} b_j\right) \leq x \leq b_n \\ 0 & si\ x > b_n \end{cases} \tag{15}$$

iii) Suppose we have only decreased the $t > 0$ latest predictions. Then the characteristic function can be calculated using the following expression:

$$\mu_{\tilde{P}_{IQR_{\uparrow}^{\downarrow}}}(x) = \begin{cases} 0 & si\ x < b_1 \\ \dfrac{x - b_1}{\dfrac{1}{n}\left(t \cdot (q_3 + 1,5 \cdot RI) + \sum\limits_{j=1}^{n-t} b_j\right) - b_1} & si\ b_1 \leq x \leq \dfrac{1}{n}\left(t \cdot (q_3 + 1,5 \cdot RI) + \sum\limits_{j=1}^{n-t} b_j\right) \\ \dfrac{(q_3 + 1,5 \cdot RI) - x}{(q_3 + 1,5 \cdot RI) - \dfrac{1}{n}\left(t \cdot (q_3 + 1,5 \cdot RI) + \sum\limits_{j=1}^{n-t} b_j\right)} & si\ \dfrac{1}{n}\left(t \cdot (q_3 + 1,5 \cdot RI) + \sum\limits_{j=1}^{n-t} b_j\right) \leq x \leq q_3 + 1,5 \cdot RI \\ 0 & si\ x > q_3 + 1,5 \cdot RI \end{cases} \tag{16}$$

iv) Suppose that we have increased the $s > 0$ earlier predictions and decreased the $t > 0$ latest predictions. Then the characteristic function can be calculated using the following expression:

$$\mu_{\tilde{P}_{IQR}\downarrow\atop t}(x) = \begin{cases} 0 & si \ x < q_1 - 1{,}5 \cdot RI \\[2mm] \dfrac{x - (q_1 - 1{,}5 \cdot RI)}{\frac{1}{n}\left(s \cdot (q_1 - 1{,}5 \cdot RI) + t \cdot (q_3 + 1{,}5 \cdot RI) + \sum\limits_{j=s+1}^{n-t} b_j\right) - (q_1 - 1{,}5 \cdot RI)} & si \ q_1 - 1{,}5 \cdot RI \le x \le \frac{1}{n}\left(s \cdot (q_1 - 1{,}5 \cdot RI) + t \cdot (q_3 + 1{,}5 \cdot RI) + \sum\limits_{j=s+1}^{n-t} b_j\right) \\[2mm] \dfrac{(q_3 + 1{,}5 \cdot RI) - x}{(q_3 + 1{,}5 \cdot RI) - \frac{1}{n}\left(s \cdot (q_1 - 1{,}5 \cdot RI) + t \cdot (q_3 + 1{,}5 \cdot RI) + \sum\limits_{j=s+1}^{n-t} b_j\right)} & si \ \frac{1}{n}\left(s \cdot (q_1 - 1{,}5 \cdot RI) + t \cdot (q_3 + 1{,}5 \cdot RI) + \sum\limits_{j=s+1}^{n-t} b_j\right) \le x \le p_{n-t} \\[2mm] 0 & si \ x > q_3 + 1{,}5 \cdot RI \end{cases} \qquad (17)$$

Finally, we have an interesting case, namely the option of allowing entrepreneurs or business owners to impose some bounds in the values. This option lets us enter their subjectivity to eliminate some expert opinion, and allows us to control measures in situations where there is hidden business information, such as, for example, the maximum production capacity.

For example, imagine the situation of Figure 8. Since our estimate of maximum production is less than that forecasted by some of the experts, it does not makes sense to include such values in our sales forecast.

Like the previous cases, we again have two possible ways to work with these extreme values: either removing or approaching the minimum and maximum bounds.

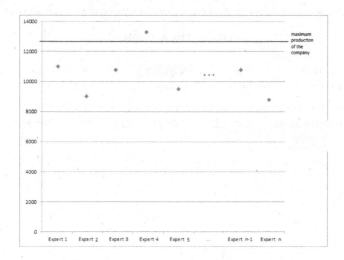

Fig. 8. In this figure shows an unattainable sales value for our production

Suppose that first s and last t of the ordered series are extreme values. Below or above the values we find the greatest lower bound and the least upper bound as determined by a business owner or entrepreneur, and we will denote them glb and LUB.

In the case of eliminating extreme values and in the case of not trusting the experts' prediction of extreme values, we propose the triangular fuzzy number obtained by applying the minimum, average, and maximum of the series obtained after removing outliers:

$$\tilde{P}_{minMAXx} = (OWA_*(\ p_{s+1}, p_{s+2},..., p_{n-t-1}, p_{n-t}),\ OWA_{ave}(p_{s+1}, p_{s+2},..., p_{n-t-1}, p_{n-t}),$$
$$OWA^*(\ p_{s+1}, p_{s+2},..., p_{n-t-1}, p_{n-t})) \qquad (18)$$

The characteristic function can be obtained from the following expression:

$$
\mu_{\tilde{P}_{glbLUBx}}(x) = \begin{cases} 0 & si \ x < b_{s+1} \\[2ex] \dfrac{x - b_{s+1}}{\dfrac{1}{n-s-t}\sum\limits_{j=s+1}^{n-t}b_j - b_{s+1}} & si \ b_{s+1} \leq x \leq \dfrac{1}{n-s-t}\sum\limits_{j=s+1}^{n-t}b_j \\[2ex] \dfrac{b_{n-t} - x}{b_{n-t} - \dfrac{1}{n-s-t}\sum\limits_{j=s+1}^{n-t}b_j} & si \ \dfrac{1}{n-s-t}\sum\limits_{j=s+1}^{n-t}b_j \leq x \leq b_{n-t} \\[2ex] 0 & si \ x > b_{n-t} \end{cases} \tag{19}
$$

However, if we rely on expert opinion and we want their views to be reflected in the sales forecast, we can move according to the extreme values corresponding to the glb or LUB values, thus obtaining a new series of cardinal set like the first but in which the first s values are now the greatest lower bound and the past t are now the least upper bound.

From the new series we propose the creation of a triangular fuzzy number:

$$
\tilde{P}_{glbLUB\updownarrow} = (\ OWA_*, OWA_{ave}, OWA^*\ ) \tag{20}
$$

whose characteristic function, just as in the previous case, also depends on the number of extreme values modified. So, we will distinguish four cases:

i)   If we have not already modified any value, those extreme values are located within the zone of no modification, the characteristic function has the form:

$$
\mu_{\tilde{P}_{glbLUB\updownarrow}}(x) = \begin{cases} 0 & si \ x < b_1 \\[2ex] \dfrac{x - b_1}{\left(\dfrac{1}{n}\sum\limits_{j=1}^{n}b_j\right) - b_1} & si \ b_1 \leq x \leq \dfrac{1}{n}\sum\limits_{j=1}^{n}b_j \\[2ex] \dfrac{b_n - x}{b_n - \left(\dfrac{1}{n}\sum\limits_{j=1}^{n}b_j\right)} & si \ \dfrac{1}{n}\sum\limits_{j=1}^{n}b_j \leq x \leq b_n \\[2ex] 0 & si \ x > b_n \end{cases} \tag{21}
$$

ii) Suppose we have only increased the $s > 0$ earlier predictions. Then the characteristic function can be calculated using the following expression:

$$
\mu_{\tilde{P}_{glbLUB\updownarrow}}(x) = \begin{cases} 0 & si \ x < glb \\[2ex] \dfrac{x - glb}{\dfrac{1}{n}\left(s\cdot glb + \sum\limits_{j=s+1}^{n}b_j\right) - glb} & si \ glb \leq x \leq \dfrac{1}{n}\left(s\cdot glb + \sum\limits_{j=s+1}^{n}b_j\right) \\[2ex] \dfrac{b_n - x}{b_n - \dfrac{1}{n}\left(s\cdot glb + \sum\limits_{j=s+1}^{n}b_j\right)} & si \ \dfrac{1}{n}\left(s\cdot glb + \sum\limits_{j=s+1}^{n}b_j\right) \leq x \leq b_n \\[2ex] 0 & si \ x > b_n \end{cases} \tag{22}
$$

iii) Suppose we have only decreased the $t > 0$ latest predictions. Then the characteristic function can be calculated using the following expression:

$$\mu_{\tilde{P}_{glbLUB\downarrow\uparrow}}(x) = \begin{cases} 0 & si\ x < b_1 \\[2ex] \dfrac{x - b_1}{\dfrac{1}{n}\left(\sum\limits_{j=1}^{n-t} b_j + t \cdot LUB\right) - b_1} & si\ b_1 \leq x \leq \dfrac{1}{n}\left(\sum\limits_{j=1}^{n-t} b_j + t \cdot LUB\right) \\[3ex] \dfrac{LUB - x}{LUB - \dfrac{1}{n}\left(\sum\limits_{j=1}^{n-t} b_j + t \cdot LUB\right)} & si\ \dfrac{1}{n}\left(\sum\limits_{j=1}^{n-t} b_j + t \cdot LUB\right) \leq x \leq LUB \\[3ex] 0 & si\ x > LUB \end{cases} \tag{23}$$

iv) Finally, suppose that we have increased the $s > 0$ earlier predictions and decreased the $t > 0$ latest predictions. Then the characteristic function can be calculated using the following expression:

$$\mu_{\tilde{P}_{glbLUB\downarrow\uparrow}}(x) = \begin{cases} 0 & si\ x < glb \\[2ex] \dfrac{x - glb}{\dfrac{1}{n}\left(s \cdot glb + \sum\limits_{j=s+1}^{n-t} b_j + t \cdot LUB\right) - glb} & si\ glb \leq x \leq \dfrac{1}{n}\left(s \cdot glb + \sum\limits_{j=s+1}^{n-t} b_j + t \cdot LUB\right) \\[3ex] \dfrac{LUB - x}{LUB - \dfrac{1}{n}\left(s \cdot glb + \sum\limits_{j=s+1}^{n-t} b_j + t \cdot LUB\right)} & si\ \dfrac{1}{n}\left(s \cdot glb + \sum\limits_{j=s+1}^{n-t} b_j + t \cdot LUB\right) \leq x \leq LUB \\[3ex] 0 & si\ x > LUB \end{cases} \tag{24}$$

5   Conclusions

This paper has developed a new method for incorporating the particular knowledge of an entrepreneur or business owner to calculate a sales forecast by experts. The method offers the option of removing or softening the extreme projections of a data set. For this, the business owner or entrepreneur must answer the following questions:

Do we consider all the results as valid? Are there upper and lower bounds for the results? Do we want to incorporate the extreme values?

The method offers the possibility of obtaining a triangular fuzzy number that represents the sales forecast, taking into consideration the response of each question according to scheme 1.

With the possibility of allowing the entrepreneurs or business owners to incorporate or not extreme values, it opens the way to allow them to finally decide the characteristic function of the sales forecast, making the prediction an absolutely personal estimate, as it should be, since the ultimate decision should be theirs alone.

We want to mention that while at work, in order to make the ideas more understandable, we used only the minimum, maximum and average functions, although there are other aggregation functions to be considered in the calculation of a triangular fuzzy number, like functions that allow averaging parts of the series, creating segments that emphasize pessimistic, most trusted, and optimistic areas.

Do we consider all the results as valid?				Propose of triangular fuzzy number
Yes			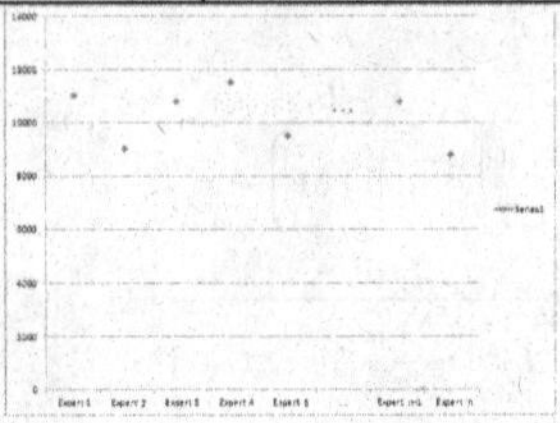	$\tilde{P}$
	Are there upper and lower bounds for the results?			
		Do we want to incorporate the extreme values?		
No	Yes	Yes	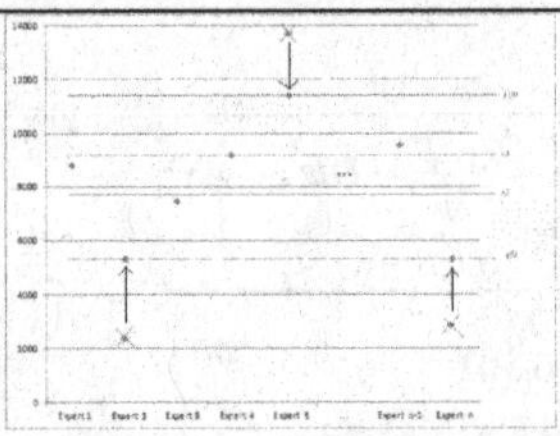	$\tilde{P}_{glbLUB\uparrow}^{\downarrow}$
		No	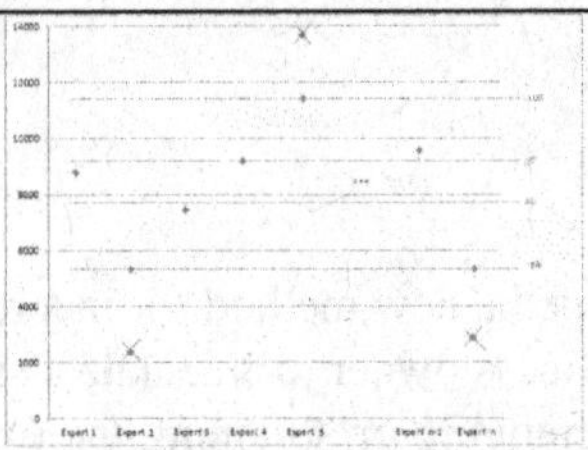	$\tilde{P}_{glbLUB\,X}$
	No	Yes	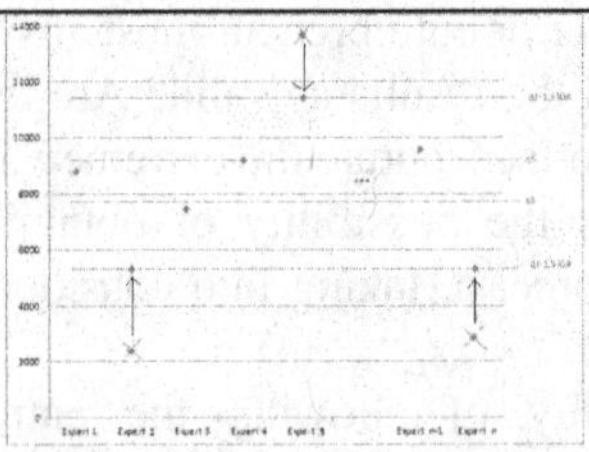	$\tilde{P}_{IQR\uparrow}^{\downarrow}$
		No	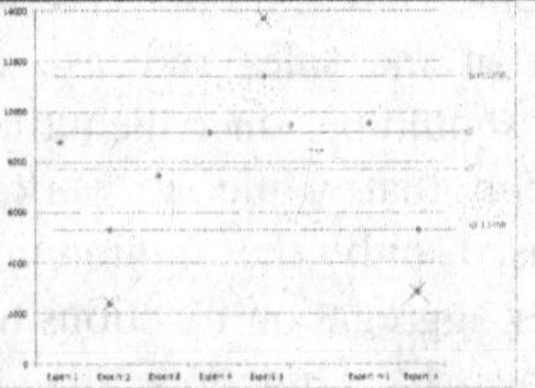	$\tilde{P}_{IQRx}$

Scheme 1. Method to soften or eliminate extreme values

References

1. Dalkey, N., Helmer, O.: An Experimental Application of the Delphi Method to the Use of Experts. Management Science 9(3), 458–467 (1963)
2. Linstone, H., Turoff, M. (eds.): The Delphi Method: Techniques and Applications. Addison-Wesley, Reading (1975)
3. Landeta, J.: El Método Delphi. Ariel, Barcelona (1999) (in Spanish)
4. Zadeh, L.A.: Fuzzy Sets. Information and Control 8, 338–353 (1965)
5. Kaufmann, A., Gil-Aluja, J.: Introducción de la Teoría de los Subconjuntos Borrosos a la Gestión de las Empresas. Milladoiro, Santiago de Compostela (1986) (in Spanish)
6. Gil-Aluja, J.: Fuzzy Sets in the Management of Uncertainty. Springer, Berlin (2004)
7. Gil-Lafuente, A.M.: Fuzzy Logic in Financial Analysis. Springer, Berlin (2005)
8. Keighobadi, J., Yazdanpanah, M., Kabganian, M.: An Enhanced Fuzzy H8 Estimator Applied to Low-Cost Attitude-Heading Reference System. Kybernetes 40(4), 300–326 (2011)
9. Merigó, J.M.: Fuzzy Decision Making Using Immediate Probabilities. Computers & Industrial Engineering 58(4), 651–657 (2010)
10. Merigó, J.M.: Fuzzy Multi-Person Decision Making With Fuzzy Probabilistic Aggregation Operators. International Journal of Fuzzy Systems 13(3), 163–174 (2011)
11. Merigó, J.M., Gil-Lafuente, A.M.: Fuzzy Induced Generalized Aggregation Operators and its Application in Multi-Person Decision Making. Expert Systems with Applications 38(8), 9761–9772 (2011)
12. Kaufmann, A., Gil-Aluja, J., Terceño, A.: Matemáticas para la Èconomía y la Gestión de Empresas. Aritmética de la incertidumbre, vol. 1. Foro Científico, Barcelona (1994) (in Spanish)
13. Yager, R.R.: On Ordered Weighted Averaging Aggregation Operators in Multi-Criteria Decision Making. IEEE Transactions on Systems, Man and Cybernetics 18, 183–190 (1988)
14. Beliakov, G., Pradera, A., Calvo, T.: Aggregation Functions: A Guide for Practitioners. Springer, Berlin (2007)
15. Karayiannis, N.: Soft Learning Vector Quantization and Clustering Algorithms Based on Ordered Weighted Aggregation Operators. IEEE Transactions on Neural Networks 11, 1093–1105 (2000)
16. Merigó, J.M.: New Extensions to the OWA Operator and its Application in Decision Making. PhD Thesis (2008) (in Spanish)
17. Xu, Z.S.: An Overview of Methods for Determining OWA Weights. International Journal of Intelligent Systems 20, 843–865 (2005)
18. Yager, R.R.: Families of OWA Operators. Fuzzy Sets and Systems 59, 125–148 (1993)
19. Yager, R.R.: Generalized OWA Aggregation Operators. Fuzzy Optimization and Decision Making 3, 93–107 (2004)
20. Yager, R.R., Kacprzyk, J.: The Ordered Weighted Averaging Operators: Theory and Applications. Kluwer Academic Publishers, Norwell (1997)

A Metaheuristic Approach for Supply Chain Network Design Problems

Uday Venkatadri[1], Soumen Bose[2], and Amir Azaron[3,4,*]

[1] Department of Industrial Engineering, Dalhousie University, Halifax, Nova Scotia, Canada
[2] Department of Industrial Engineering and Management, Indian Institute of Technology Kharagpur, Kharagpur, West Bengal, India
[3] Department of Financial Engineering and Engineering Management, Reykjavik University, Reykjavik, Iceland
[4] Michael Smurfit Graduate School of Business, University College Dublin, Carysfort Avenue, Blackrock, Co. Dublin, Ireland
amir@ru.is

Abstract. We develop a multi-objective stochastic programming model for supply chain design under uncertainty using a metaheuristic approach. This is a comprehensive model, which includes both the strategic and tactical levels. The uncertainty regarding demands, supplies, processing and transportation costs is captured by generating discrete scenarios with given probabilities of occurrence. To solve the problem, we use multi-objective simulated annealing and compare the results against the goal attainment technique. Numerical results show that the proposed metaheuristic approach is a very practical solution technique.

Keywords: Supply chain management, Metaheuristics, Stochastic programming, Multiple objective programming.

1   Introduction

In this paper, we focus on the strategic level of supply chain management. At this level, decisions are made on the number and location of facilities, their storage and handling technology, means of transportation, and decisions on subcontracting (Martel [1]). An interesting high-level problem at the strategic level is the multi-commodity distribution system design problem. It was first proposed by Geoffrion and Graves [2]. Many extensions to this problem have been considered (Martel [1], Martel and Venkatadri [3], Vidal and Goetschalckx [4]) and many versions have been solved in practice (Arntzen et al. [5] and Shapiro et al. [6]).

In this paper, we discuss the following extensions to the basic supply chain design problem:

1. The multi-objective nature of the problem.
2. Stochasticity in the supply chain modeled by scenarios which can take into account supplier reliability and production uncertainty.

---

* Corresponding author.

K.J. Engemann, A.M. Gil Lafuente, and J.M. Merigó (Eds.): MS 2012, LNBIP 115, pp. 114–122, 2012.
© Springer-Verlag Berlin Heidelberg 2012

These extensions were first proposed by Azaron et al. [7] who use the goal attainment method to solve a formulation of the extended problem. There have been other attempts in the literature to model stochastic aspects of the supply chain. For example, MirHassani et al. [8] present a two-stage multi-period stochastic model and solve their problem using Bender's decomposition. Tsiakis et al. [9] also considered a two-stage stochastic programming model for supply chain network design under demand uncertainty. The authors developed a large-scale mixed-integer linear programming model for this problem. Santoso et al. [10] integrated a sampling strategy with an accelerated Benders decomposition to solve supply chain design problems with continuous distributions for the uncertain parameters. Alonso-Ayuso et al. [11] proposed a branch-and-fix heuristic for solving two-stage stochastic supply chain design problems. However, the robustness of the decision to uncertain parameters is not considered in above studies.

Since the final mathematical model proposed in [7] is a large scale mixed-integer non-linear program, it was felt that developing a meta-heuristic approach would be useful. Moreover, the goal attainment technique used in [7] gives the same importance to the first and second stage decisions, with the trade-off between the objectives made in Phase I itself. However, in practice, the network configuration decisions are much more important than the flow decisions and we can decide the trade-off between the objectives in Phase II, once the uncertain parameters are revealed. To address these issues, we develop a multi-objective simulated annealing approach to solve the supply chain design problem.

2   Problem Formulation

The sets used to formulate the problem under consideration are as follows:

> L: Set of scenarios.
> N: Set of all nodes. This set consists of supplier, intermediate and demand
>   nodes in the supply chain.
> A: Set of all arcs.
> S: Set of all supply nodes.
> D: Set of all demand nodes.
> I: Set of all intermediate (production or warehouse) nodes.
> K: Set of products in the supply chain.
> A_i : Set of all nodes diverging from node $i \in I$.
> B_i : Set of all nodes converging into node $i \in I$.

We now define the various parameters and variables used for the model formulation. The parameters are:

> m_i : Capacity of facility in node $i \in I$, if it is built.
> c_i : Fixed cost of facility in node $i \in I$.
> r_i^k : Per-unit processing requirement at node $i \in I$ for product $k \in K$.
> p_l : Probability of occurrence of scenario $l \in L$.

h_{il}^k: Unit shortage cost for node $i \in D$ in scenario $l \in L$ for product $k \in K$.

f_{il}: Unit cost of expansion of facility in node $i \in I$ in scenario $l \in L$.

q_{ijl}^k: Unit sum of transportation and production costs for $(i, j) \in A$ & $i \neq j$, in scenario $l \in L$ for product $k \in K$.

Ω_l: Available budget in scenario $l \in L$.

d_{il}^k: Demand (in units) at node $i \in D$ in scenario $l \in L$ for product $k \in K$.

s_{il}^k: Supply (in units) for product $k \in K$ at node $i \in S$ in scenario $l \in L$.

o_i: Limiting expansion capacity (in units) of facility in node $i \in I$.

V: A very large positive value.

The model variables are:

$y_i = 1$, if the facility in node $i \in I$ is built; 0, otherwise.

x_{ijl}^k: Flow (in units) from node $i \in N$ to node $j \in N$ in scenario $l \in L$ for product $k \in K$.

z_{il}^k: Shortage (in units) at node $i \in D$ in scenario $l \in L$ for product $k \in K$.

e_{il}: Expansion of node $i \in I$ in scenario $l \in L$.

$u_l = 1$, if total cost in scenario $l \in L$ is greater than Ω_l. 0, otherwise.

$u_l' = 1$, if total cost in scenario $l \in L$ is less than Ω_l. 0, otherwise.

The three objectives to be minimized are as follows:

Objective 1: Expected total cost represented by expression 0.1.

$$\sum_{i \in I} c_i y_i + \sum_{k \in K} \sum_{i \in N} \sum_{j \in N} \sum_{l \in L} p_l \left(q_{ijl}^k x_{ijl}^k + h_{il}^k z_{il}^k + f_{il} e_{il} \right) \tag{0.1}$$

Objective 2: Variance represented by expression 0.2

$$\sum_{k \in K} \sum_{i \in N} \sum_{j \in N} \sum_{l \in L} p_l \left(A_{ijl}^k - \sum_{k \in K} \sum_{i \in N} \sum_{j \in N} \sum_{l \in L} p_l A_{ijl}^k \right)^2 \tag{0.2}$$

where, $q_{ijl}^k x_{ijl}^k + h_{il}^k z_{il}^k + f_{il} e_{il} = A_{ijl}^k$

Objective 3: Financial Risk represented by expression 0.3.

$$\sum_{l\in L} p_l u_l \qquad (0.3)$$

The set of constraints for our model is as follows (refer to Azaron et al. [7] for more details about the mathematical model):

$$\sum_{i\in B_j} x_{ijl}^k - \sum_{p\in A_j} x_{jpl}^k = 0 \;\; \forall \;\; j\in I,\, l\in L\, \&\, k\in K \qquad (1)$$

$$\sum_{i\in B_j} x_{ijl}^k + z_{jl}^k \geq d_{jl}^k \;\; \forall \;\; j\in D,\, l\in L\, \&\, k\in K \qquad (2)$$

$$\sum_{j\in A_i} x_{ijl}^k \leq s_{il}^k \;\; \forall \;\; i\in S,\, l\in L\, \&\, k\in K \qquad (3)$$

$$\sum_{k\in K} r_j^k \left(\sum_{i\in B_j} x_{ijl}^k\right) \leq m_j y_j + e_{jl} \;\; \forall \;\; j\in I\, \&\, l\in L \qquad (4)$$

$$e_{il} \leq o_i y_i \;\; \forall \;\; i\in I\, \&\, l\in L \qquad (5)$$

$$\sum_{i\in I} c_i y_i + \sum_{k\in K}\sum_{i\in N}\sum_{j\in N}\left(q_{ijl}^k x_{ijl}^k + h_{il}^k z_{il}^k + f_{il}e_{il}\right) - \Omega_l \leq Vu_l \;\; \forall 1 \qquad (6)$$

$$\sum_{i\in I} c_i y_i + \sum_{k\in K}\sum_{i\in N}\sum_{j\in N}\left(q_{ijl}^k x_{ijl}^k + h_{il}^k z_{il}^k + f_{il}e_{il}\right) - \Omega_l \geq Vu_l' \;\; \forall 1 \qquad (7)$$

$$u_l + u_l' \leq 1 \;\; \forall\, l\in L \qquad (8)$$

$$x_{ijl}^k \geq 0 \;\; \forall \;\; (i,j)\in A,\, l\in L\, \&\, k\in K \qquad (9)$$

$$z_{il}^k \geq 0 \;\; \forall \;\; i\in D,\, l\in L\, \&\, k\in K \qquad (10)$$

$$y_i \in \{0,1\} \;\; \forall \;\; i\in I \qquad (11)$$

$$u_l \in \{0,1\} \;\; \forall \;\; l\in L \qquad (12)$$

$$u_l' \in \{0,1\} \;\; \forall \;\; l\in L \qquad (13)$$

3   Solution Strategy

3.1   Single Objective Solution

If each objective is treated separately, the problems with objectives 0.1 and 0.3 are binary-integer programs. In these cases, small problems may be directly solved using mixed-integer solvers such as Cplex and Lindo. For bigger problems, the alternatives are to use either exact methods such as decompositions or approximate methods such as Simulated Annealing (SA) or TABU search. The single objective problem with objective 0.2 is non-linear. It may be solved using a non-linear solver such as Lingo or Minos. This becomes a challenge for larger problems because of the binary integer variables.

3.2   Multi-objective Solution Using Goal Attainment

We are interested in finding the Pareto frontier for the problem across the three objectives. Azaron et al. [7] used goal attainment technique to solve the problem. The main idea with goal attainment is to set goals and weights for the three objective functions. Using this method, smaller problems can be solved using Lingo assuming that a goal and weight has been specified for each value. It is our computational experience that Lingo does not work well for large problems because of the binary-integer variables. Another problem with goal attainment is that the designer has to solve several problems with different goals and weights in order to develop the Pareto frontier for the problem, and this can be quite cumbersome. Finally, the trade-off between the objectives is done in Phase I, using goal attainment technique. So, there will be no easy separation of the first and second stage decisions and we will end up with a large set configuration decisions each connected to an associated set of flow decisions and we still don't know which configuration to select.

3.3   Multi-objective Solution Using Multi-Objective Simulated Annealing

In order to overcome the problems we faced with goal attainment, we propose using the multi-objective simulated annealing (MOSA) algorithm to solve the problem. MOSA works similarly to SA but is designed for problems with multiple objectives and automatically return an approximation to the Pareto frontier.

MOSA was first proposed by Suppaptinarm et al. [12]. The multi-objective simulated annealing methodology maintains an archive of mutually non-dominated solutions A_s found during the course of the search process. The algorithm starts with a single solution with high individual temperatures assigned to each of the objectives. A new solution is then generated by performing a neighbourhood search. In our case, the neighbourhood is defined by a string of predetermined binary integer variables for the build variables. Once the binary integer variables are set to either 0 or 1, we solve the constraint set for each of the three separate objectives. The problem with objectives 0.1 and 0.3 in this case are simply linear programs that can be solved easily for very large cases using commercial packages such as Cplex or Lindo. For the second objective, we use Lingo to solve the non-linear problem with linear constraints. We found that this approach is feasible for fairly large problems. The three objective function values for each predetermined string of 0/1 variables (first-stage decision variables) is the solution for the string. A neighbouring string has one 0 value set to 1 or vice versa. The solution generated is archived in A_s if it is not dominated by any of the solutions in A_s. If it dominates any solution in A_s, then that solution is removed from the list. If the solution is not archived, then we still move to it with probability P, where P is defined as below:

$$P = \min\left\{1, \prod_{i=1}^{3} \exp\left(\frac{-\Delta_i}{T_i}\right)\right\} \qquad (14)$$

In the above equation, the probability of acceptance of the solution for dominated solutions is dependent on the difference of the current and the new solution objective function values (Δ_i), and on the current temperature of the objectives (T_i). The first update of the temperatures takes place after N_{T1} number of pre-specified iterations. During the first temperature update, the temperature T_i associated with objective i is set equal to the standard deviation of the accepted solution values for that objective.

First Temperature Update

$$T_i = \sigma_i, \qquad \forall i \qquad (15)$$

Further temperature updates are performed every N_{T2} number of iterations or after N_A number of solutions have been accepted since the last update.

Subsequent Temperature Updates

$$T_i^{'} = \alpha_i T_i, \qquad \forall i \qquad (16)$$

where $\alpha_i = \max\left\{0.5, exp\left[\frac{-0.7 T_i}{\sigma_i}\right]\right\}$.

The use of the probability function P and the above mentioned temperature criteria ensure that no scaling of objectives is required.

The algorithm also utilizes a return to base strategy, in which the current solution is replaced with a solution from the mutually non-dominant solution set. The use of the return to base strategy results in the intensification and diversification of the search process. The first return to base occurs at the same time as the first temperature update, i.e., after N_{T1} iterations. For the first return to base, all the solutions in the solution archive, A_s are in the candidate list A_{Ii} i.e., all the solutions in the archive have equal probability of being set as the current solution. As the search progresses, the size of candidate list A_{li} is reduced as $A_{li} = \phi_{li} A_s, i \geq 1$, where $\phi_{li} = r_l \phi_{l_{i-1}}$ and $0 < r_l < 1$. Also, the solutions with minimum value of objective function for each of the objective and a number of most isolated solutions are entered into the candidate list. The degree of isolation of a solution is calculated as follows:

Degree of Isolation

$$I(X_j) = \sum_{i=1}^{Card(Archive)} \sum_{k=1}^{3} \left(\frac{f_k(X_i) - f_k(X_j)}{f_{k\,max} - f_{k\,min}} \right)^2 \tag{17}$$

where, $I(X_j)$ is the degree of isolation of the solution X_j.

Card(Archive) is the total number of solutions in solution archive. $f_{k\,max}$ and $f_{k\,min}$ are the maximum and minimum value of the kth objective function term, among the solutions in the solution archive.

As the search progresses, the return to base is performed after N_{Bi} ($N_{Bi} = r_B N_{Bi-1}, i > 1, 0 < r_B \leq 1$) iterations or N_A acceptances, whichever condition is satisfied earlier. However, if the new value of N_{Bi} is less than the pre-specified lower bound (A_c), then N_{Bi} is set equal to A_c. The algorithm terminates after a fixed number of iterations.

The following are the parameter settings we chose for running our MOSA implementation (details of these parameters are given in [12]):

1. Maximum MOSA runs: 10
2. No. of Iterations per run: 15
3. Iterations for First Temperature Update: 5
4. Iterations for Further Temperature Updates (after the First Temperature Update): 5
5. Accepted Solution Number Criteria for temperature updates: 4
6. Minimum Size of the Candidate List for performing first Return to Base: 3
7. Rate of decrease of no. of iterations require to perform subsequent Return to Base: 0.9
8. Reduction Rate of Candidate List for Return to Base Operation: 0.9.

4   Computational Results

We solved three problems using MOSA and goal attainment (GA). The first problem was a small scale problem representing a three tier supply chain with 3 supply, 4 intermediate, and 3 demand nodes. The second problem was an intermediate scale problem representing a three tier supply chain with 10 supply, 10 intermediate, and 9 demand nodes. The largest problem we solved was a three tier supply chain with 24 supply, 28 intermediate, and 27 demand nodes. All problems were solved on a PC Pentium IV 2.1-GHz processor using the Lingo 10 solver.

Based on the results, for a particular setting of goals and weights, GA took on an average 3 minutes to solve the small problem. It took nearly 6 minutes to build the Pareto frontier for the problem with MOSA (10 runs of 15 iterations per run), though the final non-dominated solution set was obtained in the 3rd iteration of run 1 itself.

For the intermediate problem, with a particular setting of goals and weights, GA took on an average 14 minutes to solve the problem. It took nearly 105 minutes to solve the problem with MOSA (i.e., 10 runs of 15 iterations per run). The final non-dominated solution set was obtained in the 3rd iteration of the 6th run, approximately 60 minutes from start.

For the large problem, Lingo was unable to return a solution using the GA method due to the increased complexity of the binary-integer non-linear problem due to the non-linearity in the second objective 0.2. Nevertheless, with the MOSA algorithm, we were able to get a good non-dominated solution set by the 9th iteration of the 2nd run (658 minutes from start). When MOSA was run longer, it too had to abort due to a memory overflow error caused by Lingo.

5   Conclusion

The proposed model in this paper accounts for the minimization of the expected total cost, the variance of the total cost and the financial risk in a multi-objective scheme to design a robust supply chain network. Then, a meta-heuristic approach was developed to solve the problem.

The basic difference between MOSA and the goal attainment is that the search may turn out to be a little more efficient since the goals and weights can be applied later, in the second stage. So, MOSA seems to be a conceptually more attractive solution technique than goal attainment.

References

1. Martel, A.: The Design of Production-Distribution Networks: A Mathematical Programming Approach. In: Geunes, J., Pardalos, P.M. (eds.) Supply Chain Optimization, pp. 265–306. Springer (2005)
2. Geoffrion, A.M., Graves, G.W.: Multicommodity Distribution System Design by Benders Decomposition. Management Science 20, 822–844 (1994)
3. Martel A., Venkatadri, U.: Optimizing Supply Network Structures under Economies of Scale. In: Proceedings of International Conference on Industrial Engineering and Production Management, Glasgow, United Kingdom (1999)
4. Vidal, C., Goetschalckx, M.: Strategic Production-Distribution Models: A Critical Review with Emphasis on Global Supply Chain Models. European Journal of Operational Research 98, 1–18 (1997)
5. Arntzen, B.C., Brown, G.G., Harrison, T.P., Trafton, L.L.: Global Supply Chain Management at Digital Equipment Corporation. Interfaces 25, 69–93 (1995)

6. Shapiro, J., Singhal, V., Wagner, S.: Optimizing the Value Chain. Interfaces 23, 102–117 (1993)
7. Azaron, A., Brown, K.N., Tarim, S.A., Modarres, M.: A Multi-Objective Stochastic Programming Approach for Supply Chain Design Considering Risk. International Journal of Production Economics 116, 129–138 (2008)
8. MirHassani, S.A., Lucas, C., Mitra, G., Messina, E., Poojari, C.A.: Computational Solution of Capacity Planning Models Under Uncertainty. Parallel Computing 26, 511–538 (2000)
9. Tsiakis, P., Shah, N., Pantelides, C.C.: Design of Multiechelon Supply Chain Networks under Demand Uncertainty. Industrial and Engineering Chemistry Research 40, 3585–3604 (2001)
10. Santoso, T., Ahmed, S., Goetschalckx, M., Shapiro, A.: A Stochastic Programming Approach for Supply Chain Network Design under Uncertainty. European Journal of Operational Research 167, 96–115 (2005)
11. Alonso-Ayoso, A., Escudero, L.F., Garin, A., Ortuno, M.T., Perez, G.: An Approach for Strategic Supply Chain Planning under Uncertainty Based on Stochastic 0-1 Programming. Journal of Global Optimization 26, 97–124 (2003)
12. Suppaptinarm, A., Seffen, K.A., Parks, G.T., Clarkson, P.J.: A Simulated Annealing Algorithm for Multiobjective Optimization. Engineering Optimization 33, 59–85 (2000)

Random Forests for Uplift Modeling: An Insurance Customer Retention Case

Leo Guelman[1], Montserrat Guillén[2], and Ana M. Pérez-Marín[3]

[1] Royal Bank of Canada, RBC Insurance, 6880 Financial Drive, Mississauga, Ontario,
L5N 7Y5, Canada
leo.guelman@rbc.com
[2] Riskcenter, University of Barcelona, Diagonal, 690, E-08034 Barcelona, Spain
mguillen@ub.edu
[3] Riskcenter, University of Barcelona, Diagonal, 690, E-08034 Barcelona, Spain
amperez@ub.edu

Abstract. Models of customer churn are based on historical data and are used to predict the probability that a client switches to another company. We address customer retention in insurance. Rather than concentrating on those customers with high probability of leaving, we propose a new procedure that can be used to identify the target customers who are likely to respond positively to a retention activity. Our approach is based on random forests and can be useful to anticipate the success of marketing actions aimed at reducing customer attrition. We also discuss the type of insurance portfolio database that can be used for this purpose.

1   Introduction and Background

Customer retention is a concept that involves all efforts made by a selling company to retain its customers. It has obvious links with marketing strategies, quality, customer service and profitability. When looking at customer retention in the context of insurance products, the number of existing contributions is still scarce and mainly focuses on predictive models for the probability of a customer to switch to another company.

In this article we propose a method based on random forests to evaluate the effectiveness of a retention strategy. We use uplift modeling to present a measure that can be applied to every individual customer. This technique can predict the magnitude of a customer's reaction to a given strategic marketing campaign.

Insurance policies are non-typical goods which are based on the promise that the insurer will compensate the customer in the event of a loss. The contract usually lasts for one year and is renewed after that. Renewal requires an active action from the customer part, and this means that there is a period where he may have doubts about the suitability of the contract, its price and the type of insurance coverage. It is also important to note that if an accident has happened during the recent period previous to renewal, then the customer has had a closer relationship with the insurer, meaning that the perception of the service quality can influence the decision to renew.

K.J. Engemann, A.M. Gil Lafuente, and J.M. Merigó (Eds.): MS 2012, LNBIP 115, pp. 123–133, 2012.
© Springer-Verlag Berlin Heidelberg 2012

Customer churn, or customer attrition, refers to the loss of the customer in general, but it also implies that the profit a customer is going to generate vanishes. Customer lifetime value (CLV) is the present value of the future profit stream expected over a given time horizon of transacting with the customer [1]. This value can be forecast for insurance products assuming that the probability of policy cancellation can be predicted and it can also include the potential gain due to selling to the customer new products from the same company. This is known as cross-selling, which means that present customers of a company can be approached and encouraged to increase their engagement with the company by purchasing one or many additional products. It is one of the main tools for managers to strengthen the customer relationship [2] and it is typical in the insurance sector. In the financial sector, the customers' data base (see [3] and [4]) can be used to select preferred customers and to identify which customers are at risk of not renewing their contract (see also [5]).

The research on customer loyalty in insurance dates from the sixties, when factors associated with the increasing demand for insurance were identified. [6] and [7] mentioned household income and women entering the labour market, respectively. [8] explained that insurance demand increases with the demand of other goods, while [9] showed that the level of insurance increased as the number of insurable risks and their weight in the asset portfolio also increases.

[10] were the first to investigate the problem of customer loyalty in the insurance industry. They showed that the level of satisfaction was higher among those that renewed their contract. [11] suggested that a optimal strategy is to increase the loyalty of customers that are more profitable to the company in the long term. [12] proposed a theoretical method to calculate CLV based on expected profit and loss. [13] analysed a sample of German automobile policy holders and found that customer churn is mainly due to premium level and the reputation of the company.

[14] made a remarkable step forward, when he introduced a two-stage segmentation process to identify four different groups of health insurance policy holders. He identified age, sex, type of coverage, and seniority as suitable predictors for customer churn. Thereafter, different loyalty strategies were applied to each group based on their particular needs and as a result, customer retention was observed to increase by approximately 7%. [15] provide a very useful overview about the literature on customer loyalty and they conclude that a small increase in customer retention from 85% to 90% results in net present value profits rising from 35% to 95% amongst the business they examined. They proposed a risk-adjusted CLV, which combines the prediction of CLV with the future risk including the risk of a claim and the risk of churn.

[16], [17], [18] and [19] provide an approximation to business risk management in the insurance company. These authors consider different types of products simultaneously and provide some guidance for targeting business risk management in the insurance industry. [20] find empirical evidence of the moderating effects of inertia and switching costs on the satisfaction-retention link in the auto liability insurance context. They show that the barriers made by switching costs and the behavioral lock-in effect produced by inertia create a pull-back effect, which prevents customers from switching to another insurance provider even in the face of dissatisfaction with the quality of service by the existing provider.

2   The Uplift Problem

Let the predictive learning problem be characterized by a vector of inputs or predictor variables $\mathbf{x} = \{x_1,...,x_p\}$ and a binary response variable coded as $y \in \{0,1\}$. In the context of customer retention, the input variables may represent a collection of quantitative and qualitative attributes of the customer, and the response is the actual churn outcome (1 if churn, and 0 otherwise). In addition, assume the data is randomly partitioned into treatment ($t = 1$) and control ($t = 0$) subgroups.

Given a collection of M instances $\{(y_i, \mathbf{x}_i, t_i) \; ; \; i = 1,..., M\}$ of known $(y, \mathbf{x}, t)$ values, churn models typically found in the literature and applied work ([21], [22] and [23]), start by using non-treatment data to identify the characteristics of customers who are more likely to leave. That is, they seek to find a prediction function $\hat{f}(\mathbf{x})$ that estimates

$$\hat{f}(\mathbf{x}_i) = E(y_i \mid \mathbf{x}_i; t_i = 0) \tag{1}$$

This model is then used so retention efforts are focused on customers with highest $\hat{f}(\mathbf{x}_i)$.

A moment of thought should reveal that estimating (1) is not the right objective: the emphasis should not be on targeting customers with high probability of leaving, but those likely to respond positively to a retention activity. This is the idea behind *uplift models*, which estimate

$$\hat{f}^{uplift}(\mathbf{x}_i) = E(y_i \mid \mathbf{x}_i; t_i = 0) - E(y_i \mid \mathbf{x}_i; t_i = 1) \tag{2}$$

That is, in the uplift context we seek to estimate, for each customer, the reduction in churn probability as a result of the treatment. The retention efforts can then be centered on those customers whose decisions are likely to be positively influenced by the treatment (i.e., those with highest $\hat{f}^{uplift}(\mathbf{x}_i)$). A similar logic holds for marketing *response* and *purchase* models. In that context, the objective should not be to estimate the probability that a customer will purchase a product or service if we treat them, but to estimate the *change* in the class probabilities caused by the treatment. The problem can also be extended to the n-treatment case, where the objective is to identify the optimal treatment for each customer.

3   State of Art Solutions to the Uplift Problem

In this section, we briefly discuss two overall approaches to the uplift problem. Surprisingly, despite its importance on practical applications, the problem has received very little attention from the machine learning and statistical communities.

3.1   Subtraction of Two Models

The most obvious approach to estimate (2) is to build two separate models, one for the control and one for the treatment data. The predicted uplift for an instance where only **x** is observed, is measured by subtracting the class probabilities from the two models. This method benefits from its simplicity and software is readily available for implementation. Unfortunately, this method suffers from many drawbacks and does not work well in practice [24]. The most relevant aspect to note is that the difference between two independent `accurate' models, does not necessarily lead to an accurate model itself. Instead of giving a predictable outcome in the two groups separately, a second class of models attempt to fit the difference in behavior between the two groups directly.

3.2   Modeling Uplift Directly

Modeling uplift directly is a challenging task, as it generally requires designing novel methods/algorithms. The solution may also require domain expertise in a relevant programming language to implement the ideas.

[25] proposes building a single model by explicitly creating an interaction term between each predictor in $\mathbf{x} = \{x_1,\ldots,x_p\}$ and the treatment t. The model is then fitted using standard *logistic regression*. The estimated parameters of the interaction terms measure the additional effect of each predictor due to treatment. The model is then applied by subtracting the predicted probabilities by setting, in turn, $t = 1$ and $t = 0$[1].

[26] argues that the most fundamental element for building good uplift models is variable selection. The proposed method is based on what he calls a *net weight of evidence*, which is used to compute a *net information value* for each variable. Once variable selection is performed, the model is trained using *naïve Bayes* classifiers and the so called *bifurcated regression* models.

[24] build a CHAID-type decision tree [27], but with modified split criteria to model uplift directly. At each split, the difference between treatment and control probabilities is computed for the left (Δp_l) and right (Δp_r) branches of the tree. The split that results in the largest difference $\Delta\Delta P = |\Delta p_l - \Delta p_r|$ is selected. As already noted in [28], the main drawback of this approach is that the relative sample size between the left and right node is not taken into account. As it is always easy to find small subgroups with extreme outcome rates, this method overemphasize the importance of small subgroups.

A more recent and rigorous approach was developed by [29]. They present tree-based classifiers and pruning methods with the purpose of maximizing the difference between the class distributions in treatment and control groups. To this end, they design split criteria based on information theory. In particular, they propose split

---

[1] Most of the uplift modeling literature centers on the context of marketing purchase models. If purchase is the outcome variable, the objective in (2) must be re-expressed as the difference between the treatment and the control group (as opposed to the difference between the control and treatment group).

methods based on conditional divergence measures, including Kullback-Leibler divergence, squared Euclidean distance, and chi-squared divergence. The biggest problem faced by the authors is the lack of a suitable data set to test their method, and thus they resorted to publicly available data from which they created artificial treatment and control groups.

Finally, [28] have a long-standing interest in uplift models. Their method is similar in principle to the one by [29], but based on trees with modified split criteria and includes a variance-based pruning method.

4   Random Forests for Uplift Modeling

4.1   A Proposed Algorithm

Much of the work on uplift models up to date has been reported independently, in the sense that no single article has been published comparing the predictive performance of the various methods. This is probably due to the fact that testing each of the most prominent solutions to the uplift problem would require writing their software implementation (or alternatively buying software, in the case of [28]). However, a closer look into these methods anticipates their individual merits and potential pitfalls. Plausible enhancements could be made by borrowing some of the most promising ideas while adding some minor, but important, modifications.

Tree-based models are a natural approach to solve (2), as they partition the input space into segments, and appropriate split criteria can be designed to model uplift directly. From all split criteria proposed in the uplift literature, the one described by [29] seems to be the most appropriate, as it is more in the style with modern algorithms, which are based on information theory. However, a major concern with their approach is that their model is based on a single tree. A key problem with trees is their high variance as a result of the hierarchical nature of the splitting process: the effect of an error in the top split is propagated down to all of the splits below. [28] propose using *bagging*[2] [30] for stability. The idea behind bagging is to average many noisy models (such as trees) to reduce the variance. However, the correlation between pair of bagged trees limits the benefits of averaging (see [31]). The idea in *random forest* [32] is to improve the variance reduction of bagging by reducing the correlation between the trees, without increasing the variance of the individual trees too much.

A proposed algorithm for uplift based on random forest is shown in Algorithm 1. In short, an ensemble of uplift trees are grown on bootstrap samples of the training data (which should include both treatment and control records). The tree-growing process involves selecting $n \leq p$ input variables at random as candidates for splitting. This reduces the correlation between trees, and hence reduces the variance of the ensemble. The split criteria is based on a conditional divergence measure, as defined in [29]. The predicted uplift is obtained by averaging the uplift predictions of the individual trees in the ensemble.

---

[2] Short for *bootstrap aggregation*.

The model only has two tuning parameters: the number of variables in the random subset at each node (typical values are between 1 and $\sqrt{p}$) and the number of trees in the forest. Model performance tends not to be very sensitive to the values of these parameters[3], but this should be tested for the uplift case. Another important aspect to note about this algorithm, is that trees are grown to maximal depth (i.e., no *pruning* is done). However, it has been shown that small gains can be made by limiting the number of splits [33]. This also remains to be tested in the context of uplift.

```
Algorithm 1 Random Forest for Uplift
1: for b = 1 to B do
2:   Draw a bootstrap sample of size M with replacement
     from the train data
3:   Grow an uplift decision tree UT_b to the bootstrap
     data:
4:   for each terminal node do
5:     repeat
6:       Select n variables at random from the p variables
7:       Select the best variable/split-point among the n.
         The split criterion should be based on a
         conditional divergence measure.
8:       Split the node into two branches
9:     until a minimum node size m_min is reached
10:  end for
11: end for
12: Output the ensemble of uplift trees UT_b; b = {1,…,B}
```

13: The predicted uplift for a new data point x, is obtained by averaging the uplift predictions of the individual trees in the ensemble: $\hat{f}^{uplift}(\mathbf{x}) = \frac{1}{B}\sum_{b=1}^{B} UT_b(\mathbf{x})$

4.2   Variable Importance

An additional piece of information, generally desirable from a predictive learning algorithm, is a measure of the relative importance of the input variables on the

---

[3] However, [33] shows that when the number of variables is very large and the signal-to-noise ratio is very low, then a low value of n tends to give poor performance.

response. In the case of random forest, one option is to first compute the improvement in the split-criterion attributed to each variable over all non-terminal nodes of the tree. This is then averaged over all trees in the forest. In principle, a similar idea can be used in the uplift case, as long as the appropriate split criteria designed to address the uplift problem directly is used.

4.3   Assessing Model Performance

In conventional statistical learning classifiers, *model assessment* refers to estimating the prediction error (generalization error) of the model. This is generally accomplished by choosing an appropriate loss function to define lack of fit on an independent test sample. This is more complex in the case of uplift models, as loss functions cannot be evaluated at the individual observational unit (since a subject cannot be simultaneously treated and not-treated). However, assessing performance for uplift models is still possible. The method commonly used in the literature consists in first scoring both the treatment and control groups based on the estimated model. The churn rate for the highest p percent scored treated subjects can then be subtracted from the highest p percent scored control subjects to obtain an estimate of the model uplift at each p.

   If we are in a data-rich situation, an independent test sample can be used for model assessment. Alternatively, *out-of-bag* samples can be used. In this case, the random forest predictor for each observation is constructed by only using those trees corresponding to bootstrap samples in which that observation did not appear. This is similar to performing *N-fold* cross-validation.

5   The Data

A data set from a major Canadian insurer is at our disposal to test the algorithm described in section 4. In anticipation for a large rate increase to be experienced by most of the existing books of Auto insurance clients, this company was interested in designing retention strategies to minimize its policyholders' attrition rate. For that purpose, during Feb--Apr 2011, an experimental retention program was implemented, from which policies coming up for renewal were randomly allocated into one of the following three groups: a) a *Letter only* group. Policyholders under this group received a letter in the mail, which notified them about the rate increase and explained the motives for it. In addition, the letter encouraged the customer to call the company to ensure that all applicable discounts were in place for the policy (such as discounts resulting from also insuring their property with the company and more than one vehicle), b) a *Letter + Outbound Courtesy Call* group. In addition to the letter, this group also received a courtesy call made by one of the company's licensed insurance advisors. The purpose of the call was to verbally reinforce the items described on the letter. The advisors were also trained to deal with situations of customer dissatisfaction with the purpose of maximizing customer retention, and c) a *Control* group. No retention efforts were applied to this group. As the campaign was designed with randomized assignment, the results observed on this group represent a baseline for the impact of other two treatment types on the retention outcome.

The company offers a renewal package to its existing book of business 45 days prior to the policy renewal date. This was also the timing for the retention program discussed above. If the status of the policy was active at renewal date, it was considered as a renewal policy (and terminated otherwise). Table 1 shows the retention results. The observed difference in retention rates between each treated group and the control group is small, but there is some evidence of a slightly positive impact of the outbound call.

Although the treatment had almost a neutral impact on retention for the entire sample, it may be the case that the campaign had positive retention effects on some subgroup of customers, but they were offset by negative effects on other subgroups. Adverse effects can happen in retention programs [34], for example, if the customer is already dissatisfied and perceives the call as intrusive, or if it triggers a behavior to shop for better pricing among other insurers.

Table 2 shows most of the available variables to be used as model inputs. They were categorized into policy, driver and vehicle characteristics.

Table 1. Retention rates by group

	Overall	Letter only	Letter + Call	Control
Retained policies	22,231	11,143	7,649	3,439
Canceled policies	2,131	1,113	690	328
Retention rate	91.3%	90.9%	91.7%	91.3%

Table 2. Model input variables

Policy characteristics	Driver characteristics	Vehicle characteristics
Change in premium	At-Fault accidents	Type of vehicle
Current premium	Not-At-Fault accidents	Age of vehicle
Endorsements	Accident surcharge	Price of vehicle
Coverage	Accident surcharge change	Lease vehicle flag
Multi-Vehicle disc.	Convictions	Rate group differential
Time since inception	Convictions surcharge	
Home policy	Convictions surcharge change	
Prior Carrier	Driver's age	
Postal code u/w score	Gender	
Territory	Marital status	
Territory change	Years licensed	
Prior insurance	Age licensed	
Payment type	Occ. driver under 25	
NSF activity	Driver record	
Channel	Driver record change	
Prior carrier	Insurance lapses	
Transaction frequency	Insurance suspensions	

6   Conclusions

The results are promising because there is evidence that, at least when customers receive the outbound courtesy call, the retention rate increases for this group compared to the other customers. However, with the proposed method, we will be able to disentangle whether all customers react in the same way to the campaigns. The method allows for the prediction of how much an individual reacts to the company's persuasion.

The main advantage of the method that has been proposed here is that it allows to predict the expected change in the probability that a customer switches to another company when he is actively approached by the company before renewal. The expected change is obtained individually and it can be compared with the actual predicted probability of policy cancellation. This approach leads to an efficient selection of the customers to whom the company must devote the marketing efforts.

We also see many parallelism with previous research on customer loyalty in insurance. There is a strong empirical evidence that policy characteristics such as time since inception and no change in the territory affect positively the probability to stay in the same company. Moreover, from previous papers, we do expect that age and claim history will affect the probability of customer churn, and we also expect that they will be affecting the inertia of customer-specific retention strategies.

Acknowledgments. M. Guillén and A. M. Pérez-Marín thank the support by the Spanish Ministry of Science grant ECO2010-21787-C03-01.

References

1. Kotler, P.: Marketing during periods of shortage. J. Marketing 38(3), 20–29 (1974)
2. Kamakura, W.A., Ramaswami, S., Srivastava, R.: Applying latent trait analysis in the evaluation of prospects for cross-selling of financial services. Int. J. Res. Mark. 8, 329–349 (1991)
3. Seng, J.-L., Chen, T.C.: An analytical approach to select data mining for business decision. Expert Syst. Appl. 37(9), 8042–8057 (2010)
4. Liao, S.-H., Chen, Y.-J., Hsieh, H.-H.: Mining customer knowledge for direct selling and marketing. Expert Syst. Appl. 38(5), 6059–6069 (2011)
5. Larivière, B., Van den Poel, D.: Predicting customer retention and profitability by using random forest and regression techniques. Expert Syst. Appl. 29(2), 472–484 (2005)
6. Hammond, J.D., Houston, D.B., Melander, E.R.: Determinants of household life insurance premium expenditures: an empirical investigation. J. Risk Insur. 34(3), 397–408 (1967)
7. Duker, J.M.: Expenditures for life insurance among working-wife families. J. Risk Insur. 36(5), 525–533 (1969)
8. Mayers, D., Smith Jr., C.W.: The interdependence of individual portfolio decisions and the demand for insurance. J. Polit. Econ. 91(2), 304–311 (1983)
9. Doherty, N.A.: Portfolio efficient insurance buying strategies. J. Risk Insur. 51(2), 205–224 (1984)

10. Crosby, L.A., Stephens, N.: Effects of relationship marketing on satisfaction, retention, and prices in the life insurance industry. J. Marketing Res. 24(4), 404–411 (1987)
11. Jackson, D.: Determining a customers lifetime value, part three. Direct Marketing 52(4), 28–30 (1989)
12. Berger, P.D., Nasr, N.: Customer lifetime value: marketing models and applications. J. Interact. Mark. 12, 17–30 (1998)
13. Schlesinger, H., Schulenburg, J.M.: Customer information and decisions to switch insurers. J. Risk Insur. 60(4), 591–615 (1993)
14. Cooley, S.: Loyalty strategy development using applied member-cohort segmentation. J. Consum. Mark. 19(7), 550–563 (2002)
15. Ryals, L.J., Knox, S.: Measuring risk-adjusted customer lifetime value and its impact on relationship marketing strategies and shareholder value. Eur. J. Marketing 39(5/6), 456–472 (2005)
16. Brockett, P.L., Golden, L., Guillén, M., Nielsen, J.P., Parner, J., Pérez-Marín, A.M.: Survival analysis of household insurance policies: How much time do you have to stop total customer defection? J. Risk Insur. 75(3), 713–737 (2008)
17. Guillén, M., Nielsen, J.P., Pérez-Marín, A.M.: The need to monitor customer loyalty and business risk in the European insurance industry. Geneva Pap. R. I. – Iss. P. 33, 207–218 (2008)
18. Guillén, M., Pérez-Marín, A.M., Alcaniz, M.: A logistic regression approach to estimating customer profit loss due to lapses in insurance. Insurance Markets and Companies: Analyses and Actuarial Computations 2(2), 42–54 (2011)
19. Guillén, M., Nielsen, J.P., Scheike, T.H., Pérez-Marín, A.M.: Time-varying effects in the analysis of customer loyalty: A case study in insurance. Expert Syst. Appl. 39, 3551–3558 (2012)
20. Lai, L.-H., Liu, C.T., Lin, J.T.: The moderating effects of switching costs and inertia on the customer satisfaction-retention link: auto liability insurance service in Taiwan. Insurance Markets and Companies: Analyses and Actuarial Computations 2(1), 69–78 (2011)
21. Neslin, S., Gupta, S., Kamakura, W., Lu, J., Mason, C.: Defection detection: Measuring and understanding the predictive accuracy of customer churn model. J. Marketing Res. 43, 204–211 (2006)
22. Lemmens, A., Croux, C.: Bagging and boosting classification trees to predict churn. J. Marketing Res. 43, 276–286 (2006)
23. Coussement, K., Van den Poel, D.: Churn prediction in subscription services: An application of support vector machines while comparing two parameter-selection techniques. Expert Syst. Appl. 34, 313–327 (2008)
24. Hansotia, B., Rukstales, B.: Incremental value modeling. J. Interact. Mark. 16(3), 35–46 (2002)
25. Lo, V.: The true lift model. ACM SIGKDD Explorations Newsletter 4(2), 78–86 (2002)
26. Larsen, K.: Net models. In: M2009 - Annual SAS Data Mining Conference (2009)
27. Kass, G.: An exploratory technique for investigating large quantities of categorical data. Appl. Stat. – J. Roy. St. C 29(2), 119–127 (1980)
28. Radcliffe, N., Surry, P.: Real-World Uplift Modelling with Significance-Based Uplift Trees. Portrait Technical Report, TR-2011-1 (2011)
29. Rzepakowski, P., Jaroszewicz, S.: Decision trees for uplift modeling with single and multiple treatments. Knowl. Inf. Syst. (2011), doi: 10.1007/s10115-011-0434-0
30. Breiman, L.: Bagging predictors. Mach. Learn. 26, 123–140 (1996)
31. Hastie, T., Tibshirani, R., Friedman, J.: The elements of statistical learning, 2nd edn. Springer, New York (2009)

32. Breiman, L.: Random forests. Mach. Learn. 45, 5–32 (2001)
33. Segal, M.: Machine learning benchmarks and random forest regression, Technical report, eScholarship Repository, University of California (2004), http://escholarship.org/uc/item/35x3v9t4
34. Stauss, B., Schmidt, M., Schoeler, A.: Customer frustration in loyalty programs. Int. J. Serv. Ind. Manag. 16(3), 229–252 (2005)

Age Diversity in the Boardroom: Measures and Implications

Idoya Ferrero-Ferrero, María Ángeles Fernández-Izquierdo,
and María Jesús Muñoz-Torres

Finance and Accounting Department, Universitat Jaume I, Av. de Vicent Sos Baynat s/n,
12071 Castelló de la Plana, Spain
{ferrero,afernand,munoz}@uji.es

Abstract. In light of the recent corporate governance developments after the global financial crisis, this study aims to explore how board diversity affects corporate performance. Focusing on age diversity as separation, variation and disparity and using board of directors as unit of analysis, this study empirically tests the effects of each type of age diversity on corporate performance in a sample of 205 European listed firms for the year 2009. The results reveal that age diversity as variety positively impacts on corporate performance and suggests to increase board age variety to adapt different views and make more effective decisions,in board of directors. However, this study does not find clear evidence on the impact of age diversity as separation and disparity on corporate performance. This study advances the understanding of board behaviour and its relationships with corporate results and presents a new approach to study age diversity as an integrated view.

Keywords: Age diversity, board of directors, separation, variety, disparity.

1   Introduction

Board of directors is one of the most significant governance issues under review by the corporate governance initiatives, since the most recent financial crisis has revealed serious weaknesses of this body to fulfill its duties. A common recommendation to improve board effectiveness is concerning diversity of the boards because diversity may broaden the debate within the boards and help to avoid the danger of narrow "group think" [1]. However, in the academia world, the results of research on the association between top management team diversity on corporate performance have been inconclusive [2]. In this sense, this study expects to contribute to the understanding of board diversity and examine how it might affect corporate performance.

Following [3], this study uses the term diversity to describe the distribution of differences among the members of a unit with respect to a common attribute. These authors point out that diversity can be defined as three different ways: diversity as separation, variety, and disparity; and each type of diversity might have different effects on corporate performance. In particular, they explore the typology's

K.J. Engemann, A.M. Gil Lafuente, and J.M. Merigó (Eds.): MS 2012, LNBIP 115, pp. 134–143, 2012.
© Springer-Verlag Berlin Heidelberg 2012

implications for the special case of demographic diversity, showing that the same demographic attribute within units may be conceptualized as separation, variety or disparity. In this vein, the attribute "age" remains as one of the most important demographic variable to examine and there is an active promotion of age diversity in board to encourage the different perspectives of different age groups, and as an integral part of succession planning [4].

This study aims to test empirically the theoretical consequences build on Harrison and Klein's [3] diversity typology. In particular, the main purpose is to examine how board age diversity, in terms of separation, variety and disparity, affects corporate performance. This paper expects to contribute to the existing management literature, business practice and public policy in several ways. First, this study contributes to the theoretical understanding of board diversity and their consequences on corporate performance, since it integrates psychologists, sociologists theories with an management and economic orientation. Second, it empirically explores novel measures of age diversity of the boards that reflects different types of diversity. Third, this study also examines the impact of the three forms of diversity simultaneously with the aim to isolate, and, hence, analyse the opposite effects of different diversity types on strategic behavior and performance. Fourth, in light of the corporate governance guidelines that recommend increase board diversity, this paper brings new evidence about the three types of diversity and the finding suggests to increase board age diversity as variety to adapt different views and make more deliberate decisions in the board, which improve corporate performance.

This paper is divided into five sections. After this introduction, a review of the theoretical framework is provided. The third section includes information on the methodology used in estimating the model. The fourth section presents the findings and empirical analysis. The final section summarizes and concludes the study.

2   Theoretical Framework

The upper echelon theory has received wide interest in the field of organization behaviour. A seminal paper of this theory is the publication of Hambrick and Mason in 1984 titled "Upper echelons: The organization as a reflection of its top managers". Hambrick and Mason [5] propose a model of how upper echelon characteristics may become reflected in organizational outcomes. However, recent research on understanding of the complex roles played by top managers and top management teams requires applying alternative theories in combination with the upper echelons perspective in order to find the answer to the fundamental questions of whether heterogeneity in top management team composition is contributing to firm strategy and performance [2]. In this vein, the role of individual psychological factors and team processes on executive decision-making have led to wider application of group psychology and sociology theories combined with upper echelons theory.

The research question in this paper is to examine empirically the effect of diversity as separation, variety, and disparity on corporate performance. To that end, three hypotheses are developed based on upper echelon theory, which states that the aggregate characteristics of top management team have influence over corporate performance, and the new directions for diversity theory proposed by Harrison and Klein [3].

Diversity as Separation. It refers to differences in position or opinion among unit members, is closely related to theories of similarity-attraction, social identity and self-categorization. These theories posit that individuals are attracted to others with similar attributes to themselves and greater similarity presumably lead to shared results, fewer disagreements and conflicts, higher levels of cooperation and cohesion, trust, and social integration. Therefore, relationships with similar others make possible to reach a consensus easier and make decisions in an efficient way. Consequently, the following hypothesis is proposed:

Hypothesis 1: A greater separation leads to a lower level of corporate performance.

Diversity as Variety. It indicates differences in kind or category, knowledge or experience among unit members. In this case, information processing theory and human cognition theory assume that teams whose members draw from different pools can translate greater information richness within a unit into better choices, plans, or products, deliver from different views and, thus, make more effective decisions. Based on these arguments, the following hypothesis is expected:

Hypothesis 2: A greater variety leads to a higher level of corporate performance.

Diversity as Disparity. It represents differences in concentration of valued social assets of resources such as pay, power, prestige and status among unit members. This third perspective builds on distributive justice theory, tournament theory, and stratification, status hierarchy or characteristics theories. The basic idea of this perspective is that in teams where few members have a marked influence over the group decision, they control the flow of information, impose their views and limit a democratic participation in the team. Likewise, low-status members tend to be conformist and contribute less to the team performance. Consequently, the decisions are made in worse conditions and it negatively impact on corporate performance. Therefore, the following hypothesis is presented:

Hypothesis 3: A greater disparity leads to a lower level of corporate performance.

Harrison and Klein [3] argue that some team attributes are strongly related to a particular type of variety. For instance, pay is a good proxy of diversity as disparity, since its structure captures the differences in the power of the members in a team. However, these authors also remark that demographic variables most frequently included in diversity studies, such as age, sex, race/ethnicity, organization and team tenure, may be meaningfully conceptualised as separation or as variety, or as disparity.

In this context, this study focuses on age diversity. As remark Shore et al. [6] the research on age diversity is much less developed than that on race and gender, suggesting the need for new paradigms and new approaches to studying age diversity.

Moreover, in spite of the growing number of international initiatives that encourage age diversity to improve the overall level of knowledge on the top management team its potential effects on performance are not yet fully understood [7], and the limited empirical studies show inconclusive results. As Nielsen [2] suggests, it is possible that the inconclusive findings of previous research result from the fact that interactions between different aspects of diversity are omitted. This author also highlights that the distinction between diversity as variety, separation and disparity [3] needs to be applied to future research. In this context, this study expects to contribute to theoretical and empirical understanding of board diversity by means of test the different effects of age diversity as separation, variety and disparity on corporate performance.

3   Data and Methodology

In this section it is presented the sample, the model estimated and the variables used to test the three hypotheses developed.

3.1   Sample

The sample selection process starts with firms listed in FTSE 100, FTSE SMALL CAP, DAX 30, and CAC 40 for year 2009. Data on board characteristics are from BoardEx and data on financial items are from Worldscope. Given the limitation of available data, the final sample consists of 2,152 individual observations of director's characteristics. These directors are members of 205 boards. Therefore, this empirical study uses data from 205 European listed firms.

3.2   Model

In order to test the hypotheses, this study estimates the linear regression model presented in equation (1).

$$PERFORMANCE_i = \beta_0 + \alpha_1 \cdot DIVERSITY_i + \beta_1 \cdot SIZE_i + \beta_2 \cdot CAPEX_i + \beta_3 \cdot LEVERAGE_i$$
$$+ \beta_4 \cdot GROWTH_i + \beta_5 \cdot LIQUIDITY_i + \sum_{J=1}^{2} \lambda_J \cdot COUNTRY_i + \sum_{J=0}^{7} \lambda_J \cdot INDUSTRY_i + \mu_i \tag{1}$$

Consistent with previous research [8] and [9], this equation contains corporate performance as the dependent variable which is explained by board diversity and control variables. The board diversity variable specified in equation 1 is divided in the three types of diversity – separation, variety and disparity – with the aim to test their effects on corporate performance.

Given the feature of the sample, the equation 1 is regressed by means of OLS. The estimator process uses a robust variance matrix, in particular, White-corrected standard errors in presence of heteroskedasticity. Additionally, the issue of collinearity is explored by means of the variance inflation factors for the independent variables.

3.3   Variables

In this section, the different variables used in the equation 1 are presented.

Dependent Variable. This study aims to examine how age diversity of the board directors affect corporate performance. Consistent with previous studies, [10] and [11], the corporate performance variable (PERFORMANCE) is estimated by an accounting performance measure, specifically, earnings before interests and taxes divided by book value of total assets (EBITA). This study also considers a proxy for accounting performance removing the influence of the home country and industry performance. Therefore, the country- and industry-adjusted EBITA is defined as the difference between a firm's EBITA and the average EBITA across all listed firms in the same two-digit SIC and from the country in which the company is registered.

Independent Variables. Harrison and Klein [3] explain implications for research design; in particular, they suggest appropriate operationalisation for each type of diversity (DIVERSITY). Following their suggestions, this study uses measures of age diversity as separation, variety and disparity. Table 1 shows the graphic illustration of the empirical minimum and maximum levels of the variables used as proxies of the three types of diversity, using the sample of this study.

Diversity as separation is considered as a continuous variable. Maximum occurs when there are two subgroups divided but balanced within a team and each subgroup takes though stances but both subgroups show disagreement and opposition between themselves. Age diversity as separation (AGE SEPARATION) is measured by standard deviation of the age of directors that are members of the board.

Diversity as variety is a categorical attribute. Focusing on age, this variable represents differences in personality, traits, skills, attitudes, mental health, work values and behaviours. These differences may be categorized according to the generations, since the social and historical experiences and circumstances from a respective generation have influenced the individuals' behaviour. There is a strong consensus about the four major generations of the twentieth century: the Greatest Generation (1922-1945), Boomers (1946-1964), Xers (1965-1983), and Generation Y (1984-2002). This study uses these generations as a qualitative distinction to define the different categories. Note that any member belongs to the Generation Y, thus age diversity as variety is measured by three categories based on generations - the Greatest Generation (64 - 87), Boomers (45 - 63), and Xers (26 - 44).

Following previous studies, [8] and [12], and according to the properties of this variable [3], this study uses to Blau's Index calculate age diversity as variety (AGE VARIETY).

Blau's Index is calculated by equation 2, where k is a particular category (generation) and P_k is the proportion of directors of a particular category within the board. This index has been divided by its theoretical maximum with the aim of standardising the results and making the interpretation of the index easier.

$$Variety = 1 - \sum_{K=1}^{3} P_K^2 \qquad (2)$$

Table 1. Illustrations of the empirical levels of age diversity as separation, variety and disparity

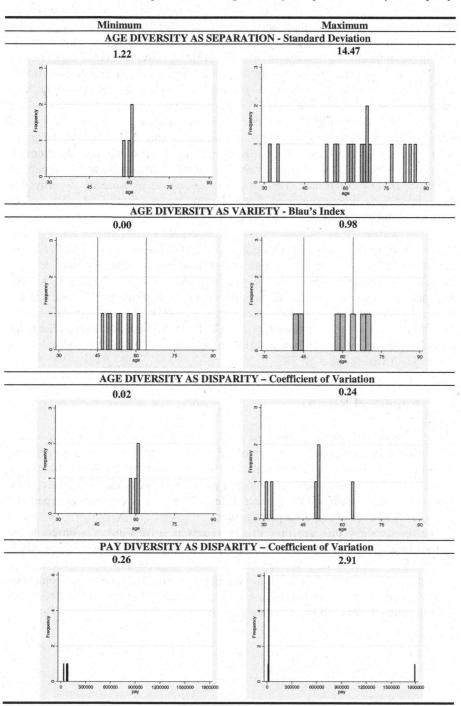

Diversity as disparity is a continuous variable which represents the differences in concentration of power and status in a board. Disparity reflects both the distances between members and the dominance of who have higher amounts of a particular attribute. This asymmetry is captured by the coefficient of variation, which as been used in previous studies to measure disparity [13]. Age diversity could be treated as disparity since age may be positively associated with authority and empowerment, since older members might be seen as possessing higher levels of task-relevant experience and tacit knowledge [3]. Therefore, the proxy proposed to measure age disparity (AGE DISPARITY) is the coefficient of variation of the age of the board members. However, focusing on board of directors as unit of analysis, maybe age is not a good proxy of power. As suggest Kang et al. [4], most of the older directors are ex-managers from various corporations that enjoy their retirement by sitting on various boards of companies. This study uses a second measure of disparity that seems to fit better the distribution of power on the board. That is the coefficient of variation of director's pay (PAY DISPARITY).

Control Variables. Consistent with previous empirical research, [11] and [12], the control variables used are: the natural log of total assets as an indicator for size (SIZE); capital expenditures divided by sales as proxy for investment ratio (CAPEX); total debt per unit of total assets as a proxy for capital structure (LEVERAGE); annual growth rate of sales as indicator of growth (GROWTH); and current assets to current liabilities as proxy for liquidity (LIQUIDITY). Additionally, dummy variables are considered to reflect differences between countries (COUNTRY), and industries (INDUSTRY) using one-digit SIC.

4   Empirical Results

Table 2 reports the estimations that test the hypotheses based on accounting performance. Focusing on the effect of age diversity of separation on corporate performance, the results indicate that the coefficient of this variable is not statistically significant. This finding does not support hypothesis 1 presented in the theoretical framework, which predicted that greater differences in age of directors as separation lead to a lower level of corporate performance.

The results show that age diversity as variety positively impacts on corporate performance. Therefore, this study finds empirical evidence to support hypothesis 2, that is, a greater age diversity as variety leads to a higher level of corporate performance. This finding is in line with Harrison and Klein [3] who argue that variety broadens the cognitive, behavioural repertoire and views of the board and leads to better choices and improvements in performance.

Table 2. Regression of the relationship between corporate performance and board age diversity

Dependent Variable	PERFORMANCE (EBITA)			PERFORMANCE (country- and industry-adjusted EBITA)		
Independent Variables	**(a)**	**(b)**	**(c)**	**(a)**	**(b)**	**(c)**
AGE SEPARATION	-0.04		0.02	-1.13		-1.05
	(0.30)		(0.32)	(0.84)		(1.00)
AGE VARIETY	6.70*	6.98*	7.92**	13.24*	13.78*	16.17*
	(2.88)	(2.84)	(2.96)	(6.41)	(6.33)	(7.22)
AGE DISPARITY		-5.58			-72.90	
		(18.27)			(48.49)	
PAY DISPARITY			0.94			5.76
			(1.70)			(4.32)
FIRM SIZE	0.78*	0.76*	0.94*	1.07	0.94	1.22
	(0.35)	(0.36)	(0.39)	(0.79)	(0.81)	(0.90)
CAPEX/SALES	0.02	0.023	0.00	$0.28^{\dagger}$	$0.28^{\dagger}$	0.18
	(0.06)	(0.06)	(0.07)	(0.15)	(0.15)	(0.18)
LEVERAGE	-0.11*	-0.11*	-0.11*	-0.20*	-0.19*	$-0.13^{\dagger}$
	(0.04)	(0.04)	(0.04)	(0.07)	(0.07)	(0.07)
SALES GROWTH	0.14***	0.14***	0.11**	0.21**	0.21**	0.21*
	(0.03)	(0.03)	(0.03)	(0.07)	(0.07)	(0.08)
LIQUIDITY	0.03***	0.03***	0.03**	0.08***	0.08***	0.07**
	(0.01)	(0.01)	(0.01)	(0.02)	(0.02)	(0.02)
CONSTANT	-4.12	-3.73	-6.50	0.06	1.93	-10.11
	(4.23)	(4.33)	(4.72)	(9.52)	(9.83)	(10.29)
	Country and Industry Dummies Included					
R^2	0.23	0.23	0.30	0.26	0.27	0.26
F- test	4.70***	4.71***	3.83***	7.13***	7.21***	6.25***
N. obs.	205	205	173	205	205	173

Robust standard errors are in brackets. †p<0.10; *p<0.05; **p<0.01; ***p<0.001.

Regarding diversity as disparity, this study does not find evidence supporting hypothesis 3. Therefore, the sample data does not support the theoretical assumption that inequality in terms of power and status of the board directors leads a lower level of corporate performance. This relationship is not significant using age disparity or pay disparity. In the case of age disparity, one possible explanation of the finding consistent with Kang et al. [4] could be that the age is not an attribute that reflects the distribution of power of the board. Whit respect to the unexpected results related to pay disparity, one explanation could be that larger difference in pay also lead to directors to elicit stronger individual efforts. In fact, Henderson and Fredrickson [14] find a balance between the arguments that "more equal pay" promotes collaboration, greater coordination and the opposite view that suggest that "larger pay differences" create a tournament-like incentives that better address the monitoring difficulties that arise with joint decision making.

5   Conclusion and Discussion

This study aims to explore how board age diversity affects corporate performance. Despite recent corporate governance initiatives recommend to increase board age diversity, theories predict differing effects of board diversity to corporate performance, and previous research find inconclusive results. In response of these disappointing cumulative results, Harrison and Klein [3] present a diversity typology which involves differences in the meanings of diversity, maxima, and theoretical relationships with corporate performance. The first type is diversity as separation which refers to differences in position or opinion. The second type, the variety, represents differences in kind, category, or knowledge. Finally, diversity as disparity indicates differences in power or status among group members.

This study develops three hypotheses build on upper echelons perspective [5] and Harrison and Klein's [3] diversity typology. Focusing on age diversity, which is one of the demographic diversity less developed, and using board of directors as unit of analysis, this study empirically tests the effects of each type of age diversity on corporate performance in a sample of 205 European listed firms for the year 2009.

The main results reveal that age diversity as variety positively impacts on corporate performance. This is consistent with Harrison and Klein's [3] view, who argue that high levels of diversity as variety enhance information resources and broaden the cognitive and behavioral range of the unit. However, the age diversity as separation and disparity can not be ascertained in the same way as age diversity as variety, since this study does not find clear evidence on the impact of these types of diversity on corporate performance.

The results have important implications for theory, business practice and public policy. First, this study contributes to the theoretical understanding of board diversity and their consequences on corporate performance, since it integrates psychologists, sociologists theories with an management and economic orientation and this theoretical pluralism better captures the complexity of the behavior of boards of directors. Second, it empirically explores novel measures of age diversity of the boards that reflects different types of diversity. Third, this study also examines the impact of the three forms of diversity simultaneously with the aim to isolate, and, hence, analyse the opposite effects of different diversity types on strategic behavior and performance. Fourth, the finding suggests that corporate governance guidelines encourage board age diversity as variety to adapt different views and make more deliberate decisions in the board, which improve corporate performance.

To sum up, this study offers new and interesting insights on the consequence of different types of age diversity in board of directors and encourages to future research to consider integrated views and multiple dimensions of diversity to advance in the understanding of board behaviour and its relationship with corporate results.

Acknowledgment. The authors wish acknowledge the financial support received from projects P1•1B2010-13 and P1•1B2010-04 through the Universidad Jaume I and the Master in Sustainability and Corporate Social Responsibility offered by Universitat Jaume I of Castellon, Spain.

References

1. European Commission: Corporate Governance in Financial Institutitons: Lessons to be drawn from the current financial crisis, best practices. COM 284 (2010)
2. Nielsen, S.: Top management team diversity: a review of theories and methodologies. International Journal of Management Reviews 12(3), 301–316 (2010)
3. Harrison, D., Klein, K.: What's the difference? Diversity constructs as separation, variety, or disparity in organizations. Academy of Management Review 32, 1199–1228 (2007)
4. Kang, H., Cheng, M., Gray, J.: Corporate Governance and Board Composition: diversity and independence of Australian boards. Corporate Governance: An Interview Review 15(2), 194–207 (2007)
5. Hambrick, D.C., Mason, P.A.: Upper echelons: the organization as a reflection of its top managers. Academy of Management Review 9, 193–206 (1984)
6. Shore, L.M., Chung-Herrera, B.G., Dean, M.A., Ehrhart, K.H., Jung, D.I., Randel, A.E., Singh, G.: Diversity in organizations: Where are we now and where are we going? Human Resource Management Review 19, 117–133 (2009)
7. Kunce, F., Boehm, S.A., Bruch, H.: Age diversity, age discrimination climate and performance consequences – a cross organizational study. Journal of Organizational Behavior 32, 264–290 (2011)
8. Campbell, K., Minguez-Vera, A.: Gender Diversity in the Boardroom and Firm Financial Performance. Journal of Business Ethics 83, 435–451 (2008)
9. Mahadeo, D., Soobaroyen, T., Hanuman, V.O.: Board Composition and Financial Performance: Uncovering the Effects of Diversity in an Emerging Economy. Journal of Business Ethics (2011), doi:10.1007/s10551-011-0973-z
10. Cornett, M.M., McNutt, J.J., Tehranian, H.: Corporate governance and earnings management at large US bank holding companies. Journal of Corporate Finance 15, 412–430 (2009)
11. Cheng, S.: Board size and the variability of corporate governance. Journal of Financial Economics 87, 157–176 (2008)
12. Miller, T., Triana, M.C.: Demographic Diversity in the Boardroom: Mediators of the Board Diversity-Firm Performance Relationship. Journal of Management Studies 46(5), 755–786 (2009)
13. Siegel, P.A., Hambrick, D.C.: Pay Disparities Within Top Management Groups: Evidence of Harmful Effects on Performance of High-Technology Firms. Organization Science 16(3), 259–274 (2005)
14. Henderson, A.D., Fredrickson, J.W.: Top Management Team Coordination Needs and the CEO Pay Gap: A Competitive Test of Economic and Behavioral Views. The Academy of Management Journal 44(1), 96–117 (2001)

Hierarchical Structure of Variables in Export Agribusiness: The Case of Michoacan[*]

Joel Bonales Valencia and Odette Virginia Delfín Ortega

Institute of Economic and Enterprise Investigations
Universidad Michoacana de San Nicolás de Hidalgo
Morelia, Michoacán, México

Abstract. The present investigation has as its aim, to determine the ways in which are affected the quality, the price, the technological innovation, the environmental management, the market and the public agroindustrial policies in the international competitiveness of the agroindustrial sector of Michoacan's situation. A very exhaustive theoretical review was carried out. In order to determine the relevance between the variables considered in this investigation, the Technique of Analytical Hierarchial Structuring was used, which is utilized to evaluate the importance of the problems posed or the causes of the same. The method consists of realising comparisons by pairs between each one of the alternatives that in this case are variable and, by means of a previously specific scale, to evaluate the magnitudes of preference, among them, based on the objective that is persued to make the comparison.

Keywords: Agroindustrial sector, competitiveness, Analytical Hierarchial Structuring.

1   Background

Agriculture is a vulnerable sector in Mexico and in the State of Michoacan, so that their development is an economic and social balance. To the extent that the quality, price, technological innovation, environmental management, market and public policy impact on the competitive development of the state, agribusiness companies are more profitable and the industry is constantly growing. International markets daily become more demanding and regulations coupled with quality are a challenge for local supply.

Market forces directly influence the export supply and the agricultural sector is one example through technological development and has been able to improve its various stages of production and industrialization of its production. The FAO states that "the industrialization of agriculture and agribusiness development is a common process that is creating an entirely new industrial sector" (FAO, 2007).The state of Michoacan is the largest producer of fruits at the national level, and according to figures from SAGARPA, produces about 2 million tons (SAGARPA, 2009). But agribusiness has

---

[*] This work investigated the international competitiveness of the agroindustrial sector of Michoacan. Information was obtained from 51 companies.

K.J. Engemann, A.M. Gil Lafuente, and J.M. Merigó (Eds.): MS 2012, LNBIP 115, pp. 144–150, 2012.

not developed in the same vein and is nationally ranked 18, a situation that makes us direct our attention to promote this development and achieve greater competitiveness, directed not only at national but also international markets.

This investigation, to determine the performance of the agro-export competitiveness in the state of Michoacan, will detect the points of greatest strength to continue with the same kind of support and detect weaknesses in order to provide the basis for the structuring of public policies that help to reinforce specific points, using Variable Hierarchical Structure.

Therefore, the main objective is to describe the impact of quality, price, technological innovation, environmental management, market and public policies on the competitiveness of the agribusiness sector of the state of Michoacan, and foster the development of the state. The results obtained can be used for decision making in the private sector and government, so government programs will be aimed in that direction. This leads to an increase in industrialized agricultural enterprises and will strengthen the sector.

2   Theory and Hypotheses

International trade is of critical importance in theoretical models of competitiveness. According to Porter's (2007) theory of comparative advantage, a nation need not produce all its own goods and services and can increase its competitiveness by only producing goods and services for which it holds a comparative advantage. Although there is a lack of consensus on the definition of "competitiveness", international competitiveness is most commonly associated with greater productivity as measured by penetration in international markets, investment flows, and workforce unit costs. Although a literature review revealed over 100 difference competitiveness models, the following eight are the most relevant to this study.

In agreement with Worldwide Report on Competitiveness, competitiveness exists at the country level, the sector level, and the company level (Sala-i-Martin et al, 2009). The extent to which national sociopolitical and economic conditions are favorable to business is the primary determinant of country level competitiveness. Sector level competitiveness is the degree to which an industrial sector has the potential to grow and to produce attractive yields on investment. Company level competitiveness is determined by an organization's ability to design, produce, and commercialize goods and services that are more attractive than those of market competitors.

Competitiveness is a relative concept with not all nations, sectors or companies being able to be equally competitive in global markets. Widely competitive a national atmosphere for a business can not be it for another one. The Mexican Secretariat of Commerce (SECOFI) identified the following factors in their competitiveness model: (1) atmosphere with clear and permanent rules; (2) economies of scale; (3) economies of specialization (particularly important for small and medium companies; (4) flexible and fast adoption of the most appropriate technology (including processes of shared production); and (5) markets operating correctly. Nevertheless, this model has been criticized because it does not consider factors such as the development of qualified human resources, communications and transportation infrastructures, regulatory clarity, and the presence of a comprehensive and high priority industrial policy for long term investment in science and technology.

1. Porter (2007: 855) defined competitiveness as the production of goods and services that are of higher quality and lower prices than those of domestic and international competitors.   National competitiveness results in real income growth for a country's population.   Porter's model identified four generic attributes of a nation that determine its competitive environment: (1) Related conditions of the factors, 2) market demand conditions; (3) industries and government support; and (4) the strategy, structure and rivalry of the company.   These attributes measure the degree to which national environments are supportive of sector competition. Porter's "Diamond" reflects the many diverse elements of a nation, measures the manner by which nations create competitive forces and influence organizations, as well as identifying the need strategies and instruments for enhancing competitive advantage.

2. The Presidential Commission on Industrial Competitiveness of the United States (1985) concluded that national competitiveness is the basis for a standard of life that maintains and increases real income. Reason why it is from fundamental importance for the expansion of the use and the international fulfillment of obligations.

3. The World Economic Forum (2009) of the OECD used 330 indicators to measure eight basic factors that distinguish between the "soft side" (entry distribution, quality of the environment, and cultural values) and "hard facts" (GDP, economic growth, inflation, and balance of trade) of competitiveness

Quality

The quality is a significant variable that it influences in the competitiveness of the companies. According to the models of competitiveness of: The European Union, the Technological Institute of Massachussets, the OECD, BANCOMEXT, Michael And Porter, Carlos Wagner, Alexander Serralde, Sergio Hernandez, Alexander Lerma, Ricardo Arachavela and Vicente Felgueres. From the industrial point of view, the word quality means: lo better to satisfy desires and tastes with the consuming public (Feigenbaun, 1990). All the products have quality and, depending if they satisfy or not the consuming public, traditionally the systems of quality control have settled down to assure minimum norms quality related to the necessities and tastes of the consumers.

The norms or specifications of quality are the pattern against which the characteristics of quality of the products are moderate that make or produce. These ideas take us hypothesis to our first of work:

Hypothesis 1: A higher degree of quality agro-industrial products for export, results in greaterinternational competitiveness.

Price

This variable form leaves from the models of the OECD, the Technological Institute of Massachussets, Michael And Porter, Thomas J. Peters, Carlos Wagner, Safe Julio, Ricardo Arachavela and Vicente Felgueres.

The price is the only element of the marketing mixture that is generating of income; all the others are costs: therefore, it must be an active instrument of the strategy in the main areas of the decision making of marketing. The price is an important competitive tool to face and to overcome to the rivals and products near substitutes. The prices next to the costs, will determine the long term viability of the company (Chawla, A.2003)

The function of the analysis of the market in relation to the determination of the prices of export consists of establishing a maximum limit to the corresponding decision, from the demand of the product and the characteristics of the competitors. In the countries developing it is the situation of the market that determines the range of prices of exportation.Reason why, one sets out:

> Hypothesis 2: The better the export price of agro-industrial
> products, increasing the international competitiveness.

Environmental Management

Minimize intrusion into the various ecosystems, maximizing the chances of survival of all life forms, however small and insignificant that result from the point of view, not a kind of magnanimity by the weaker creatures, but for true intellectual humility, recognizing that not really know what the loss of any living species could mean forthe biological balance.

> Hypothesis 3: The implementation of environmental management in
> the agro business processesresults in an improvement in international
> competitiveness.

Technological Innovation

Form leaves from the models of the OECD, the Technological Institute of Massachussets, Bancomext, Michael And Porter, Alfonso Cebreros, Henry Mintzberg, Ricardo Arachavela and Vicente. It is a determining variable that influences in the competitiveness of the companies. A thorough study of the variable technology was made to deduce its real definition, its dimensions and indicators.

The technology has been always point used in the speeches of the Mexican Government of an opportunistic and deceptive way. To thus they demonstrate to the National Plan of Desarrollo (PND) and the National Program of Technological Development and Científico it (PRONDETYC). The cost in science and technology is the plus under the emergent countries and comparativily inferior to Economies like the similar Spanish and Korean economic powers (Marcovitch, 1995)

The technology comprises of trinomio science-technology-production. Due to it, the following hypothesis sets out:

> Hypothesis 4: Through the development of technological innovation,
> we can achieve international competitiveness in the agribusiness sector,
> linking scientific research, encouraging the development of patents.

Market

The market variable is also an important point of analysis because many companies now limit their target market Michoacan, to attracting potential customers in international trade fairs or are the same customers who come in search of the product, without having actually performed an analytical study on it, Pedraza mentions the benefit of a market segment is achieved when it meets the following characteristics: grade which can measure the size and buying power of a segment; degree to which the segments are the big enough or profitable, Accessibility (how easy to reach and serve effectively segments) actionability (degree to which effective programs can be formulated to bring and serve the segments (Pedraza Rendon, 2002).

> hypothesis 5: The international competitiveness of the agribusiness sector is given from the strengthening of the market, market segmentation, demand analysis, marketing strategies and appropriate distribution channels.

Public Policies

Public policy is the set of activities of the institutions of government, acting directly or through agents and that are intended to have a certain influence on the lives of citizens (Pallares, 1988), the author also mentions that must be considered as a decision-making process and make it as a set of decisions that arise to take place in a given period of time. To be considered public, must have been created or processed through institutions or government agencies.

The instruments used by government institutions that make public policy, are the actions of public policy. These instruments are legal standards, services, financial resources and persuasion.

> Hypothesis 6: The agro-industrial policies through government programs promoting international competitiveness of the agribusiness sector.

Hierarchial Structuring Variables

They were six variables that emerge from the theoretical framework of competitiveness, to interact and determining the assumptions to prove or disprove as the result of the collection, processing and interpretation of data obtained from the field study.

To determine the relevance between variables considered in this investigation, we used the analytic hierarchy technique, which is used to assess the significance of the problems or the cause thereof. The method involves making pairwise comparisons between each of the alternatives in this case are variable and, on a scale previously specifically assess the magnitude of preference between them, depending on the purpose intended to make the comparison.

Market mentions that the problem statement assumes that all variables and objectives can be divided into different subsets and that there is order among them, so that the elements of a set have higher or lower priority than another. So, the problem is summed to determine the importance of each of the variables considered, so one must know the scale of values associated with each of them, ie what is the best or worst in terms of variable how they contribute to a greater or lesser degree to the achievement of each of the objectives set for analysis.

The structure of the problem requires the division level objectives, which in this case the general aims of competitiveness and both can be divided into the layers of processing the product, marketing and business as a whole. The method first compares the importance of each of the variables together, based on comparison to one of the objectives of the next higher level. Thus, comparing in each stratum, the items were identified as targets. In summary, the method provides the relative importance of each variable in each of the levels using the following theorem (Saaty,1980):

$$W = \left[V_1^5 V_2^5 V_3^5 V_4^5 \middle| V_1^4 V_2^4 V_3^4 \middle| V_1^3 V_2^3 \middle| V_1^2 \right] \tag{1}$$

Subject:
W = Total relevance
V = Independent variable

Table 1. Hierarchy of objetives and variables, based in the results of the theoretical framework of documentary research

International Competitiveness					
Product		Company	Distribution		
Quality	Price	Environmental Management	Technological Innovation	International Market	Public Policies

The fact that the evaluation of each variable is made by a person implies that there is some bias opinion occurred, but to subdivide the original problem in a large number of subproblems and the error is minimized.

To solve the problem, first raised the different strata in which the targets were located and the variables considered, and relations between them according to the model mentioned above. Thus the following table is constructed with three levels where the first is the main objective of competitiveness study, in the second level are three objectives for the application of this technique are considered the product, marketing and business. Finally, the third layer, are the variables considered in the study.

Table 2. Significance of the independen variable, based on the resultsof field research

Objetives	Quality	Price	Environmental Management	Technological Innovation	International Market	Public Policies	Total
Product	33.4%	30.2%	8.5%	13.2%	11.9%	2.9%	100%
Company	33.2%	41.4%	5.2%	9.5%	8.4%	2.4%	100%
Distribution	32.9%	32.4%	6.6%	10.8%	14.4%	2.9%	100%
Total	33.2%	34.4%	6.9%	11.3%	11.4%	2.7%	100%

Solving the matrix system as required by Theorem "Saaty", and applying the estimates of importance of objectives and variables whose scale is presented in the rating scale of activities, obtain the results shown in the table below, which shows that the total sum to relevances hundred percent.

According to this technique, it appears that the overall importance of the variables is given in the following order: quality, price, international market, technological innovation, environmental management,and public policies. The results indicate the main objective with 40.9% product, company with 32.8% and the distribution with 26.3%. This is the importance given to the analysis of competitiveness through the variables considered.

3   Conclusions

The results obtained by processing data show a complete picture of how you are doing export agribusiness in the state. We start analyzing the results of the ranking of the variables obtained through the methodology AHP Saaty T, which shows the weight to give each of the companies to independent variables, where first is the variable quality followed by the price variable, third market, technological innovation fourth, fifth environmental management and ultimately public policy environment.

This weighting was based on their perception of how this variable impacts on competitiveness, by giving companies more value to the variable price, means that its objectives, goals and projections will be based on a better structuring of the export price on the other side, public policies have not impacted on the competitiveness of their products, because it really has not been very significant their contribution, have benefited through programs of export promotion, tax refund, but much needed support in a matter of modernization and technology transfer.

References

1. FAO. El estado mundial de la agricultura y la alimentación. Alimentación, O. d. (ed.) Recuperado el febrero de 2009 (2007),
   http://www.rlc.fao.org/prior/desrural/alianza.htm
2. Feigenbaun, A.: Total Quality Control (vol. Cuarta Edición). Mc Graw Hill, USA (1990)
3. Marcovitchy, J., Silber, S.D.: Innovación tecnológica, competitividad y comercio internacional. (Universidad de San Pablo) Recuperado el 3 de septiembre de 2009 (1995),
   http://www.science.oas.org/espanol/redes/part2_re.pdf
4. Pallares, F.: Las políticas públicas: El sistema político en acción. Revista de Estudios Políticos (Nueva Época) (62), 141 (1988)
5. Pedraza Rendón, O.H.: Modelo de plan de negocios para micro y pequeña empresa. In: ININNE (ed.) Universidad Michoacana de San Nicolás de Hidalgo, Morelia, Michoacán (2002)
6. Porter, M.: Estrategia Competitiva: Los conceptos centrales. Recuperado el noviembre de 2011 (2007),
   http://www.fing.edu.uy/iimpi/academica/grado/adminop/
   Teorico/AO_7porter1.pdf
7. Saaty, T.L.: The Analytic Hierarchy Process. McGraw Hill, New York (1980)
8. Sala-i-Martin, X., Blanke, J., Geiger, T., Drzeniek Hanouz, M., Mia, I.: The Global Competitiveness Report 2009-2010. Recuperado el noviembre de 2009, World Economic Forum (2009), http://www.weforum.org/documents/GCR09/index.html
9. SAGARPA. Monitor agroeconómico 2009 del estado de Michoacán. Recuperado el mayo de (abril de 2009),
   http://www.sagarpa.gob.mx/agronegocios/Estadisticas/Document
   s/MICHOACAN.pdf?Mobile=1&Source=%2Fagronegocios%2FEstadistic
   as%2F_layouts%2Fmobile%2Fview.aspx%3FList%3Dea4191c6-15b5-
   4625-afe9-be7e6cce2216%26View%3Df5c8d175-3fb9-49f2-86e6-
   c9db05b29bfb%26Curren
10. WEF. Finding from the Global Competitiveness Index 2007-2008. Austin, E. L., Mia, I. (eds.) Recuperado el septiembre de (2009),
    http://www.weforum.org/pdf/Global_Competitiveness_Reports/
    Mexico.pdf

Division of Labour and Outsourcing: Mutually Exclusive Concepts?

Wilfred I. Ukpere[1,*] and Mohammed S. Bayat[2]

[1] Department of Industrial Psychology and People Management, Faculty of Management,
University of Johannesburg, RSA
wiukpere@uj.ac.za
[2] Management College of South Africa (MANCOSA), RSA

Abstract. Division of labour has been recognised as a source of wealth to nations, by some experts without considering its effects on labour. However, some experts are of the view that increasing division of labour contracts the range of choice of ways of making a living for the working class. Thus the outcome of division of labour is the lowering of value of individual workers. Workers' activities become increasingly narrow and monotonous, through division of labour which mars instead of developing their creativity. Outsourcing on the other hand, has been considered a requirement for corporations to earn higher profits and to respond to competition. Thus, companies that refuse to look outward for lower-cost inputs may lose their competitive advantages under a global setting. This paper contends that the logic of outsourcing cannot be mutually exclusive from the logic division of labour. Outsourcing simply implies the internationalisation of division of labour. Whereas division of labour was localised, outsourcing has been globalised to provide capital with additional leverage to exploit labour globally.

Keywords: Division of labour, Globalisation, Machinery, Outsourcing, Technology.

1   Introduction

Management is the act of doing things through people in an orderly, organised manner. This act of doing things through people has come in different forms at different times over the years. Since the inception of the production function to the current period of globalisation, people have utilised other people to achieve some desired results. However, this process of utilising people over the years has not been an easy task. Different methods of utilising people have emerged over the years. However, for this discourse, we shall consider the mutuality between logics of division of labour and outsourcing.

---

* Corresponding author.

K.J. Engemann, A.M. Gil Lafuente, and J.M. Merigó (Eds.): MS 2012, LNBIP 115, pp. 151–165, 2012.

The Logic of Division of Labour

The renowned treatise of Adam Smith [41]; [8] on 'division of labour', has been considered as a major step in the productive and optimal utilisation of labour. Adam Smith, cited in [20]; [17]; [8] presented one of the clearest and oldest illustrations of the application of division of labour. While analysing the pin-making process in his famous book, 'Wealth of Nation' (WN), Smith explained: *"A person working alone could perhaps make twenty pins per day. But by breaking down the task into a number of simple operations, however, ten workers carrying out specialised job in collaboration with one another could produce 48,000 pins per day. The rate of production per worker, in other words, is increased from 20 to 4800 pins, each specialist operator producing 240 times as much as he or she was working in isolation"*. Thus, Smith has suggested that the increase in productivity is itself a result of three processes: an increased dexterity that accompanies the confinement of workers to a specialised task, a saving of time for workers from passing from one tool to another and the possibility for the creation and introduction of technology that the division of labour stimulates.

In concordance with the above thought, Charles Babbage cited in [17] has posited: *"Technological progress in production can be measured by the degree to which the tasks of each worker are simplified and integrated with those of other workers. This process reduces the price employers have to pay for hiring workers and the time needed to learn each job, as well as weakening the workers' bargaining power and thus keep wage costs down"*.

Taylorism seems to agree with Smithan views by explicating how industrial processes could be broken down into simple operations that could be precisely timed and organised [43]. Taylor was only concerned with improving industrial efficiency, but gave little consideration to how products should be marketed. *'Mass production necessitates mass markets'* and the industrialist, Henry Ford, was one of the first to take notice of this and exploited its possibilities. Fordism was designated to the system of mass production, which is tied to the cultivation of mass markets.

Henry Ford established his first plant at Highland Park, Michigan, in 1913 and made only one product- the Model T Ford- thereby allowing the introduction of specialised tools and machinery that was designed for speed, precision and simplicity of operation. The most spectacular innovation of Ford was the construction of a *'moving assembly line'*, which was inspired by Chicago slaughterhouses, where animals were disassembled section-by-section on a moving line. Each worker on Ford's assembly line had a specific task, such as fitting the right-side door handles, while the car bodies moved along the line. The result of this system was astronomical, since before 1929, when production of the model T ceased, fifteen million cars were made and 80 per cent of the cars in the world were registered in the United States of America [17].

The French sociologist, Emile Durkheim, in wrote towards the end of the 19th century, that the greater the division of labour, the more people would have to depend on each other, and the closer they would become [21].

Thus, Smith and his associate, without considering the negative effect of division of labour on the working class, believed it leads to efficiency and growth, thereby tracing the wealth of the nation to the interaction between a growing division of labour and the scope of market relations.

Contrary to the above views, Marx cited in [21] has opined: *"division of labour impoverishes the worker and makes him a machine...for as soon as labour is distributed, each man has a particular exclusive sphere of activity, which is forced upon him and from which he cannot escape"*. It was on this grounds that Marx cited in [14] further rebuff Smith and his supporter: *"Now it is quite possible to imagine, with Adams Smith that the difference between the above social division of labour and the division in manufacture, is merely subjective, exists for the observer who in the case of manufacture can see at a glance all the numerous operations being performed on one spot, while... the spreading-out of the work over great areas and the great number of people employed in each branch of labour obscured the connection. But what is it that forms the bond between the independent labour of the cattle-breeder, the tanner and the shoemaker? It is the fact that their respective products are commodities. What, on the other hand, characterises the division of labour in manufacture? It is the fact that the specialised worker produces no commodities"*. Division of labour in the workplace may be acceptable to the capitalists in their respective sweatshops. However, the social organisation of division of labour is totally unacceptable. Based on this, Marx cited in [14] posited: *"division within the workshop implies the undisputed authority of the capitalist over the workers, who are merely members of the total mechanism, which belong to him"*. In that case, Marx perceived division of labour (most especially in a capitalist workplace) as ceaseless exploitation of the workers [13].

The outcome of capitalist system of division of labour is 'surplus value' through productivity increases at deplorable wages, which reduces the value of labour [32]; [23]. Marx cited in [18] has further affirmed: *"The source of man's immediate difficulty is the division of labour. Division of labour was the very essence of all that was wrong with the world. It is contrary to man's real essence"*.

The effect of division of labour in the workplace is the creation of a hierarchy of skills and wages in correspondence to the increasing specialised tasks that are undertaken. In many instances, however, the detailed labour becomes increasingly simple even if certain dexterity is gained with practice. Hence, in addition to the creation of skilled specialised job, there is an overwhelming amount of simple, unskilled work activities that are formed to which a large section of the working class is assigned [43]. Marx traced the introduction of machinery that eroded jobs to the increasing division of labour and went further to show the extent to which machinery production utilises the division of labour to diminish the role of workers in the field of production.

[32], maintained that *"whereas manufacture adopted existing methods of production and transformed them through the utilization of the co-operation and division of labour, machinery took on much greater significance and transformed the role played by labour as a whole in the production process. In manufacture, the division of labour brings a range of specialist tools for the workman to use in his detailed task. Machinery production brings the displacement of the worker from the handling of his own tools and instead he becomes a tool of the machine. He becomes robbed even of the simple and specialised task that has been left by the manufacture. The pace of the machine dictates the pace of work. In short, machinery seizes the division of labour created by manufacture, intensifies it and transforms it into a division of tasks between the parts of the machine to which labour becomes an*

appendage". In other words, it is no longer the labourer that employs the means of production, but it is the means of production that employs the labourer, since labour has taken the position as one of the limbs of the machine, which he has created [27].

Moreover, as the division of labour intensified, the source of value and surplus value become more deeply obscured. Firstly, increasing productivity is associated with the power of collective labour organised in co-operation with division of labour that mars the skills of many workers. Therefore, it is capital that increasingly appears to be the source of wealth, since what is a gain for the productive power of capital through collective labour, is a loss to the labourer in terms of dilapidated skills, functions and control. With the growing use of fixed and constant capital and the displacement of the labourer by machine, the significance of labour, as a source of value is increasingly denied, and regarded as one source of value amongst other things [32].

Adam Smith himself, even later wrote, with discernment, about the intellectual degradation of the worker in a society in which the division of labour has proceeded exceedingly far. Smith cited in [22] remarked: *"for by comparison with the alert intelligence of the husbandman, the specialised worker generally becomes as stupid and ignorant as it is possible for a human being to become"*. Smith went further to assert that there is a tendency in commercial society, owing to advanced division of labour, to corrupt the 'intellectual, social and martial virtues' of its citizens [2]. This *'leprosy'* (division of labour) is so great a public evil that it leaves people *'mutilated'* and deformed in their character [2].

Therefore, increasing division of labour (specialisation) narrows the range of choice of ways of making a livelihood for workers [37]. The outcome of division of labour is the lowering of value (in terms of dexterity and remuneration) of individual workers (Hooker, 1999:2). By way of increasing division of labour, workers' activities become more and more narrow and monotonous and instead of developing man's creative powers, it evaporates it, degenerating people into 'idiocy' and 'cretinism' [27]; [22].

It may be easy to show how the growing international division of labour helps to boost world economic performance, while at the same time ignoring its effect on the working class, namely the distribution of this performance. In this regard, [30] opined: *"World market integration is economically very efficient. But, in the absence of state intervention, the global economic machine (division of labour) is anything but efficient in distributing the wealth so produced; the number of losers far exceeds the number of winners"*. Division of labour introduces inequality between occupations and generates disunity amongst workers, which results in social inequality that divides society into haves and have-nots, rulers and the ruled, exploiters and exploited [27]. In Marx's view, division of labour pits a man against his fellow man; creates class differences and destroys the unity of the human race [18].

Division of labour under capitalism therefore amounts to the creation of a class of wage-labourers dispossessed of means of production and forced to become appendages of the machine [32]; [14]; [27]; [33]. Marx observed that the introduction of machinery (a by-product of division of labour) is a stage in the development of capitalism. In his view, manufacture compelled different capitalists to accumulate and this force was strengthened with the introduction of machinery that necessitates huge funding costs, which was once beyond the power of gathering funds through savings and capital accumulation. The reorganisation of capital through liquidation, acquisition and amalgamation/merger, became the new trend and credit system through banking, which was utilised as another instrument of such accumulation [14]. In addition, the greatest stimulus to production by huge automation is only achieved by eradicating the possibility for competition from capitals, which is always achieved through backward methods, namely retrenchment, downsizing, re-engineering and of course through the logic of outsourcing. At this juncture, the discussion will shift towards the logic of outsourcing in order to explicate its mutuality with division of labour in the workplace.

The Logic of Outsourcing
Outsourcing or the foreign 'sourcing' of inputs is one important aspect of globalisation in production, which is the reason why outsourcing is at times referred to as globalisation of production. As a matter of fact, there are barely any products these days that does not have some foreign components in it. Thus, this method of production has developed so rapidly, to the extent that it has become difficult to determine the nationality of most products. Indeed, [39] has stated: *"should Honda Accord produced in Ohio, be considered American? What about a Chrysler mini-van produced in Canada, especially now that Chrysler has been acquired by Daimler-Benz (Mercedes)? Is a Kentucky Toyota or Mazda that uses 50% imported Japanese parts American? It is clearly becoming more and more difficult to define what is American and opinions differ widely. One could legitimately even ask if this question is relevant in a world growing more and more interdependent and globalised. Today, the ideal corporation is strongly decentralised to allow local units to develop products that fit into local cultures, and, yet, it is much more centralised at its core to coordinate activities around the globe".*

Indeed, outsourcing has become a requirement for corporations to earn higher profits and to respond to import competition [15]. In this case, companies that refuse to look outward for lower-cost inputs may loose their local and international competitiveness under a global setting. [39] further maintained that the need to be competitive, *"...is the reason that $625 of the $860 total cost of producing an IBM PC was incurred for parts and components manufactured by IBM outside the United States or purchased from foreign producers during the mid-1980s. Such low-cost offshore purchase of inputs is likely to continue to expand rapidly in the future and is being fostered by joint ventures, licensing arrangements, and other non-equity collaborative arrangements. Indeed, this represents one of the most dynamic aspects of the global business environment of today".* According to Roger Herman, of The Herman Group (Greensboro, North Carolina, USA) cited in [34], *"One area that will continue on its current part, is outsourcing.* Indeed, outsourcing has greatly increased over the last two decades. For example, between 1972 and 1990, imported intermediate inputs to Britain alone increased from 5.3% of materials purchased to 11.6% of materials purchased [15]. In fact, substantial evidence points to the fact that outsourcing has become widespread among modern producers [28]. For example, Nike only employs 2,500 persons in the USA for marketing and other headquarters services, whereas about 75,000 persons are employed in Asia to produce shoes that are purchased back by Nike. Also, currently, General Electric in the US imports from Samsung in Korea all the microwaves that are marketed under their brand-name [29]; [25]. Outsourcing is also claimed as an important activity in industries such as footwear [47]; [1], textiles [16]; [18] and electronic [3]. These aforementioned examples demonstrate that outsourcing applies to both finished goods and intermediate inputs [3].

Reflecting a bit on the meaning of outsourcing, [3] have stated that *"outsourcing takes place where companies take the benefit of both the low wage costs of relatively labour-abundant countries and modern production techniques, and whereby the process of producing a product can be broken into a number of discrete activities, by shifting the low-skill intensive section of production abroad, but continue to carry out the high-skill intensive activities themselves, and, once the low-skill activities have been completed, the goods are then imported back from the low-wage countries and either used as intermediate inputs or sold as finished goods".* Hence, outsourcing includes parts and components arrangement from offshore and contract work done by others. Moreover, another category of outsourcing includes goods that are produced entirely by subcontractors, where the outsourcing manufacturer attaches its brand-name to the completed product. An example of such outsourcing is reflected in the statement of the president of the American division of Levi Strauss, John Ermatinger cited in [25]: *"Our strategic plan in North America is to focus intensely on brand management, marketing and product design as a means to meet the casual clothing wants and needs of consumers. Shifting a significant portion of our manufacturing from the U.S. and Canadian markets to contractors throughout the world will give the company greater flexibility to allocate resources and capital to its brands. These steps are crucial if we are to remain competitive".* Similarly [25] noted: *"From El Paso to Beijing, San Francisco to Jakarta, Munich to Tijuana, the global brands are sloughing the responsibility of production into their contractors; they just tell them to make the damn thing, and make it cheap, so there's lots of money left over for branding. Make it really cheap".* Indeed, contracting a foreign firm to manufacture goods that have been designed and distributed by companies in the advanced country, has become an important form of outsourcing.

Another type of outsourcing includes contract work done by others through the use of foreign plants for product assembly. Currently, assembly services represent large shares of U.S. imports from low-wage countries. For example, imports from offshore assembly plants accounted for 42.2% of U.S. imports from Mexico (Feenstra and Hanson, 1996:4-5). Therefore, within the context of globalisation and the need to respond to competition, organisations are reverting to their core functions, consolidating these functions and casting off or outsourcing peripheral activities. [5]. Hence, [24] observed that *"The merits of large highly integrated corporations occupying multiple stages in the value chain were diminished in favour of 'focus, focus, focus'. This new conventional wisdom decreed that organisations should concentrate on those activities in which they have significant, competitive advantage. Any 'non-core' activities should be outsourced".*

Indeed, with respect to the global economy, outsourcing has become the manufacturer's new international economies of scale. [39] noted that *"just as companies were forced to rationalise operations within each country in the 1980s, they now face the challenge of integrating their operations for their entire system of manufacturing around the world to take advantage of the new international economies of scale. What is important is for the firm to focus on those components that are indispensable to the company's competitive position over subsequent product generation and 'outsource' all the rest from outside suppliers in order to have a distinctive productive advantage"*. On this keynote, [25] stated: *"Many companies now bypass production completely and instead of making the products themselves, in their own factories, they 'source' them, much as corporations in the natural-resource industries source uranium, copper or logs. They close existing factories, shifting to contracted-out, mostly offshore, manufacturing. And, as the jobs fly offshore, something else is flying away with them: the old-fashion idea that a manufacturer is responsible for its own workforce"*.

In fact, the origin of outsourcing could be traced to two important revolutions that occurred in business, namely the Multi National Corporation (MNC) and the *'Retail'* business revolutions. In the first instance, the MNC revolution enabled business to learn how to render high-tech innovations that complements with globally mobile production systems. Moreover, MNC activities offered a first scope within which capital was able to put labour in international competition, and this competition, indeed, has had significant, negative effects on manufacturing wages, employment and union membership [9]; [10]. Then, the second revolution, namely the retail business revolution, was linked to some new sourcing model based on big-box discount stores. The first phase of this revolution could be traced back 40 years ago with the surfacing of big-volume discount stores like Wart-Mart, which was founded in 1962 [35]. During this phase, the business model was mainly based on national sourcing, and the big-box stores bought from the cheapest, national manufacturers. These stores created competition for producers nationally, so that for example, companies in California were forced to compete with those in New York. This national rivalry provided lower prices, and it was beneficial because all suppliers were located and operated within the same territorial jurisdiction. However, to some extent, it also led to cut-throat competition as it pressured some manufacturers to move southward to non-union *'right-to-work'* states where organising workers was much more difficult and labour costs were lower [35]

However, the most contested phase of this revolution commenced in the 1980s, when the big-box discount stores began to move out of their territorial boundaries to outsource goods and services. The consequence of this new trend was that US suppliers were no longer merely in national competition, but they were in an international bidding competition. Thus, California was no longer only competing with New York, but US producers were now forced to compete with companies in China, Indonesia and Mexico. According to [35], *"The economic logic of this global sourcing model is simple. Scour the world for the cheapest supplier and lowest cost- the so-called 'China price'- and then require US manufacturers and workers to match it if they wish to keep their business"*.

However, today outsourcing is not only a game of the retail sector, but has drifted to the production and service sectors [40]. In fact, production and service companies, in the advanced countries, are now busy shifting their activities offshore (have become transnationals and mobile) in order to compete with low cost companies, in the developing countries. [15] reported that more and more TNCs are engaged in a substantial amount of outsourcing. For example, the Compaq Computer Corporation purchases parts for personal computers from its foreign branches and from outside foreign suppliers. In both cases Compaq imports components that it could have previously produced domestically. Hence, both forms of outsourcing will definitely affect the range of activities that Compaq would perform in its local production operations. Another example is when Texas Instruments, which set up an impressive software programming operation in Bangalore in India some few years ago, was emulated by other American MNCs. Presently, Motorola, IBM, AT&T, and many other high-tech firms have currently shifted a lot of their basic research abroad and, in 2004, IBM indicated that it would shift about 7,500 high-tech jobs abroad to lower costs [39].

Thus, outsourcing can be viewed as an application of the retail sector's sourcing model to production and service sectors, and this development has been accelerated by technological innovations and improvements in computing, electronic communication and the internet [42]. [12] reiterated: *"In a way, the latest outsourcing phase is simply a result of the internet bubble. Thousands of kilometres of fibre optic cable and high bandwidth connections, laid during the boom years, have united much of the world in high speed connectivity. Relentless growth in storage capacity and high-speed transmission (digital scanning is currently at 200 pages a minute), has meant that anything can be digitised and sent anywhere for processing".* Similarly, [35] remarked: *"Owing to improvements in electronic communication and the internet, many services that were previously non-tradable, have become tradable. These include basic computer maintenance and software programming, tax preparation and accounting, architectural planning, and telephone call centres. Even retail sales are potentially tradable, as indicated by the success of the Amazon.com business model".* In effect, all types of companies are now engaged in outsourcing, and also want suppliers to meet the so called 'China price'. These dynamics, though originating in the retail sector, have eroded manufacturing and service jobs and wages [36].

Outsourcing does, indeed, deliver low prices but at the expense of workers [35]. In this regard, with American workers, [12] observed: *"Blue-collar workers, long wary of outsourcing, have been joined by programmers, engineers and office workers. The media is covering the story more than ever before. One CNN program has begun campaigning against outsourcing, compiling a web based list of companies (so far totalling 326) that it accuses of 'exporting America' by 'either sending American jobs overseas, or choosing to employ cheap overseas labour, instead of American workers".* [3] felt that trade with the low-wage countries via outsourcing, will surely shift employment away from less-skilled towards skilled workers in the advanced nations and put a downward pressure on the relative wages and employment of low-skilled workers within industries.

Although it was previously believed that outsourcing was companies' response to global competition, however, from the 1980s onward, it became clear that most TNCs have resorted to it, in order to counter some of the rigidities necessitated by government regulations in some countries. A majority of the TNCs adopted it because it provided some leeway from a legal point-of-view against labour. However, there remain innumerable pitfalls with such modes of outsourcing, particularly where skilled jobs are sent away. In such situations, employers may find in the long-run, that the decision to outsource is not as sagacious as envisaged [5]. For example, the negative repercussion of this style of outsourcing, can be exemplified by the two US auto part companies Visteon and Delphi, former subsidiaries of Ford and General Motors, respectively. Visteon and Delphi had initially competed nationally. However, as both Ford and General Motors announced in 2005 their intentions to outsource from low cost companies in order to meet the *'China price'* of being globally competitive, both Visteon and Delphi, owing to higher union wages and benefits in America, joined the race by shedding jobs and moving production offshore, including to China. However, in the long run, both came to realise the difficulties that are associated with this mode of outsourcing and in October 2005, Delphi became bankrupt [35].

In fact, this increasing drive to outsource began after the 1987 recession, when companies became desperate for cost-cutting measures to boost profits. Thus, with manufacturing transferred overseas, high-speed imaging and communication technology helped to reduce costs in software applications, data processing, accounting and customer service. In addition to the high-tech innovations, there was also an increase in the number of English-speaking accountants, engineers and business students who came from low cost country universities, such as India and, for many of these new graduates, the call centres are a first step into the job market, while indeed, many of these graduates are willing to obtain a little portion of what their counterparts earn in America and Europe. This is the main reason why most TNCs have rushed into India's cyber-office space [12]. Presently, most TNCs have become *'virtual manufacturers'*. Although their product design and marketing is done in their parent countries, the actual production work is carried out in lower cost locations, such as China or Mexico [12]. [24] commented: *"Much was written about the advent of the 'virtual corporation' that would only exist as a brand. Labour market problems were to be vanquished in the electronic lightening of new technology"*.

The positive side of this story is that countries like India, have leapfrogged into the 21st century by setting up high speed networks, effectively turning their cyberspace into virtual office space for the West. In that case, an employee sitting in Chennai, in India, can examine the image of a medical insurance claim in the West on his computer screen and complete the form for processing. In that light, Andrew Grove, CEO of Intel Corporation cited in [12] exclaimed: *"From a technical and productivity standpoint, the engineer sitting 6,000 miles away, might as well be in the next cubicle and on the local area network"*.   Certainly, it is this imperceptible worker that is willing to work for a little fraction of the average US wage that has eroded America jobs. Indeed, McKinsey's cited in [12] predicted that by 2008 IT services and back-office work in India will grow fivefold and will employ four million people. This is indeed a sign that the upward trend in service outsourcing to low cost nations, will continue unabated to the detriment of service workers in advanced countries. In this respect, [12] retorted: *"There is a gnawing fear that, given the cost advantages and unlimited supply of competent workers, jobs now leaving the US may not come back. The savings that corporations achieved through outsourcing will reduce consumer prices and raise shareholders' profits, but without necessarily creating any jobs at home. The classic solution to the problem of job loss, created by technology, has been to promote education and retraining programs. But, if an unlimited supply of workers with similar skills is available at the end of a broadband wire for a tenth of the salary, the textbook economics remedy may not work. There will obviously be many office jobs requiring direct client and team contact, but those jobs that can be done in isolation, are increasingly up for grabs in a global labour market"*.   Indeed, work that was previously done in the United States and other industrialised countries, is now done at a lower cost in some developing countries. This is not only for low-skilled assembly-line jobs but includes job requiring high computer and engineering knowledge.

Previously, it was a general belief in the US that only the unskilled job would fly away, leaving American workers with the highly paid white-collar jobs. However, recently IBM has moved millions of white-collar jobs to countries like India and China, and contended that such moves enhance their competitive advantage, hold costs down for American consumers, and help to develop poorer nations while supporting overall employment in the United States, by improving productivity and the nation's global reach. Nevertheless, this rationale does little to calm the growing concerns of many politicians and employees in the US and other advance countries that are affected. This is the reason why it was predicted that outsourcing will result in unemployment, resentment of foreigners training in the US and retaliatory unionisation efforts [19]. However, in spite being concerned about the repercussions, IBM continued with their plans to outsource thousands of jobs overseas. According to the company's employee relations officer, Tom Lynch (cited in Raynor, 2003:4), *"This challenge really hits us squarely between the eyes. We don't want to sit back and say 'don't do it' because it's going to be a real problem. Our competitors are doing it and we have to do it"*. Similarly, [39] commented: *"Globalisation in production and labour markets is important and inevitable- important because it increases efficiency; inevitable because international competition requires it. Besides the well-known static gains from specialisation in production and trade, globalisation leads to even more important dynamic gains from extending the scale of operation to the entire world and from leading to the more efficient utilisation of capital and technology wherever they are more productive. Otherwise, competitors would do so and the firm would lose its markets and might even be forced to shut down. For the same reason, firms must outsource labour services or employ labour off-shore where it is cheaper or more convenient"*. However, one computer company executive worried that *"Once those jobs leave the country, they will never come back"* [46].

John Ermatinger, while explaining the decision to shut down twenty-two Levi plants and lay-off more than 16,000 workers, stated: *"As far as the company is concerned, those 16,310 jobs are off the payrolls for good, replaced by contractors throughout the world. Those contractors will perform the same tasks as the old Levi's- owned factories- but the workers inside will never be employed by Levi Strauss"* [25]. Indeed, most Americans have only now come to fully realise that there is a truly competitive labour force around the world that is willing and able to do their jobs more efficiently at much lower costs [6]. Thus, service industries are not immune to global competition and outsourcing. For example, more than 3,500 workers in the island of Jamaica, are connected to the United States by satellite dishes to make airline reservations, process tickets, answer calls to toll-free numbers, and do data entry for US airlines at a much lower cost than could be done in the United States. Therefore, even highly skilled and professional people are not spared from the competition which is triggered by outsourcing [30]; [31].

Outsourcing in production is the cause of the decline in the demand for, and the wages of semi-skilled and unskilled labour in the advanced countries [42]. For example, in the age of outsourcing, companies throughout the Silicon Valley have abolished many permanent jobs, and contracted work to agencies for temporary staff, where they have few or no benefits [4]; [11]. Regular employment in major computer companies, like Sun, Hewlett-Packard and Apple, is stagnant or declining, while these same companies subcontract for many of their components with manufacturers who pay 30% less to their employees [6]. Indeed, most workers will turn to free agents, and a number of them will sell their services through an international network of brokers [34].

Therefore, outsourcing could be one the main culprits for the widening inequality in the advanced nations. In this regard, [15] pointed out that since the late 1970s, the wages of less-skilled U.S. workers have decreased dramatically, both in real terms and relative to the wages of more-skilled U.S. workers. Indeed, two main explanations frequently offered for the apparent shift in demand away from low-skilled workers in the UK and other industrial countries, are that skill-biased labour-saving technical progress has reduced the relative demand for unskilled workers and that increased international trade with nations that have an abundant supply of low-skilled and low-wage labour, has decreased the demand for low-skilled workers in the advanced, industrialised countries [3]. Therefore, ignoring outsourcing misses an important channel through which trade affects the demand for labour of different skill types [15]. Hence, when firms outsource, they narrow the range of activities that the domestic industry performs, which can reduce the industry unit demand for less-skilled labour[15]. In that case, outsourcing can have a damaging effect on the economic fortune of the less-skilled within the advance countries [3], and, indeed, that could explain the reason why workers who were retrenched in the US during the 1980s, were not rehired [15].

Ensuing job losses appear more unnerving for three additional reasons. Free trade theorists, Stolper and Samuelson cited in [35] have long established that when a rich capital-abundant country engages in free trade with a poor labour-abundant country, wages in the rich country decrease. Therefore, by combining global sourcing with globalisation of production, the new system places the Stolper-Samuelson effect on motion [35]. Hence, one explanation of how trade with low wage countries may push down the relative wages and employment of unskilled workers within industries, is provided by the notion of 'outsourcing' [35].

Outsourcing can account fully for 51.3% of the increase in the non-productive wage share. Therefore, outsourcing performs substantially better in accounting for the increase in the relative demand for non-production labour in advanced nations. This should not be too surprising, since the new measure of outsourcing is a much more direct estimate of the extent to which industries move production activities offshore [15]. For example, the price of imports from low-wage countries, relative to the price of UK products, explains some of the rise in UK inequality. It may be the case that this relative price term captures the *'threat'* of increased competition from low-wage countries, arising from appreciation of the UK currency. Therefore, as an alternative to reducing labour costs by outsourcing, may have also encouraged other firms to implement measures, which restrain the wages and, perhaps terminate the employment of less-skilled workers in order to remain competitive against low-wage countries.

At the same time, this *'threat'* may have made it considerably easier for firms to obtain the agreement of their workforce for the implementation of such measures [3]. In the UK, the relative wages and employment of the low-skilled, have indeed fallen dramatically during the 1980s. For example, the real earnings of the top tenth of male earners in the UK, rose at a rate five times faster than that of the earnings of the bottom tenth of male earners [7]. The unemployment rate of less-skilled males in the UK rose from 6.4% in the mid-1970s to 18.2% by the mid 1980s, whereas over the same period, the unemployment rate of skilled males only rose from 2.0% to 4.7% [3].

Furthermore, outsourcing is deeply embedded in the service sectors and has led to a spate of retrenchments in this sector. This development will continue to be a cause for renewed concern in the advanced countries. For example, in the US today, the discordance towards outsourcing is not like that of the past decades when manufacturing jobs left America for cheaper shores, and then, some opposition were calmed by the prospect of US workers moving *'up the economic ladder'*, as the US transitioned to a service economy. However, currently, the upper end of that very growing sector of the US economy, is under threat. Unlike in the 1990s, when a period of mass lay-off was more than offset by the net creation of 22 million new jobs, the current job creation machine seems to be sluggish [12]. Mandel, cited in [38] reiterated: *"Two decades ago, the loss of auto jobs and other high paying manufacturing jobs sparked fears of a hollowing- out of the US economy. Yet, painful as the loss of those positions were, strong economic growth and innovation created far more- and better- jobs to replace them. Now, the same process, many economists argue, is going on in services. Yes, some individuals are losing out as well-educated programmers or engineers can do the same jobs for far less halfway across the globe. But, as the US economy evolves, innovation will create new high-paying jobs. Others, though, argue that the outsourcing of highly skilled service jobs is fundamentally different- and poses greater risks for the US economy"*.

According to a private IT research firm, Forrester, about 400,000 American service jobs have been moved overseas since 2000. Over the last decade, the top 25 British companies have destroyed more than 200,000 British manufacturing jobs, and acquired, or created, a similar number of jobs overseas [31]. Forrester further maintained that by 2015, about 3.3 million American service jobs will be off-shored. Disturbed about the current events, an Illinois congressman, who previously chaired the US House Committee on Small Business, Don Manzullo, opined: *"What do you tell the Ph.D., or professional engineer, or architect, or accountant, or computer scientist to do next? Where do you tell them to go"?* [12].

Indeed, global outsourcing poses some new economic challenges and its solution requires a new set of institutions. The task is complicated by problems associated with a lack of global, regulatory institutions and changes in the balance of power between the government and the TNCs that makes it difficult to enact needed reforms. The eminent researcher and writer, Dr Craig Roberts, expressed his remorse towards the current logic and maintained that: *"Trade implies reciprocity. It is a two way street. There is no reciprocity in outsourcing, only the export of domestic jobs...if there are no given endowments because of business know-how, capital and technology are globally mobile, the advantages lies with countries with untapped pools of educated and skilled low-wage labour"* [38]. In a similar frame of mind, the former US Democratic Party Presidential candidate, Senator John Kerry, criticised companies outsourcing American jobs, and nicknamed them *'Benedict Arnold companies'*, after the most reviled traitor of the American War of Independence. During his presidential campaign, Kerry stated: *"Companies will no longer be able to surprise their workers with pink slips instead of pay check; they will be required to give workers three months notice if their jobs are being exported offshore"* [12]. Dr. Alfie Kohn is of the opinion that outsourcing is essentially detrimental, and believes that even productivity would all be improved if this pattern of relentless competition is abandoned [38].

Presently, a plethora of anti-outsourcing legislation has been introduced to US state legislature, and the US Senate has considered banning the outsourcing of government-funded projects [12]. Though, previously, tough immigration policies worked to keep millions of Third World workers out of the developed country shores. Nevertheless, the challenge of a global, visual labour force will necessitate a new global development strategy. Of course, trade protectionism may seem plausible under the current dispensation. However, the solutions of another era may not measure well in the post industrial age [12]. In fact, sooner or later, firms may come to realise that they are trading off long-term interest, in favour of short-term gains for immediate returns. In fact, most firms that outsource, may eventually find that other alternatives would have been wiser to protect their knowledge assets and workers, rather than search the globe for the next low-wage earners. They may come to realise that their real strategic advantage comes from maximising knowledge and leveraging ideas in the long run, instead of maximising employee costs now for short-term gain [38].

Moreover, Dapice's cited in [45] contended that the argument that outsourcing allows hundreds of thousands of people in developing countries, like Vietnam, the chance to earn wages, pull themselves out of poverty and, in turn buy goods that are produced overseas is unrealistic. In fact, most of the workers in those poor nations cannot afford the goods, which they have produced for the TNCs in their countries, much less the ones produced overseas [44]. In addition, most of these jobs sent to the Third World nations are of low quality, cheap, and do not arrive in the Third World in the same form that it left the First World. This, indeed, amounts to a global depletion of the value of labour. Furthermore, countries in the Third World do not need cheap or devalued jobs because these have not brought development and growth to the region. It simply drains their human resources. What are most needed in the poor countries are the cancellation of their debts, reparation and developmental aids and overheads. These would enable them to create jobs for their people rather than being exploited by the so-called job providers who pay mere subsistence wages.

Analysis and Conclusion
Division of labour in production obscures the role of labour in the production of value and surplus value. In the social division of labour, the concealment is reinforced. The confinement of each worker to a particular task, within a particular sector, renders impossible a direct vision of the performance of surplus labour. Therefore, as the social division of labour develops, the source of profit in surplus labour will continue to be less transparent. Outsourcing as well obscures the role of labour in the production of surplus value. Therefore, outsourcing, off-shoring, foreign-sourcing, globalisation of production and whatever other name it may be called, is nothing but division of labour (discussed above), which has assumed an international dimension. Whereas previous division of labour was localised, the current one has been internationalised or globalised to provide capital with additional leverage to exploit labour globally. Hence, the logic of outsourcing cannot be mutually exclusive from the logic division of labour.

References

1. Adams, M., et al.: The relationship between Globalisation, Internationalisation and unemployment. Unpublished article prepared for CPUT (2005)
2. Alvey, J.E.: Adam Smith's three strikes against commercial society. International Journal of Social Economics 25(9), 1425–1441 (1998)
3. Anderton, B., Brenton, P.: Outsourcing and low-skilled workers in the UK. Centre for the Study of Globalisation and Regionalisation (CSGR) working Paper no.12 (1998)
4. Bacon, D.: The new face of union busting. dbacon@igc.org (1996) (accessed June 10, 2004)
5. Bendix, S.: Industrial relations in South Africa. Juta, Cape Town (2005)
6. Benner, C.: Growing together or drifting apart: A status report on social and economic well-being in Silicon Valley. Working partnerships & Economic policy institute, Washington (1998)
7. OECD Employment Outlook, pp. 157-184 (July 1993)
8. Boonzaier, B.: Revision of job characteristics Model. Cape Technikon (D.Tech Thesis), Cape Town (2001)
9. Bronfenbrenner, K.: Uneasy terrain: The impact of capital mobility on workers, wages, and union organising. Report prepared for the US Trade Deficit Review Commission, Washington (2000)
10. Bronfenbrenner, K., Luce, S.: The changing nature of global restructuring: The impact of production shifts on jobs in the US, China, and around the globe. Report prepared for the US-China Economic and Security Review Commission, Washington (2004)
11. Burbach, R.: Globalisation and postmodern politics: from Zapatistas to high tech robber barons. Plato, London (2001)
12. Chanda, N.: Outsourcing debate-Part II: Challenge of virtual office-space needs long-term solutions, not election year quick fix. YaleGlobal (February 27, 2004)
13. Dowd, D. (ed.): Understanding capitalism: critical analysis from Karl Marx to Amartya Sen. Pluto, London (2002)
14. Fine, B.: Theories of the capitalist economy. E. Arnold, London (1982)
15. Feenstra, R.C., Hanson, G.H.: Globalization, outsourcing and wage inequality. NBER Publications, Cambridge (1996)
16. Gereffi, G.: The role of big buyers in global commodity chain: How US retail networks affects overseas production patterns. In: Gereffi, G., Korzeniewicz (eds.) Commodity Chain and Global Capitalism. Praeger, Westport (1993)
17. Giddens, A.: Sociology, 2nd edn. Polity Press, Cambridge (1993)
18. Greaves, B.B. (ed.): Free market economics: a basic reader. Foundation for Economic Education, Irvington on Hudson (1975)
19. Greenhouse, S.: IBM explores shift of white-collar jobs overseas. The New York Times, 1 (July 22, 2003)
20. Hackman, J.R., Oldham, G.R.: Work redesign. Addison-Wesley, Philippines (1980)
21. Hawthorn, R.: Asking about society. Cambridge University Press, Cambridge (1981)
22. Heilbroner, R.: Adam Smith (2002), http://www.kat/history/Eco/SmithAdam.html (September 5, 2004)
23. Hooker, R.: Capitalism (January 19, 2004), http://www.wsu.edu
24. Kirkbride, P. (ed.): Globalization: the external pressures. John Wiley, Chichester (2001)
25. Klein, N.: No logo: no space, no choice, no job. Flamingo, London (2001)
26. Kumar, P.: Personnel management and industrial relations. Kadar Nath Ram Nath, Meerut (2000)

27. Leatt, J., Kneifel, T., Nurnburger, K.B. (eds.): Contending ideologies in South Africa. David Philip, Cape Town (1986)
28. Legrain, P.: Open World: The truth about globalisation. Abacus, London (2002)
29. Magaziner, K., Patinkin, M.: Fast heat: How Korea won the microwave war. Harvard Business Review, 83–92 (January-February 1989)
30. Martin, H.-P., Schumann, H.: The global trap: globalisation and the assault on prosperity and democracy. HSRC, Pretoria (1997)
31. Matthews, R.: The myth of global competition and the nature of work. Journal of Organisational Change Management 11(5), 378–398 (1998)
32. Marx, K.: Capital: a critique of political economy. Penguin, Harmondsworth (1976)
33. Miles, R.: Capitalism and unfree labour: anomaly or necessity. Tavistock, New York (1987)
34. Nel, P.S., et al.: Human resources management. Oxford University Press, Cape Town (2004)
35. Palley, T.: The economies of outsourcing: How should policy respond? FPIF
36. Policy Report (November 16, 2006), http://www.fpif.org/fpifxt/3134
37. Pasricha, A.: WTO, self-reliance and globalisation. Deep & Deep, New Delhi (2005)
38. Raynor, W.: Outsourcing Jobs Off-shore: Short and long-term consequences (2003), http://www.network.com/Pages/Opinion/Raynor/Outsourcing%20Consequences.html (January 16, 2007)
39. Salvatore, D.: Growth and poverty in a globalizing world. Journal of Policy Modeling 26, 543–551 (2004)
40. Slabbert, A.: Capitalism at the crossroads. International Journal of Social Economics 23(9), 41–50 (1996)
41. Stoner, J.A.F., Freeman, R.E.: Management, 5th edn. Practice Hall, Englewood Cliffs (1992)
42. Streeten, P.: Globalisation: Threat or opportunity? Copenhagen Business School, Copenhagen (2001)
43. Worsfold, A.: Scientific management: arguing for maximum profitability, or Taylorism (2004), http://www.change.freeuk.com/learning/business/taylorism.html (March 10, 2004)
44. YaleGlobal. Outsourcing debate- Part II (2004), http://www.yaleglobal.yale.edu/display.article?id=3422 (December 20, 2006)
45. YaleGlobal. Outsourcing debate- Part III (2004), http://www.yaleglobal.yale.edu/display.article?id=3442
46. YaleGlobal. IBM explores shift of white-collar jobs overseas (2007), http://www.yaleglobal.yale.edu/display.article?id=2142 (February 17, 2007)
47. Yoffie, D.B., Gomes-Casseres, B.: International trade and competition. McGraw-Hill, New York (1994)

OWA Operators in the Assignment Process: The Case of the Hungarian Algorithm

Emili Vizuete Luciano, José M. Merigó, Anna M. Gil-Lafuente,
and Sefa Boria Reverté

Department of Business Administration, University of Barcelona,
Av. Diagonal 690, 08034 Barcelona, Spain
{evizuetel,jmerigo,amgil,jboriar}@ub.edu

Abstract. We develop a new assignment algorithm by using a wide range of aggregation operators in the Hungarian algorithm. We introduce a new process based on the use of the ordered weighted averaging distance (OWAD) operator in the Hungarian algorithm. We refer to it as the Hungarian algorithm with the OWAD operator (HAOWAD). The main advantage of this approach is that we can provide a parameterized family of aggregation operators between the minimum and the maximum. Thus, we can represent the information in a more complete way. Furthermore, we also present a general framework by using generalized and quasi-arithmetic means. We end the paper with a practical application of the new approach in a financial decision making problem regarding the assignment of investments.

Keywords: Uncertainty modelling, ordered weighted average, Hungarian algorithm, assignment theory, distance measure.

1   Introduction

In real life, we have observed that in several cases, the distance measure is usually used to calculate the deviations between different arguments. After doing an extensive literature review, we have found a great variety of distance measures. Among the existing ones, we observe that the Hamming distance is widely used by researchers in various fields of science. For example, De Luca and Termini (1973) and Kaufman (1975) used it for measuring the entropy of fuzzy sets. Gil-Aluja (1999) applied it in human resource management. Note that a very useful survey about different decision-making methods can be found in Figueira et al. (2005). Usually when dealing with distance measures it is necessary to aggregate them by using an aggregation operator.

One of the most common aggregation operators is the weighted average. It aggregates the information giving different degrees of importance to the elements. It has been applied in an incredibly wide range of aggregation operators. Another contribution that we can consider, very useful for the realization of our study, is the ordered weighted average (OWA) operator (Yager, 1988; Yager and Kacprzyk, 1997). It provides a parameterized family of aggregation operators between the minimum and the maximum. Since its introduction it has been studied by different

K.J. Engemann, A.M. Gil Lafuente, and J.M. Merigó (Eds.): MS 2012, LNBIP 115, pp. 166–177, 2012.
© Springer-Verlag Berlin Heidelberg 2012

authors. For example, Canós and Liern (2008) and Merigó and Gil-Lafuente (2011b) developed a flexible decision support system in human resource management, Xu and Chen (2008b) studied priority weights from interval fuzzy preference relations, Yager (1993) developed different families of OWA operators and Yager et al. (2011) presented an updated overview concerning the main trends in this field.

In recent years, it has appeared an interesting alternative which represents a generalization of OWA operators called the ordered weighted averaging distance (OWAD) operator (Merigó and Gil-Lafuente, 2010; Xu and Chen, 2008a). It is a distance measure that provides a parameterized family of distance aggregation operators between the minimum distance and the maximum distance.

In this paper, we present a new model for the assignment process by using the Hungarian method (Kuhn, 1955). We introduce the use of the OWAD operator in this framework. The main advantage is that we can provide a more complete representation of this process considering results from the minimum distance to the maximum one.

We also develop an application in financial management. We analyze the optimal assignment of financial products for each decision maker according to his attitudinal character.

This work is structured as follows: In section 2 we present the preliminary concepts. In Section 3, we suggest a new approach for dealing with the Hungarian Algorithm by using the OWAD operator. Section 4 analyzes the applicability of this framework in financial management and in Section 5 we summarize the main results of this paper.

2   Theoretical Foundations

In this section, we present the theoretical definitions and theorems related to the OWA operator and the Hungarian method.

2.1   The OWA Operator

The OWA operator was introduced by Yager (1988). It provides a parametrized family of aggregation operators that has been used by many authors (Cheng et al. (2009); Dong et al. (2010); Karayiannis, 2000). The principal advantage of the OWA operator is that it reorders arguments based on their values. The weights are associated with a particular position in the ordering. This reordering process introduces nonlinearity into an otherwise linear process. It can be defined as follows:

Definition 1. An OWA operator of dimension n is a mapping OWA:$R^n \to R$ that has an associated weighting vector $w = (w_1, w_2, ..., w_n)$ with $w_j \in [0,1]$ and $\sum_{j=1}^{n} w_j = 1$ such that

$$OWA (a_1, a_2, ..., a_n) = \sum_{j=1}^{n} w_j b_j , \qquad (1)$$

where b_j is the j th largest of $(a_1, a_2, ..., a_n)$

2.2   The OWA Distance Operator

The OWAD operator (Merigó and Gil Lafuente, 2010; Xu and Chen, 2008a) is an aggregation operator and it is an extension of the traditional normalized Hamming distance by using OWA operators. The purpose of this operator is the reordering of the individual distances according to their values. An interesting advantage of this operator is the possibility of calculating the distance between two fuzzy sets modifying the results according to the interests of the decision maker. It can be defined as follows:

Definition 3. An OWAD operator of dimension n is a mapping OWAD: $[0,1]^n \times [0,1]^n \rightarrow [0,1]$ an associated weighting vector w, with $\sum_{j=1}^{n} w_j = 1$ and $w_j \in [0,1]$ such that:

$$OWAD\left(\langle u_1, u_1^{(k)}\rangle, \langle u_2\, u_2^{(k)}\rangle, \dots, \langle u_n, u_n^{(k)}\rangle\right) = \sum_{j=1}^{n} w_j D_j , \qquad (3)$$

where D_j represents the j th largest of the individual distances $\left|u_i - u_i^{(k)}\right|$, with u_i and $u_i^{(k)} \in [0,1]$, and $k = 1, 2, \dots, m$.

It is necessary to remark that this operator can be generalized to all the real numbers R by using OWAD: $R^n \times R^n \rightarrow R$. Because it is possible to distinguish between ascending and descending orders, the weights of these operators are related by $w_j = w_{n-j+1}^*$, where w_j is the jth weight of the descending OWAD (DOWAD) operator and w_{n-j+1}^* the j th weight of the ascending OWAD (AOWAD) operator.

2.3   The Hungarian Method

The Hungarian method (Kuhn, 1955) was established as an assignment algorithm used in various fields of science by the diversity of applications that can be performed. In recent years we can highlight the contribution made by Golderberg (2008), which sets out the main features of this algorithm.

Let A be a $n \times n$ matrix. The following algorithm finds a permutation $\pi \in S_n$ that minimizes the expression $\sum_i A_{i,\pi(i)}$. In this algorithm, the entries of the matrix A are being modified repeatedly. Zero entries in the modified matrix may be either marked, by a star or by a prime, or unmarked. In addition, each row or column in the matrix may be either covered or uncovered. Initially, there are no starred or primed entries in the matrix and none of the rows or columns is covered:

1.   For each row in the matrix A find its minimal entry and subtract it from all entries in that row.
2.   For all $1 \leq i, j \leq n$ if $A_{ij} = 0$ then star that zero entry, unless there is already a starred zero in the same row or in the same column.
3.   Cover each column that contains a starred zero. If all columns are covered, go to Step 7.
4.   Repeat the following procedure until there are no uncovered zeros left and then go to Step 6: find an uncovered zero and prime it. If there are no starred

zeros in the same row as this primed zero, go to Step 5. Otherwise, cover this row and uncover the column containing the starred zero.

5. Construct a series of alternating primed and starred zeros as follows: Let Z_0 be the uncovered primed zero that was found in Step 4. Let Z_1 be the starred zero in the column of Z_1 (if any). Let Z_2 be the primed zero in the row of Z_1 (there will always be one). Continue to construct this series of alternating primed and starred zeros until it terminates with a primed zero that has no starred zero in its column. Unstar each starred zero of the series, star each primed zero of the series, erase all primes and uncover all rows and columns in the matrix. Go to Step 3.

6. Find the smallest uncovered value, add it to every entry in each covered row, and subtract it from every entry in each uncovered column. Go to Step 4.

7. At this stage, in each row of the matrix, as well as in each column, there is exactly one starred zero. The positions of the starred zeros describe an optimal permutation $\pi \in S_n$. Output this permutation and stop.

2.4   Fuzzy Hungarian Algorithm

The Hungarian algorithm can be used with fuzzy information. For example, Gil Aluja (1999) developed the fuzzy Hungarian algorithm for the efficient assignment of some products based on certain characteristics in different markets.

Developing the Hungarian fuzzy method, based on a distance matrix that we call $[\tilde{\tilde{Q}}]$ or its complementary matrix $[\tilde{\tilde{R}}]$; in order to find the optimal assignment we will start by using a minimum principle.

In most cases these fuzzy relations are not always the same number of rows and columns so they would have to operate with rectangular matrices. For operational reasons, it will be transformed into a square matrix by adding rows or columns needed for introducing fictious elements, as well as p_{ij} denotes the elements of the matrix, such as if we consider the matrix $[\tilde{\tilde{Q}}]$ or $[\tilde{\tilde{R}}]$.

The algorithm consists of the following steps:

1. Subtract the smallest value in each row or column based on what we have added to make the matrix square. In the case of the rows, we will have $u = \min_j p_{ij}$, obtaining $p_{ij} - u_i = p_{ij} - \min_j p_{ij}$; or $u_j = \min_j p_{ij}$ in the case of columns getting $p_{ij} - u_j = p_{ij} - \min_i p_{ij}$

   We apply the same process in each column, $v_j = \min_i(p_{ij} - u_i)$ or row $v_i = \min_j(p_{ij} - u_j)$. This means that at least we have one 0 in each column and row in a matrix whose elements take values $p_{ij} - (u_i - v_j)$ or $p_{ij} - (u_j - v_i)$.

2. We analyze if it is possible to proceed with an assignment in the case that the p_{ij} values of the solution are all zero. If so, we get an optimum. Otherwise is necessary to continue with the process as follows:

   a) We search for the row of the matrix that contains less zeros.
   b) Mark one of the zeros of each row and delete the other zeros that appear in the row and column to which it belongs the zero we have marked.

c)   Repeat this process as many times as necessary in the rows that has more zeros to be marked.

3.   We will get the least number of rows and columns that contain all zeros and we will continue the following process:

a)   Mark with an arrow ← the rows that does not exist any zero marked.

b)   Mark with an arrow ↑ the columns in which there exists a zero deleted in a row marked with an arrow.

c)   Indicate with an arrow ← the rows where there is a zero marked in a column shown with an arrow.

d)   Repeat steps b) and c) as many times as necessary until we cannot form more rows or columns.

e)   Draw a line in the rows not marked by arrows and a line in the columns marked by arrows. The result we get constitutes the minimum number of rows and columns with zeros marked or deleted.

4.   Eventually, move some zeros. We choose the smallest value of non-marked elements at the matrix. This number is subtracted from the non-marked elements and we sum them to the elements of the marked rows. We obtain a matrix with the elements p_{ij}.

5.   With this new matrix whose elements are p_{ij}, return to Step 2 following the same process shown above.

The solution obtained is not unique and it is important to note that we may find other solutions.

3   OWA Operators in the Hungarian Algorithm

In this section we introduce the new approach. First we analyze the case with the OWAD operator. We end the section studying a general framework that uses generalized aggregation operators.

3.1   Using the OWAD Operator in the Hungarian Algorithm

The Hungarian algorithm is an efficient assignment process. This algorithm is based on the similarity (or dissimilarity) between the elements considered. However, in real world problems, when calculating the similarity, we need to use a technique that permits to do so such as a distance measure. A very common one used in the Hungarian algorithm is the Hamming distance (Hamming, 1950). When dealing with similarities, we have to normalize the distance. The most common way for doing so is by using the arithmetic mean where we give the same importance to all the elements obtaining the normalized (or relative) Hamming distance.

In this paper, we suggest the use of the OWA operator in the normalized process of the Hamming distance. Thus, we use the OWAD operator (Merigó and Gil-Lafuente, 2010; Xu and Chen, 2008a) in the assignment process. Its main advantage is that we can provide a parameterized family of distance aggregation operators between the minimum and the maximum. Thus, we can analyze several similarity relations from the minimum to the maximum according to our particular attitude in the specific problem considered. It is very useful because we can consider our normal position when forming the similarity relations but at the same time we can under or overestimate this position and see if the results are the same or some important changes may appear when dealing with the assignment process. That is, the assignment process may be different depending on the particular type of OWAD operator used.

When dealing with the OWAD operator, we have to use the following formulation as it has been explained in Definition 3 for two sets $X = \{X_1, X_2, ..., X_n\}$ and $Y = \{Y_1, Y_2, ..., Y_n\}$:

$$d(X, Y) = OWAD\left(\langle x_1, y_1 \rangle, \langle x_2, y_2 \rangle, ..., \langle x_n, y_n \rangle\right) = \sum_{j=1}^{n} w_j D_j \,, \qquad (5)$$

where D_j represents the j th largest of the individual distances $|x_i - y_i|$, with x_i and $y_i \in [0, 1]$, $\sum_{j=1}^{n} w_j = 1$ and $w_j \in [0, 1]$.

Thus, we have to use it to calculate all the similarity relations between the elements of the set $T = \{T_1, T_2, ..., Tm\}$ and $Z = \{Z_1, Z_2, ..., Z_p\}$ in order to form the matrix $[R]$ as it is shown in Table 1.

Table 1. Similarity relations between T and Z

		Z_1	Z_2	...	Z_k	...	Z_p
	T_1	$d(T_1, Z_1)$	$d(T_1, Z_1)$	...	$d(T_1, Z_k)$	...	$d(T_1, Z_p)$
	T_2	$d(T_2, Z_1)$	$d(T_2, Z_2)$	...	$d(T_2, Z_k)$	...	$d(T_2, Z_p)$
$[R]$	...	...	...	...	...	...	...
	T_h	$d(T_h, Z_1)$	$d(T_h, Z_2)$	...	$d(T_h, Z_k)$	...	$d(T_h, Z_p)$
	...	...	...	...	...	...	...
	T_m	$d(T_m, Z_1)$	$d(T_m, Z_2)$	...	$d(T_m, Z_k)$	...	$d(T_m, Z_p)$

As can see, we calculate the distance by using the OWAD operator between each T_h and Z_k. Note that with the OWAD operator we can obtain a wide range of results between the minimum and the maximum distance according to our attitude in the aggregation process.

Once we have the similarity relations, it is straightforward to solve the assignment process with the Hungarian algorithm following the Steps 1 – 7 given in Section 2.5.

Remark 1. It is possible to distinguish between ascending and descending orders in the OWAD operator. The weights of these operators are related by $w_j = w*_{n-j+1}$, where w_j is the jth weight of the descending OWAD (DOWAD) operator and $w*_{n-j+1}$ the jth weight of the ascending OWAD (AOWAD) operator.

Remark 2. Another interesting transformation is possible by using $w_i* = (1 + w_i) / (n - 1)$. Furthermore, we can also analyze situations with buoyancy measures (Yager, 1993). In this case, we assume that $w_i \geq w_j$, for $i < j$. Note that it is also possible to consider a stronger case known as extensive buoyancy measure where $w_i > w_j$, for

$i < j$. Additionally, we can also consider the contrary case, that is, $w_i \leq w_j$, for $i < j$, and the contrary case of the extensive measure $w_i < w_j$, for $i < j$.

Remark 3. The maximum distance is found when $w_1 = 1$ and $w_j = 0$ for all $j \neq 1$ and the minimum distance when $w_n = 1$ and $w_j = 0$ for all $j \neq n$. A generalization of the previous ones is the step-OWAD. It sets $w_k = 1$ and $w_j = 0$ for all $j \neq k$. Note that if $k = 1$, we get the maximum and if $k = n$, the minimum distance.

Remark 4. The normalized Hamming distance is obtained when $w_j = 1/n$ for all j, and the weighted Hamming distance is obtained when the ordered position of i is the same as the ordered position of j.

Remark 5. The olympic-OWAD is generated when $w_1 = w_n = 0$, and for all others $w_{j*} = 1/(n - 2)$. Following Merigó and Gil-Lafuente (2010), it is possible to develop a general form of the olympic-OWAD by considering that $w_j = 0$ for $j = 1, 2, ..., k, n, n - 1, ..., n - k + 1$, and for all others $w_{j*} = 1/(n - 2k)$, where $k < n/2$.

Remark 6. Moreover, we can develop the contrary case of the previous one by using $w_j = (1/2k)$ for $j = 1, 2, ..., k, n, n - 1, ..., n - k + 1$, and $w_j = 0$, for all other values, where $k < n/2$.

Remark 7. Another interesting family is the S-OWAD operator. It can be subdivided into three classes: the "or-like," the "and-like" and the generalized S-OWAD operators.

- The generalized S-OWAD operator is obtained if $w_1 = (1/n)(1 - (\alpha + \beta)) + \alpha$, $w_n = (1/n)(1 - (\alpha + \beta)) + \beta$, and $w_j = (1/n)(1 - (\alpha + \beta))$ for $j = 2$ to $n - 1$, where $\alpha, \beta \in [0, 1]$ and $\alpha + \beta \leq 1$.
- If $\alpha = 0$, the generalized S-OWAD operator becomes the "and-like" S-OWAD operator.
- If $\beta = 0$, it becomes the "or-like" S-OWAD operator.

In order to analyze the aggregation, we can use several measures for characterizing the weighting vector. A very common technique for doing so is the orness measure $\alpha(W)$. It can be defined as follows:

$$\alpha(W) = \sum_{j=1}^{n} \left(\frac{n - j}{n - 1} \right) w_j, \tag{6}$$

As we can see, $\alpha(W) \in [0, 1]$. Note that the more of $\alpha(W)$ closer to 1 the more of the weight we locate at the top of W and vice versa.

Another useful measure is the balance operator. It permits to analyze if the balance of the aggregations closer to the minimum or to the maximum. It is defined as follows:

$$Bal(W) = \sum_{j=1}^{n} \left(\frac{n + 1 - 2j}{n - 1} \right) w_j, \tag{7}$$

In this case, $Bal\ (W) \in [-1, 1]$. Note that if $Bal\ (W)$ is positive, the aggregation tends to the maximum and if it is negative, it tends to the minimum.

4   Illustrative Example

In the following, we develop a simple numerical example of the new approach. We focus on a multi-person decision-making problem regarding the selection of investments.

Assume that an enterprise wants to invest some money in one product. After careful analysis of the different possibilities that the markets offer, the group of experts of the enterprise considers six possible investments:

- P_1 = Hedge Funds.
- P_2 = Investment Funds.
- P_3 = Bonds.
- P_4 = Fixed Income Notes.
- P_5 = Stocks.
- P_6 = Equity Derivatives.

When analyzing the investments, the experts have consider the following general characteristics:

- C_1 = Risks of the investment.
- C_2 = Difficulty of the investment.
- C_3 = Benefits in the long term.
- C_4 = Benefits in the mid term.
- C_5 = Benefits in the short term.
- C_6 = Social responsible investment (SRI).
- C_7 = Others aspects.

The companies involved in the following decision process that can invest in these financial products are the following:

- E_1 = Enterprise A.
- E_2 = Enterprise B.
- E_3 = Enterprise C.
- E_4 = Enterprise D.
- E_5 = Enterprise E.

With this information, the group of experts describes each financial product according to the characteristics established in aggregated form the results are shown in Table 2. We assume that each company has similar characteristics so they are more or less equally qualified for carrying out the strategic investment process and its aggregated subjective opinions are shown in Table 3. The results are valuations (numbers) between 0 and 1 being 1 the best result and 0 the worst result.

In this example, we consider the OWAD operator. In this case, the aggregated results are shown in Table 4. The third operator are presented in Table 6.

In the first case we assume that $W = (0.3, 0.2, 0.2, 0.1, 0.1, 0.1, 0)$ in the OWAD operator. With this information, it is possible to develop the Hungarian algorithm.

Table 2. Characteristics of the financial products

	C_1	C_2	C_3	C_4	C_5	C_6	C_7
P_1	0,8	0,7	0,8	0,7	0,6	0,5	0,6
P_2	0,7	0,6	0,7	0,7	0,5	0,8	0,6
P_3	0,4	0,7	0,6	0,7	0,7	0,6	0,6
P_4	0,6	0,7	0,5	0,6	0,8	0,7	0,7
P_5	0,8	0,4	0,7	0,8	0,8	0,4	0,7
P_6	0,8	0,7	0,7	0,8	0,7	0,5	0,7

Table 3. Characteristics of the enterprises

	C_1	C_2	C_3	C_4	C_5	C_6	C_7
E_1	0,6	0,5	0,8	0,6	0,5	0,5	0,7
E_2	0,7	0,6	0,7	0,8	0,9	0,8	0,6
E_3	0,8	0,9	0,8	0,7	0,5	0,7	0,5
E_4	0,6	0,8	0,9	0,8	0,7	0,5	0,8
E_5	0,9	0,8	0,7	0,8	0,9	0,6	0,7

Table 4. Aggregated distances with OWAD operator

	P_1	P_2	P_3	P_4	P_5	P_6
E_1	0,14	0,16	0,18	0,21	0,20	0,17
E_2	0,20	0,14	0,20	0,15	0,20	0,17
E_3	0,13	0,15	0,24	0,24	0,31	0,18
E_4	0,15	0,22	0,20	0,23	0,23	0,13
E_5	0,16	0,24	0,24	0,20	0,19	0,11

Next, we develop a Hungarian algorithm with the OWAD operator. We assume that $W = (0.3, 0.2, 0.2, 0.1, 0.1, 0.1, 0)$. The results are shown in Table 8. By using the same process that we use before, we observe that $0,18 + 0,15 + 0,15 + 0,15 + 0,11 = 0,74$. In this case we not add and subtract the fictitious value, and we can obtain and optimal.

Table 5. Aggregated results with the OWAD operator

	P_1	P_2	P_3	P_4	P_5	P_6
E_1	0,02	0,01	0 (1.)	0,05	0,02	0,07
E_2	0,08	0	0,03	0 (2.)	0,02	0,07
E_3	0	0 (3.)	0,07	0,08	0,13	0,08
E_4	0 (4.)	0,05	0	0,05	0,03	0
E_5	0,04	0,09	0,07	0,04	0	0 (5.)
E_0	0,07	0,05	0,02	0,04	0 (6.)	0,09

Assignment process 3
1. Enterprise 1 assigned with financial product P_1 (Hedge Funds).
2. Enterprise 2 assigned with financial product P_2 (Investment Funds).
3. Enterprise 3 assigned with financial product P_4 (Fixed Income Notes).
4. Enterprise 4 assigned with financial product P_3 (Bonds).
5. Enterprise 5 assigned with financial product P_6 (Equity Derivatives).
6. Enterprise 0 assigned with financial product P_5 (Stocks).

or
1. Enterprise 1 assigned with financial product P_5 (Stocks).
2. Enterprise 2 assigned with financial product P_2 (Investment Funds).
3. Enterprise 3 assigned with financial product P_4 (Fixed Income Notes).
4. Enterprise 4 assigned with financial product P_1 (Hedge Funds).
5. Enterprise 5 assigned with financial product P_6 (Equity Derivatives).
6. Enterprise 0 assigned with financial product P_3 (Bonds).

5   Conclusions

We have presented a new approach for decision making in an assignment process by using the OWAD operator. Its main advantage is that we can consider the information in a more complete way by using a parameterized family of aggregation operators from the minimum to the maximum distance. Thus, we can consider optimistic or pessimistic scenarios by under or over estimating the information.

The applicability of this new approach has been studied in a financial management problem regarding the assignment of financial products. We have seen that each aggregation operator may lead to different assignments since the results can be different. The use of OWAD operators permits to consider different degrees of optimism or pessimism in the analysis.

In future research, we expect to develop further developments by adding new characteristics in the problem such as the use of probabilistic information, norms, weighted averages and more complex structures.

References

1. Beliakov, A., Pradera, G., Calvo, T.: Aggregation functions: A guide for practitioners. Springer, Berlin (2007)
2. Canós, L., Liern, V.: Soft computing-based aggregation methods for human resource management. European Journal of Operational Research 189, 669–681 (2008)
3. Cheng, C.H., Wang, J.W., Wu, M.C.: OWA-weighted based clustering method for classification problem. Expert Systems with Applications 36, 4988–4995 (2009)
4. De Luca, A., Termini, S.: A definition of a nonprobabilistic entropy in the setting on fuzzy sets theory. Information and Control 20, 301–312 (1972)
5. Dong, Y., Xu, Y., Li, H., Feng, B.: The OWA-based consensus operator under linguistic representation models using position indexes. European Journal of Operational Research 203, 455–463 (2010)
6. Fodor, J., Marichal, J., Roubens, M.: Characterization of the ordered weighted averaging operators. IEEE Transactions on Fuzzy Systems 3, 236–240 (1995)
7. Figueira, J., Greco, S., Ehrgott, M.: Multiple criteria decision analysis: state of the art surveys. Springer, Boston (2005)
8. Goldberger, J., Tassa, T.: A hierarchical clustering algorithm based on the Hungarian method. Pattern Recognition Letters 29, 1632–1638 (2008)
9. Karayiannis, N.: Soft learning vector quantization and clustering algorithms based on ordered weighted aggregation operators. IEEE Transactions on Neural Networks 11, 1093–1105 (2000)
10. Kaufmann, A.: Introduction to the theory of fuzzy subsets. Academic Press, New York (1975)
11. Kuhn, H.W.: The Hungarian method for the assignment problema. Naval Research Logistics Quarterly 2, 83–97 (1955)
12. Merigó, J.M., Casanovas, M.: Decision making with distance measures and induced aggregation operators. Computers & Industrial Engineering 60, 66–76 (2011)
13. Merigó, J.M., Gil-Lafuente, A.M.: New decision making techniques and their application in the selection of financial products. Information Sciences 180, 2085–2094 (2010)
14. Merigó, J.M., Gil-Lafuente, A.M.: Decision-making in sport management based on the OWA operator. Expert Systems with Applications 38, 10408–10413 (2011a)
15. Merigó, J.M., Gil-Lafuente, A.M.: OWA operators in human resource management. Economic Computation and Economic Cybernetics Studies and Research 45, 153–168 (2011b)
16. Merigó, J.M., Gil-Lafuente, A.M., Gil-Aluja, J.: A new aggregation method for strategic decision making and its application in assignment theory. African Journal of Business Management 5, 4033–4043 (2011)
17. Xu, Z.S., Chen, J.: Ordered weighted distance measure. Journal Systems Sciences Systems Engineering 17, 432–445 (2008a)
18. Xu, Z.S., Chen, J.: Some models for deriving the priority weights form interval fuzzy preference relations. European Journal of Operational Research 184, 266–280 (2008b)

19. Xu, Z.S., Da, Q.L.: An overview of operators for aggregating information. International Journal of Intelligent Systems 18, 953–969 (2003)
20. Yager, R.R.: On ordered weighted averaging aggregation operators in multi-criteria decision making. IEEE Transactions on Systems, Man and Cybernetics B 18, 183–190 (1988)
21. Yager, R.R.: Families of OWA operators. Fuzzy Sets and Systems 59, 125–148 (1993)
22. Yager, R.R.: Heavy OWA operators. Fuzzy Optimization and Decision Making 1, 379–397 (2002)
23. Yager, R.R., Kacprzyk, J.: The ordered weighted averaging operators: Theory and applications. Kluwer Academic Publishers, Norwell (1997)
24. Yager, R.R., Kacprzyk, J., Beliakov, G.: Recent developments on the ordered weighted averaging operators: Theory and practice. Springer, Berlin (2011)

Mathematic Modeling of Reactor's Temperature Mode of Multiloop Pyrolysis Plant

Yuriy P. Kondratenko[1,2] and Oleksiy V. Kozlov[1]

[1] Admiral Makarov National University of Shipbuilding, 9 Geroiv Stalingrada Av.,
Mykolaiv, 54025, Ukraine
`kozlov_ov@ukr.net`
[2] Petro Mohyla Black Sea State University, 10, 68th Desantnykiv Str.,
Mykolaiv, 54003, Ukraine
`y_kondrat2002@yahoo.com`

Abstract. The paper considers the synthesis of mathematical model of temperature mode of multiloop pyrolysis plant's (MPP's) reactor as a complexed control object with distributed parameters. The results of computer simulation as diagrams of transient processes of different points heating of multiloop pyrolysis plant reactor's workspace are obtained on the basis of developed model. The analysis of simulation results for different temperature modes confirms the adequacy of the mathematic model of the reactor to real processes. The developed model can be used for synthesis of automatic control systems of the MPP's reactor heating temperature.

Keywords: Multiloop pyrolysis plant, mathematic model, objects with distributed parameters.

1 Introduction

The pyrolysis plants are widely used for the utilization of municipal solid wastes (MSW) in order to obtain alternative liquid fuels and pyrolysis gas. Modern samples of the MSW utilization equipment are developed on the basis of implementation of the process of multiloop circulating pyrolysis (MCP), which makes possible to obtain liquid fractions of alternative fuels that can be used in combustion engines without any extra filtration [1]. Complexity of modern multiloop pyrolysis plants (MPP) as technical objects causes necessity of complex control systems development, that are especially effective at functioning in automatic mode. The lower level of such kind of systems is an aggregate of automatic control subsystems that perform stabilization of set values of MPP's basic controlled coordinates in the conditions of disturbance action [2]. T_r (the reactor's heating temperature) is one of the important MPP's basic controlled coordinates. Mathematic and computer modeling methods are appropriate to be used for the research of the efficiency of MPP heating temperature's automatic control system (ACS) at the stage of its development. Analytical mathematic and computer modeling are quite effective and cheap instruments, comparatively with experimental and the other ones. Thus the development of mathematic model of MPP reactor's temperature mode is quite necessary.

K.J. Engemann, A.M. Gil Lafuente, and J.M. Merigó (Eds.): MS 2012, LNBIP 115, pp. 178–187, 2012.

2   Geometrical Model and Basic Heat Equations of the MPP's Reactor

The MPP's reactor is a tubular metal cylinder that is completely or partly filled with polymer wastes that are inhomogeneously distributed in space. A simplified geometrical model of the MPP's reactor as a control object by temperature is given on the fig. 1.

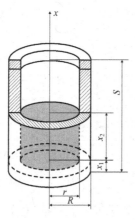

Fig. 1. Simplified geometrical model of the MPP's reactor: x – the cylinder's longitudinal axis; x_1 – the thickness of bottom flat wall; x_2 – the layer thickness of wastes that are downloaded to the reactor; S – the reactor's height; r – the internal reactor's radius; R – the external reactor's radius

Reactor is heated with the help of the flue gases that are generated as a result of burning of gaseous fuel. The principle of the MPP reactor's heating with the help of a gas burner is illustrated on the chart, shown on the fig. 2.

The amount of heat that is evolved by a gas burner can be calculated by the formula (1) [3]

$$Q_b = VQ_1^s,\qquad(1)$$

where Q_1^s – lower specific heat of combustion; V – volume of gaseous fuel. Thus gas burner's power is calculated by the formula (2)

$$P_b = \frac{dQ_b}{dt} = \frac{dV}{dt}Q_1^s = BQ_1^s,\qquad(2)$$

where B – volumetric flow rate. The heat flow rate that goes for heating of the reactor is calculated by the formula (3) [3]

$$P = P_b - \sum P_L = P_b - \left(P_{wg} + P_{icf}\right) = P_b - \left(B_{wg}c_{wg}T_{wg} + B_{icf}Q_1^s\right),\qquad(3)$$

where $\sum P_{\text{L}}$ – total loss that mainly consists of the waste gasses loss P_{wg} and loss of chemical incomplete combustion of fuel P_{icf}; B_{wg}, c_{wg}, T_{wg} – volumetric flow rate, specific heat capacity ratio and temperature of waste gasses; B_{icf} – volumetric flow rate of unburned gaseous fuel.

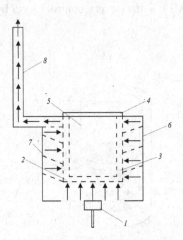

Fig. 2. MPP reactor's heating system: 1 – gas burner; 2 – reactor's bottom flat wall; 3 – reactor's side cylindrical wall; 4 – reactor's overhead lid; 5 – capacity for wastes; 6 – reactor. heating system's casing; 7 – casing's partition for turbinated way formation of flue gases motion; 8 – pipe for flue gases outlet

The heat exchange between the reactor and flue gases that pass through the heating system is mainly performed by the convection. The amount of heat that is transferred by the convection is calculated by the formula (4) that is determined by the law of Newton-Richman [4]

$$Q = \alpha \left(T_{\text{g}} - T_{\text{s}} \right) Ft, \tag{4}$$

where α – heat transfer coefficient; T_{g} – the temperature of the flue gases; T_{s} – the temperature of the reactor's surface; F – the reactor's surface area that is heating; t – heating time. The heat flow rate and specific heat flow rate that is transferred to the reactor's surface by the convection of the flue gases can be calculated by the formulas (5) and (6)

$$P = \frac{dQ}{dt} = \alpha \left(T_{\text{g}} - T_{\text{s}} \right) F, \tag{5}$$

$$q = \frac{P}{F} = \alpha \left(T_{\text{g}} - T_{\text{s}} \right). \tag{6}$$

Heat transfer coefficient α can be found [3] with the help of formulas (7)-(13)

$$Nu = \frac{\alpha l}{\lambda}, \tag{7}$$

$$Nu = f(Gr, Re, Pr), \tag{8}$$

$$Gr = \frac{g\beta l^3 \Delta T}{v^3}, \tag{9}$$

$$\beta = \frac{1}{T}, \tag{10}$$

$$Re = \frac{ul}{v}, \tag{11}$$

$$Pr = \frac{v}{a}, \tag{12}$$

$$a = \frac{\lambda}{c\rho}, \tag{13}$$

where l – linear dimension of heating surface; λ – coefficient of thermal conductivity of flue gases; Gr – Grashof criterion; Re – Reynolds criterion; Pr – Prandtl criterion; g – acceleration of gravity; β – coefficient of volumetric expansion of flue gases; ΔT – temperature difference; v – kinematic viscosity coefficient; T – temperature, K; u – velocity of the flue gases; a – thermal diffusivity coefficient; c – specific heat capacity ratio of flue gases; ρ – density of flue gases.

If the velocity of flue gases is high enough it is possible to neglect the influence of lifting force, then the equation (8) transforms into

$$Nu = f(Re, Pr). \tag{14}$$

In case of gravitational convection the velocity criterion (Re) is minor and the equation (8) transforms into

$$Nu = f(Gr, Pr) = cGr^m Pr^n, \tag{15}$$

where c, m, n – constant coefficients that could be found experimentally.

As the flue gases move between the reactor's wall and reactor heating system's casing (fig. 2), this case should be considered as a convective heat exchange in confined space. This process is usually considered as a heat exchange by the thermal conduction with introducing the concept of an equivalent coefficient of thermal conductivity λ_e, which could be found experimentally [3]. The influence of convective heat exchange and aerodynamics of a medium's motion is taken into account with the help of convection coefficient ε_c that is calculated by the formula (16)

$$\varepsilon_c = \frac{\lambda_e}{\lambda}. \tag{16}$$

The convection coefficient ε_c equals to the product $Gr Pr$, if $Gr Pr < 1000$, then $\varepsilon_c = 1$, if $Gr Pr > 1000$, then ε_c can be found by the formula (17)

$$\varepsilon_{\kappa} = 0,18(Re Pr)^{0,25}.$$ (17)

The equivalent coefficient of thermal conductivity λ_e can be found with the help of the equations (16) and (17). The specific heat flow rate can be calculated by the formula (18)

$$q = \frac{\lambda_e}{\delta}\Delta T,$$ (18)

where δ – thickness of the layer, that takes part in the heat exchange.

The process of heating of the MPP's reactor can be considered as a process of heating of a massive body [4]. The heat exchange in solids occurs by thermal conduction. Specific heat flow rate in the transmission of energy by thermal conduction through flat and cylindrical walls according to [5] is calculated by formulas (19) and (20)

$$q = \frac{\lambda}{S}\left(T_1 - T_2\right),$$ (19)

$$q = \frac{2\pi\lambda\left(T_1 - T_2\right)}{\ln\left(r_1 / r_2\right)},$$ (20)

where S – thickness of a flat wall; r_2 and r_1 – external and internal radiuses of the cylindrical wall; T_1 and T_2 – temperature values on the outer and inner surfaces of the walls.

Since the temperature at different points in the reactor will have a significantly different value, the formation of an accurate mathematical model of the reactor temperature as the object of control can be implemented by representing it as an object with distributed parameters [6]. The further formalization of the control object is based on equations with partial derivatives for the distribution of heat in solids. Analytical dependence of changes in temperature and transferred amount of heat in time for any point of the body can be obtained by solving the basic differential equation of thermal conduction Fourier-Kirchhoff [5], which in a Cartesian cylindrical coordinate systems has the following forms

$$\frac{dT}{dt} = a\left(\frac{\partial^2 T}{\partial x^2} + \frac{\partial^2 T}{\partial y^2} + \frac{\partial^2 T}{\partial z^2}\right) = a\nabla^2 T,$$ (21)

$$\frac{dT}{dt} = a\left(\frac{\partial^2 T}{\partial r^2} + \frac{1}{r}\frac{\partial T}{\partial r} + \frac{1}{r^2}\frac{\partial^2 T}{\partial \varphi^2} + \frac{\partial^2 T}{\partial z^2}\right),$$ (22)

where x, y, z – spatial coordinates of the Cartesian coordinate system; r, φ, z – spatial coordinates of the cylindrical coordinate system, ∇^2 – differential Laplace operator.

For unique solution of the equation of thermal conduction it is necessary to solve it together with boundary conditions [7], consisting of: a) initial conditions, b) surface or boundary conditions.

Boundary conditions can be represented (given) by the following ways:

1) the surface conditions of the first kind, when a temperature distribution on the body's surface is given as $T_s = f_1(t)$;

2) the surface conditions of the second kind, when a heat flow rate is given as a function of time: $q = f_2(t)$;

3) surface conditions of the third kind, when the law of change of ambient temperature is given: $T_e = f_3(t)$ and the law of heat exchange between the body surface and its ambient;

4) Surface conditions of the fourth kind characterize the conditions of heat exchange between the body and the ambient by the law of thermal conduction.

Solutions of equation of thermal conduction for a plate of infinite length and width and for a cylinder of infinite length at $\dfrac{at}{R^2} \geq 0,25$ [7] for the surface conditions of the second kind ($q_x = \text{const}$; $T_{t=0} = 0$) have the following forms (23), (24)

$$T = \frac{qS}{2\lambda}\left[\frac{2at}{S^2} + \left(\frac{x}{S}\right)^2 - \frac{1}{3} + \sum_{l=1}^{\infty}\frac{4(-1)^{l+1}}{\varepsilon_l^2}\cos\varepsilon_l\frac{x}{S}\exp\left(-\varepsilon_l^2\frac{at}{S^2}\right)\right], \qquad (23)$$

$$T = \frac{qR}{2\lambda}\left[\frac{4at}{R^2} + \left(\frac{r}{R}\right)^2 - \frac{1}{2}\right], \qquad (24)$$

where S – thickness of the plate; x – variable coordinate of the plate's thickness (on the outer surface of the plate $x = S$, on the inside – $x = 0$); ε_l – The correction factor, $= \pi l$ ($l = 1, 2, 3, ...$); l – normal range of numbers; R – radius of the cylinder, r – current radius of the point for where the temperature is calculating ($r = R$ – on the outer surface of the cylinder and $r = 0$ – on the it's longitudinal axis).

The way of the formation of mathematical models based on the equations (19), (20), (23), (24) provides high adequacy of this class of models, but in fact for the synthesis of mathematical model of the MPP's reactor it is necessary to take into account the reactor's geometry and heat flow rate's direction.

To solve this problem the authors of this article have proposed to consider the reactor's geometry as an aggregate of simple geometric forms that have partial solutions at given boundary conditions. It is necessary to introduce some assumptions for calculating of a MPP reactor's temperature mode in this way:

1) The reactor is heating with the help of flue gases from the bottom and sides, and in this case the heat flow rate is uniformly distributed over the heating surface, i.e. the value of the specific heat flow rate is the same at all points of heating space (Fig. 3, *a*);

2) The MPP's reactor can be divided into 4 main components (Fig. 3, *b*): 1 – flat bottom wall, 2 – cylindrical sidewall 3 – solid cylinder, formed of wastes loaded into the reactor 4 – flat top lid;

3) The process of the bottom wall heating of the reactor is considered as a process of the plate heating with its thickness S and the area limited by the circle with the radius R;

4) The process of heating the sidewall of the reactor is considered as a process of heating of the cylindrical wall with the outer and inner radius R, r;

5) The process of heating wastes is considered as a process of simultaneous heating of the plate and solid cylinder from the bottom and side walls of the reactor;

6) The process of heating of the reactor's lid is considered as a heating of the plate from wastes that are loaded into the reactor;

7) The calculation of the process of wastes heating is carried out for the lower and upper points lying on the longitudinal axis of the cylinder $(r = 0)$.

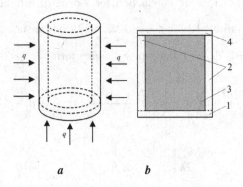

Fig. 3. The MPP reactor's model: *a* – distribution of the heat flow rate over the heating surface area; *b* – components

The mathematic model of the MPP reactor's temperature mode consists of the following formulas

$$q_{1x} = q_{1\text{inx}} - q_{1\text{outx}} = \frac{\lambda_1}{S_1}(T_{11} - T_{12}), \qquad (25)$$

$$q_{2r} = q_{2\text{inr}} - q_{2\text{outr}} = \frac{2\pi\lambda_2(T_{21} - T_{22})}{\ln(R_2 / r_2)}, \qquad (26)$$

$$q_{3x} = q_{3\text{inx}} - q_{3\text{outx}} = \frac{\lambda_3}{S_3}(T_{12} - T_{32}), \qquad (27)$$

$$q_{3r} = q_{3inr} = \frac{2\pi\lambda_3 \left(T_{22} - T_{32} \right)}{\ln\left(R_3 / r_3 \right)}, \tag{28}$$

$$q_{4x} = q_{4inx} - q_{4outx} = \frac{\lambda_4}{S_4}\left(T_{32} - T_{42} \right), \tag{29}$$

$$T_{11} = \frac{q_{1x}S_1}{2\lambda_1}\left[\frac{2a_1 t}{S_1^2} + \frac{2}{3} + \sum_{l=1}^{\infty} \frac{4(-1)^{l+1}}{\varepsilon_l^2} \cos\varepsilon_l \, \exp\left(-\varepsilon_l^2 \frac{a_1 t}{S_1^2} \right) \right], \tag{30}$$

$$T_{12} = \frac{q_{1x}S_1}{2\lambda_1}\left[\frac{2a_1 t}{S_1^2} - \frac{1}{3} \right], \tag{31}$$

$$T_{21} = \frac{q_{2r}R_2}{2\lambda_2}\left[\frac{4a_2 t}{R_2^2} + \frac{1}{2} \right], \tag{32}$$

$$T_{22} = \frac{q_{2r}R_2}{2\lambda_2}\left[\frac{4a_2 t}{R_2^2} + \left(\frac{r_2}{R_2} \right)^2 - \frac{1}{2} \right], \tag{33}$$

$$T_{32} = \frac{q_{3x}S_3}{2\lambda_3}\left[\frac{2a_3 t}{S_3^2} - \frac{1}{3} \right] + \frac{q_{3r}R}{2\lambda_3}\left[\frac{4a_3 t}{R_3^2} + \frac{1}{2} \right], \tag{34}$$

$$T_{42} = \frac{q_{4x}S_4}{2\lambda_4}\left[\frac{2a_4 t}{S_4^2} - \frac{1}{3} \right], \tag{35}$$

where q_{1x}, q_{2r}, q_{3x}, q_{3r}, q_{4x} – values of the specific heat flow rates that go to heating of the according model components; q_{in} and q_{out} – values of incoming and taken off specific heat flow rates of different components; T_{11}, T_{12}, T_{21}, T_{22}, T_{32}, T_{42} – current values of temperature of the external and internal surfaces of the according model components. Also $q_{1inx} = q_{2inx} = q$; $q_{3inx} = q_{1outx}$; $q_{3inr} = q_{2outr}$; $q_{4inx} = q_{3outx}$; q_{4outx} – the value of the specific heat flow rate that goes from the reactor's lid to the ambient by the convection and can be also calculated by the formula (6).

3   Computer Simulation of the MPP Reactor's Temperature Mode

The computer modeling of the reactor's temperature mode was carried out for the experimental MPP using data of the equations (2), (3), (16)-(18), (25)-(35). The working capacity of the MPP's reactor is 14 liters, the power of the gas burner is 24 kW. The modeling was held at constant heating power of the gas burner, at constant parameters of the coefficients of thermal conductivity and thermal diffusivity of the waste a_3 and λ_3, at ambient temperature $T = 0°C$. Modeling results in the form of diagrams of transient heating process of the lower and upper points lying on the longitudinal axis of the cylinder formed from wastes, loaded to the MPP's reactor are shown on the Fig. 4 (*a, b*).

The initial stages of transient processes of different points heating of the MPP's reactor are detailed in Fig. 5 (*a, b*) for the analysis of time-delay. The analysis of modeling results (Fig. 4, Fig. 5) shows that the characteristics of the transient heating processes of the reactor at various points differ significantly, confirming the feasibility of the MPP's reactor examination as a controlled object with distributed

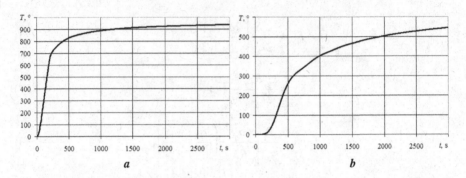

Fig. 4. Transient heating processes of the according calculated points of the MPP's reactor: *a* – the lower point laying on the longitudinal axis of the cylindrical body formed from wastes, *b* – the upper point laying on the longitudinal axis of the cylindrical body formed from wastes

parameters in its modeling. Transient heating processes of all calculated points of the MPP's reactor have aperiodic nature and significant inertia.

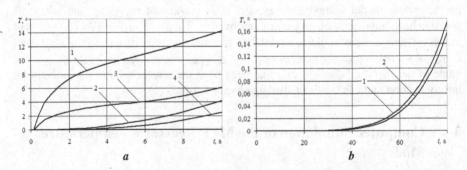

Fig. 5. Detailed initial stages of transient heating processes of the according calculated points of the MPP's reactor: *a*: *1* – outside of the bottom wall, *2* – inside of the bottom wall, *3* – outside of the sidewall, *4* – inside of the sidewall; *b*: *1* – on the surface of the cylindrical body formed from wastes at the reactor top lid, *2* – outside of the upper lid

The time of the temperature processes release on the default mode and the delay time are: for the outside of the bottom wall –2010 s and 0.2 s, for the inside bottom wall and for the lower point laying on the longitudinal axis of the cylindrical body formed from wastes – 2032 s and 2 s for the outside sidewall – 2018 s and 0.2 s, for the inside of the sidewall – 2036 s and 2.5 s, for the upper point of a cylindrical body formed of wastes at the reactor top lid – 3000 s and 30 s, for the outside of the upper lid – 3020 s and 32 s. In particular, in steady mode the heating temperature of the waste, loaded to the reactor at the bottom and the top points of the reactor differs by about 400 ° C.

The mathematical model (2), (3), (16)-(18), (25)-(35) can be used in developing and researching the effectiveness of ACS of the MPP's reactor temperature mode.

4   Conclusions

The mathematical model of the reactor temperature mode gives the opportunity to investigate the behavior of the controlled object in steady and transient modes, in particular, calculate temperatures in its various locations at a given incoming influence.

The results of modeling in the form of graphs of transient heating processes of the different points of the reactor can be used for the synthesis and researching of automatic control system of the MPP's reactor heating temperature.

Further research should be carried out in the direction of modeling of reactor's temperature mode at different disturbances as well as towards the synthesis of the complex control systems of MMP using different types of feedbacks and regulators.

References

1. Ryzhkov, S.S., Markina, L.M.: Experimental Researches of Organic Waste Recycling Method of Multiloop Circulating Pyrolysis. J. Collected Works of NUS 5, 100–106 (2007) (in Russian)
2. Kondratenko, Y.P., Kozlov, A.V.: Analysis of Complex of Tasks and Controlled Coordinates of Technological Process of Ekopirohenezisu. J. Technical News 1(33)-2(34), 13–16 (2011) (in Ukrainian)
3. Mastryukov, B.S.: Heat Engineering Calculations of Industrial Furnaces. Metallurgy, Moscow (1975) (in Russian)
4. Islamov, M.S.: Design and Exploitation of Industrial Furnaces. Chemistry, Leningrad (1986) (in Russian)
5. Fedotkin, I.M., Burlyay, I.J., Ryumshyn, M.O.: Mathematical Modeling of Processes: Reactors, Recycling and Boundary Layer Theories. Technology, Kiev (2003) (in Russian)
6. Usov, A.V., Dubrov, A.N., Dmitrishin, D.V.: Simulation of Systems with Distributed Parameters. Astroprint, Odessa (2002) (in Russian)
7. Kazantsev, E.I.: Industrial Furnaces: Reference Guide for Calculation and Design. Metallurgiya, Moscow (1975) (in Russian)

Step Size Bounds for a Class of Multiderivative Explicit Runge–Kutta Methods

Moses A. Akanbi[1], Solomon A. Okunuga[2], and Adetokunbo B. Sofoluwe[3]

[1] Department of Mathematics, Lagos State University, P.M.B. 0001 LASU Post Office
Lagos, Nigeria
akanbima@gmail.com
[2] Department of Mathematics, University of Lagos, Lagos, Nigeria
nugabola@yahoo.com
[3] Department of Computer Sciences, University of Lagos, Lagos, Nigeria
absofoluwe@yahoo.com

Abstract. A class of 3-Stage Multiderivative Explicit Runge-Kutta Methods was developed for the solution of Initial Value Problems (IVPs) in Ordinary Differential Equations. In this work, we present the bounds on the step size required for the implementation of this family of methods. This bound is one of the parameter required in the design of program codes for solving IVPs. A comparison of the step size bound was made vis-a-vis the existing Explicit Runge-Kutta Methods using some standard problems. The computation shows that the family of methods competes well with the popular methods.

Keywords: multiderivative, Rung-Kutta method, 3-stage method, step size bound.

1   Introduction

A system of first order autonomous Ordinary Differential Equations (ODEs) is of the form

$$\left.\begin{array}{l} \mathbf{y}'(x) = \mathbf{f}(\mathbf{y}(x)) \quad , \quad x \in [x_0, x_{end}], \\ \mathbf{y}(x_0) = \mathbf{y}_0. \end{array}\right\} \tag{1}$$

These systems are often associated with many problems in engineering, economics, science and even management sciences. And most times, the solution to the systems cannot be found in a closed form. This necessitated the need to continuously develop effective and efficient numerical algorithms for solving these systems of ODEs. In the design of such algorithms, the bounds on the step size are of paramount importance. In this work, the step size bounds for the 3-Stage Multiderivative Explicit Runge–Kutta (MERK) Methods recently developed by Akanbi and Okunuga [2] are presented. These errors are expected for any numerical step-by-step method for solving equation (1) because the numbers y_n furnished by the method with a fixed step length h, even if calculated to an infinite number of decimal places, will only

K.J. Engemann, A.M. Gil Lafuente, and J.M. Merigó (Eds.): MS 2012, LNBIP 115, pp. 188–197, 2012.

rarely agree with the corresponding values $y(x_n)$ of the true solution [7], where y_n denotes the exact number produced by the algorithm. This discrepancy is envisaged, since numerical algorithms are by necessity finite processes, whereas the solutions of even very simple equations such as $y' = y$ are transcendental functions and cannot therefore be calculated with a finite number of rational operations. Thus, this paper examines these errors in relation to the new 3-stage MERK methods. Results obtained by the methods are compared with other results obtained using some standard methods subsequently. In section 2 and 3, we recap the theory of Explicit Runge–Kutta (ERK) methods and development of the class of MERK methods respectively. Section 4 deals with the propagation of errors in the MERK methods while the numerical computation and conclusions are given in sections 5 and 6 respectively.

2   Explicit Runge–Kutta Methods

Explicit Runge–Kutta (ERK) methods are generally known to be one-step schemes. One advantage of ERK methods (and other one-step methods) is their self-starting nature [6, 9]. We shall stay close to the spirit of ERK methods, yet attain higher order of accuracy by incorporating higher order derivatives of f (i.e. y').

As it is well known, one-step method is of the form:

$$y_{n+1} - y_n = \Phi(x_n, y_n; h) \tag{2}$$

The Taylor's algorithm of order p is obtained from (2) by setting

$$\begin{aligned}
\Phi(x_n, y_n; h) &= \Phi_T(x_n, y_n; h) \\
&= \sum_{r=0}^{\infty} \frac{h^{r+1}}{(r+1)!} \left(\frac{\partial}{\partial x} + \frac{\partial}{\partial y} \right)^r f(x, y)
\end{aligned} \tag{3}$$

Although, a general overview of ERK algorithm has been presented in an earlier work [1], but for purpose of clarity we present a brief overview below. As it is well known, one-step method is of the form

$$y_{n+1} - y_n = \Phi(x_n, y_n; h) \tag{4}$$

The Taylor's algorithm of order p is obtained from (2) by setting

$$\begin{aligned}
\Phi(x_n, y_n; h) &= \Phi_T(x_n, y_n; h) \\
&= \sum_{r=0}^{\infty} \frac{h^{r+1}}{(r+1)!} \left(h\frac{\partial}{\partial x} + h\frac{\partial}{\partial y} \right)^r f(x, y)
\end{aligned} \tag{5}$$

and whenever f does not depend on x explicitly, we have the incremental function

$$\Phi_T(y_n; h) = \sum_{r=0}^{\infty} \frac{h^{r+1}}{(r+1)!} \left(\frac{\partial}{\partial y} \right)^r f(y) \tag{6}$$

An s-stage ERK method is of the form

$$y_{n+1} - y_n = \Phi_{RK}(x_n, y_n; h) \tag{7}$$

$$\text{where } \Phi_{RK}(x_n, y_n; h) = \sum_{r=1}^{s} b_r K_r$$

$$K_1 = hf(x, y)$$

$$K_r = hf\left(x + c_r h, y + \sum_{u=1}^{r-1} a_{ru} K_r\right), r = 2, 3, ..., s$$

$$c_r = \sum_{u=1}^{r-1} a_{ru}, r = 2, 3, ..., s.$$

The matrix representation of the ERK methods stated in equations (7) is of the form

$$
\begin{bmatrix} A & c \\ b^T & 0 \end{bmatrix} =
\begin{bmatrix}
a_{11} & a_{12} & \cdots & a_{1\,s} & a_{1\,s+1} \\
a_{21} & & & & \vdots \\
\vdots & & & & \vdots \\
a_{s\,1} & & \cdots & a_{s\,s} & a_{s\,s+1} \\
a_{s+1\,1} & & \cdots & a_{s+1\,s} & 0
\end{bmatrix} \tag{8}
$$

These coefficients are specified as follows

$$a_{r\,s+1} = c_r, \qquad r = 1(1)s$$

$$a_{s+1\,u} = b_u, \qquad u = 1(1)s$$

$$a_{s+1\,s+1} = 0,$$

$$a_{j\,u} = a_{u\,j}, \; j, u = 1(1)s$$

The basis of derivation of ERK schemes is to equate the coefficients of the incremental functions $\Phi_T(x_n, y_n; h)$ and $\Phi_{RK}(x_n, y_n; h)$ to $O(h^p)$ for a pth order method.

3   A Class of 3 – Stage MERK Methods

Traditionally, given an IVP (1), classical ERK methods are derived with the intention of performing multiple evaluations of $f(y)$ in each internal stage for a given accuracy. However, the new scheme is derived with the notion of incorporating higher order derivatives of $f(y)$. There is an appreciable improvement of the attainable order of accuracy of the method.

A 3-stage MERK method for an autonomous system is given as

$$y_{n+1} - y_n = \Phi_{MERK}(y_n, h)$$

$$\text{where } \Phi_{MERK}(y_n, h) = \sum_{r=1}^{3} b_r K_r$$

$$= b_1 K_1 + b_2 K_2 + b_3 K_3$$

$$K_1 = hf(y)$$

$$K_2 = hf\left(y + a_{21}K_1 + ha_{22}f_y K_1 + \frac{1}{2}h^2 a_{23}\left(f_y^2 + ff_{yy}\right)K_1\right)$$

$$K_3 = hf\left(y + a_{31}K_1 + a_{32}K_2 + ha_{33}f_y K_1 + ha_{34}f_y K_2 + \frac{1}{2}h^2 a_{35}\left(f_y^2 + ff_{yy}\right)K_1 + \frac{1}{2}h^2 a_{36}\left(f_y^2 + ff_{yy}\right)K_2\right)$$

Having expanded $\Phi_{MERK}(y_n, h)$ by Taylor's series, its coefficients of h as far as h^6 was compared with $\Phi_T(y_n, h)$ to generate the family of methods in Table 1. This family of 3–stage methods, which are of order 5 also demonstrates that an s–stage MERK method can attain order as high as $s + 2$. This is a significant improvement on the established maximum attainable order of $p(s) = s$ for an s–stage method by Butcher [1, 4, 5].

Table 1. Family of 3-stage MERK Methods of order 6 for Autonomous system

Parameter	$MERK_a$	$MERK_b$	$MERK_c$	$MERK_d$	$MERK_e$
b_1	$\frac{1}{6}$	$\frac{1}{6}$	$\frac{1}{6}$	$\frac{1}{6}$	$\frac{1}{10}$
b_2	$\frac{1}{6}$	$\frac{2}{3}$	$\frac{2}{3}$	$\frac{1}{6}$	$\frac{2}{5}$
b_3	$\frac{2}{3}$	$\frac{1}{6}$	$\frac{1}{6}$	$\frac{2}{3}$	$\frac{1}{2}$
a_{21}	1	$3\frac{1}{2}$	$\frac{1}{2}$	1	$\frac{5}{6}$
a_{22}	$\frac{1}{5}$	$\frac{1}{10}$	$\frac{1}{10}$	$\frac{2}{5}$	$\frac{1}{18}$
a_{23}	0	0	0	$\frac{1}{5}$	0
a_{31}	$\frac{13}{58}$	1	$\frac{3}{41}$	$\frac{3}{8}$	$-\frac{5657}{9615}$
a_{32}	$\frac{8}{29}$	0	$\frac{38}{41}$	$\frac{1}{8}$	$\frac{2954}{3205}$
a_{33}	$-\frac{281}{2320}$	$\frac{1}{2}$	$\frac{2}{205}$	$\frac{1}{40}$	$-\frac{95444}{144225}$
a_{34}	$\frac{21}{464}$	$\frac{1}{10}$	$\frac{6}{41}$	0	$-\frac{2926}{16025}$
a_{35}	$-\frac{12}{145}$	$\frac{6}{25}$	$\frac{11}{410}$	$-\frac{1}{40}$	$-\frac{368803}{1442250}$
a_{36}	$\frac{1}{145}$	$\frac{4}{25}$	$\frac{29}{205}$	0	$\frac{3763}{240375}$

From Table 1, a member of this family corresponding to $MERK_b$ is stated as follows

$$y_{n+1} = y_n + \frac{1}{6}K_1 + \frac{2}{3}K_2 + \frac{1}{6}K_3$$

$$K_1 = hf(y_n)$$

$$K_2 = hf\left(y_n + \frac{7}{2}K_1 + \frac{1}{10}hf_yK_1\right)$$

$$K_3 = hf\left(y_n + K_1 + \frac{1}{2}hf_yK_1 + \frac{1}{10}hf_yK_2 + \frac{6}{25}h^2\left(f_y^2 + ff_{yy}\right)K_1 + \frac{4}{25}h^2\left(f_y^2 + ff_{yy}\right)K_2\right)$$

4   Propagation of Errors in MERK Methods

MERK methods, like other one-step methods are based on the principle of discretization. These methods have the common feature that no attempt is made to approximate the exact solution $y(x)$ over a continuous range of the independent variable. Approximate values are sought only on a set of discrete points $x_0, x_1, x_2, \ldots$. The true solution of the differential equation at $x = x_n$ is denoted by $y(x_n)$, and the appropriate solution obtained by applying any of the MERK methods as y_n.

The objective of this work is to investigate the propagation of errors of these methods which is a crucial property of the methods. This study also helps in the selection of step length and thus, the speed of generating numerical results for IVPs.

Definition 1. *The Local Truncation Error (LTE) at x_{n+1} of the MERK methods is defined to be τ_{n+1} where*

$$\tau_{n+1} = y(x_{n+1}) - y(x_n) - \Phi_{Method}(y(x_n); h)$$

and $y(x)$ is the theoretical solution of the IVP (1).

Making the localizing assumption that no previous errors have been made (that is, that $y_n = y(x_n)$), then the LTE of MERK methods satisfies

$$\tau_{n+1} = y(x_{n+1}) - y_{n+1}$$

The study of error bounds also plays a significant role in the design of program codes for solving IVP (1). Only few codes control the LTE committed at every integration step by demanding that [6],

$$\tau_{n+1} \leq h_n \tau_n \tag{9}$$

where $h_n = x_{n+1} - x_n$ is the current step size and τ_n is the allowable error tolerance, which may depend on the independent variable x_n. Most practical codes, however, replace h_n on the right hand side of (10) by unity, thus adopting error per step criterion [6].

A user is actually interested in the true or global error specified by

$$e_{n+1} = y(x_{n+1}) - y_{n+1}, \tag{10}$$

This global truncation error e_{n+1} is defined such that it is no longer assumed that no previous truncation errors have been made. And it is well known that the variational equation

$$e\prime(x) = J(x, y(x)) e(x) \quad , \quad e(a) = \delta \tag{11}$$

(where J is the Jacobian matrix associated with the IVP (1) says how an error δ at $x = a$ propagates). The approximate equation (12) is satisfied by the error-neglecting second order terms [6].

The propagation of errors depends on two factors namely:

- the local error and
- the nature/stability of the problem.

For instance if the IVP (1) is inherently stable (that is, all the eigenvalues of J have negative real parts), then the local errors may damp out with increasing x; otherwise, the errors will be magnified with increasing x [9]. Bulirsh and Stoer [3], constructed asymptotic upper and lower bounds on the global errors emanating from extrapolation methods to IVPs. Shampine [11], generalized this idea for any one-step method endowed with an asymptotically correct LTE estimator.

There exist fundamental obstacles in the direct control of the global error, but in recent years, appreciable progress has been attained and reliable estimation of the global error [6]. The LTE and the round off errors constitute a sequence of perturbations that shift computed solutions to the neighboring integral curves. The stability of a discretization method for (1) demands that, provided the starting global error $e_0 \neq 0$, the ultimate global error e_n should be bounded [6]; that is, a finite constant K exists such that $e_n < Ke_0$,

Therefore, the LTE for MERK methods can be obtained by using the Taylor's series expansion for $y(x_{n+1})$ and y_{n+1} from Table 1 as follows,

$$\tau_{n+1} = \varphi(y(x_n)) h^5 + O(h^6) \tag{12}$$

where

$$
\begin{aligned}
\varphi(y(x_n)) = {} & -\frac{1}{120} f \left(-1 + 60 \left(a_{23}a_{32} + 2a_{22}a_{34} + a_{21}a_{36}\right) b_3\right) f_y^4 - \\
& \frac{1}{120} f^2 \Big(-11 + 60\big((a_{22}^2 + a_{21}a_{23}) b_2 + \\
& \left(a_{23}a_{32} + a_{21}^2 (a_{32}^2 + a_{34}) + (a_{33} + a_{34})^2 + \right. \\
& a_{21} \left(2a_{31}a_{34} + 2a_{32}(a_{22} + a_{33} + 2a_{34}) + a_{36}\right) + \\
& (a_{31} + a_{32})(2a_{22}a_{32} + a_{35} + a_{36}) \big) b_3 \Big) f_y^2 f_{yy} + \\
& f^3 \left(-\frac{1}{30}\left(-1 + 15a_{21}a_{23}b_2 + 15(a_{31} + a_{32})(a_{21}^2 a_{32} + a_{35} + a_{36}) b_3\right) f_{yy}^2 - \right. \\
& \frac{1}{120}\left(-7 + 60a_{21}^2 a_{22} b_2 + 20(a_{21}^3 a_{32} + 3a_{21}a_{32}(a_{31} + a_{32})^2 + \right. \\
& \left. 3(a_{31} + a_{32})^2 (a_{33} + a_{34})) b_3\right) f_y f_{yyy} \Big) - \\
& \frac{1}{120} f^4 \left(-1 + 5a_{21}^4 b_2 + 5(a_{31} + a_{32})^4 b_3\right) f_{yyyy}
\end{aligned}
$$

is called the principal error function, and $\varphi\left(y(x_n)\right)h^5$ the Principal Local Truncation Error (PLTE) [6, 8].

Suppose the following bounds for f and its partial derivatives hold for $x \in [a,b]$, $y \in (-\infty,\infty)$,

$$|f(y)| \leq Q \,, \quad \left|\frac{\partial^j f(y)}{\partial y^j}\right| < \frac{P^j}{Q^{j-1}} \,, j \leq p \tag{13}$$

where P and Q are positive constants [10], and p is the order of the method (in this case $p = 4$). Then,

$$|f_y| < P \,, \ |f_{yy}| < P^2 Q^{-1} \,, \ |f_{yyy}| < P^3 Q^{-2} \ \text{ and } \ |f_{yyyy}| < P^4 Q^{-3}.$$

The computed step size bounds for the MERK Methods in Table 1 and ERK stated earlier are shown in figures 1–4.

5   Numerical Computation and Results

In this section we compute the appropriate mesh size bounds so as to integrate the IVPs in Examples 1 – 4 above [6, 8], using the MERK methods stated in Table 3.1 with an allowable error tolerance $\varepsilon = 10^{-4}$.

Table 2. Initial Value Problems for Implementation using the MERK Methods

Example No	$y\prime(x)$	Initial Conditions	Interval of Integration	Theoretical Solution $y(x)$	P	Q
1	$-10\left(y(x)^2 - 1\right)$	$y(0) = 2$	$0 \leq x \leq 1$	$1 + \frac{1}{10x+1}$	20	10
2	$y(x)$	$y(0) = 1$	$0 \leq x \leq 1$	$exp(x)$	1	e^1
3	$\sqrt{y(x)}$	$y(0) = 1$	$0 \leq x \leq 1$	$\frac{1}{4}(x+2)^2$	$\frac{1}{2}$	1
4	$1 + (y(x))^2$	$y(0) = 1$	$0 \leq x \leq 1$	$tan\left(x + \frac{\pi}{4}\right)$	2	2

The results of these computations are shown in figures 1–4. Three other conventional methods [6] implemented and compared with the new MERK methods are stated below:

1. 2–stage scheme - Explicit Trapezoidal (ERKO).

$$\begin{bmatrix} 0 & 0 & 0 \\ 1 & 0 & 1 \\ \frac{1}{2} & \frac{1}{2} & 0 \end{bmatrix}$$

2. 3–stage scheme - Heun (ERKH).

$$\begin{bmatrix} 0 & 0 & 0 & 0 \\ \frac{1}{3} & 0 & 0 & \frac{1}{3} \\ 0 & \frac{2}{3} & 0 & \frac{2}{3} \\ \frac{1}{4} & 0 & \frac{3}{4} & 0 \end{bmatrix}$$

3.   4–stage scheme - Classical (ERK4TH).

$$\begin{bmatrix} 0 & 0 & 0 & 0 & 0 \\ \frac{1}{2} & 0 & 0 & 0 & \frac{1}{2} \\ 0 & \frac{1}{2} & 0 & 0 & \frac{1}{2} \\ 0 & 0 & 1 & 0 & 1 \\ \frac{1}{6} & \frac{2}{6} & \frac{2}{6} & \frac{1}{6} & 0 \end{bmatrix}$$

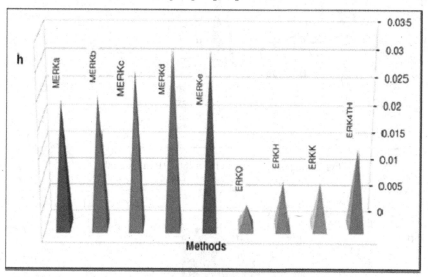

Fig. 1. Step size (h) bounds for the IVP in Example 1

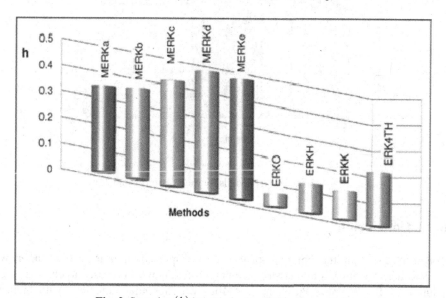

Fig. 2. Step size (h) bounds for the IVP in Example 2

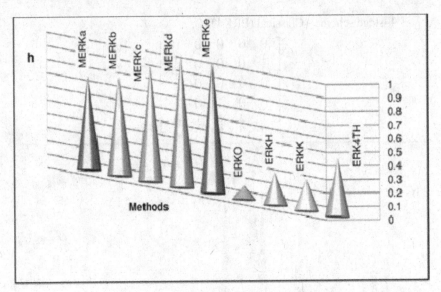

Fig. 3. Step size (h) bounds for the IVP in Example 3

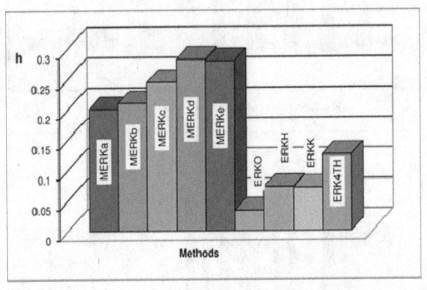

Fig. 4. Step size (h) bounds for the IVP in Example 4

6   Conclusion

Conclusively, from the computational results displayed in figures 1–4, the new MERK methods possess higher step size bounds and they can be used to compute the solution of Initial Value Problems in Ordinary Differential Equations with higher values of step size h.

References

1. Akanbi, M.A., Okunuga, S.A., Sofoluwe, A.B.: Error Bounds for 2-stage Multiderivative Explicit Runge-Kutta Methods. Adv. in Modelling & Analysis, A.M.S.E. 45(2), 57–72 (2008)
2. Akanbi, M.A., Okunuga, S.A.: On Region of Absolute Stability and Convergence of 3-stage Multiderivative Explicit Runge-Kutta Methods. Journal Sci. Res. & Dev. 10, 83–100 (2005/2006)
3. Bulirsh, R., Steor, J.: Asymptotic Upper and Lower Bounds for Results of Extrapolation Methods. Numerische Mathematik 8, 93–104 (1966)
4. Butcher, J.C.: Numerical methods for Ordinary Differential Equations in the 20th century. J. Comput. Appl. Math. 125, 1–29 (2000)
5. Butcher, J.C.: Numerical methods for Ordinary Differential Equations, 2nd edn. John Wiley & Sons Ltd., Chichester (2008)
6. Fatunla, S.O.: Numerical Methods for IVPs in ODEs. Academic Press Inc., USA (1988)
7. Henrici, P.: Discrete Variable Methods in ODE. John Wiley & Sons, New York (1962)
8. Lambert, J.D.: Computational Methods in ODEs. John Wiley & Sons, New York (1973)
9. Lambert, J.D.: Numerical Methods for Ordinary Differential Systems: The Initial Value Problem. John Wiley & Sons, England (1991)
10. Lotkin, W.: On the Accuracy of R-K Methods. MTAC 5, 128–132 (1951)
11. Shampine, L.F., Baca, L.S.: Global Error Estimates for ODEs based on Extrapolation Methods. SIAM Journal on Scientific and Statistical Computing 6, 1–14 (1985)

Tapping the Potential of Virtual World-Based Simulations in Higher Education

Donald R. Moscato and Shoshana Altschuller

Information Systems Department, Hagan School of Business, Iona College,
715 North Avenue, New Rochelle, NY 10801
{dmoscato,saltschuller}@iona.edu

Abstract. The purpose of this paper is to illustrate the pedagogical possibilities of employing virtual world simulations across many disciplines of higher education. We first present a discussion of the varied approaches to pedagogy along with the identification of the pros and cons for each one. Next, we present an introduction to virtual world concepts and key issues surrounding their use for educational purposes. A specific discussion of the virtual world Second Life is presented. Following these preliminary sections, we present by subject areas, the many potential and actual implementations of virtual world simulations in higher education.

Keywords: Virtual world, education, pedagogy.

1   Introduction

Although the concept of a virtual world tends to be defined along its various characteristic dimensions [1] (e.g. presence [2] and persistence and representation [3, 4], in essence, a virtual world is a simulation of the real world. It should come as no surprise, therefore, that participants of higher education have found numerous ways to incorporate virtual worlds into their programs of learning [5, 6]. What better way to examine and probe the actual world than by traversing its virtual replica? In fact, the characteristics that define virtual worlds create an environment that lends itself, in many ways, to a learning experience that is unparalleled in traditional forms of education or in other modes of online learning.

2   Pedagogy in Higher Business Education

Anyone who has participated in higher education knows that the student can be exposed to a wide variety of educational approaches to teach the subject matter of a course. In this section, we identify these methods and discuss briefly some of the advantages and disadvantages of each one. These educational methodologies represent a range of levels of interactivity. In recent years there has been a major thrust to promote active learning in the university environment [7]. The literature also

K.J. Engemann, A.M. Gil Lafuente, and J.M. Merigó (Eds.): MS 2012, LNBIP 115, pp. 198–209, 2012.

discusses several forms of learning styles that people utilize. These are visual learners, audio learners, linear learners and experiential learners [8, 9]. The approaches that follow can and should incorporate a mix of these approaches so that students can benefit from their preferred learning modality.

2.1   Lecture Method

This traditional approach follows the model of the sage on the stage. The approach is independent on class size, however, in smaller venues a Socratic method can be utilized by the instructor. This will avoid the trap of the passive student just listening and taking notes and not participating in any interactive manner. There is little or no immersive student experience.

2.2   Video

This passive approach to education has the student viewing a pre-recorded session and responding afterward according to the learning objective of the exercise [10]. It is passive in nature in that there is no dynamic student involvement. An advantage is that it can be viewed on the student's own time schedule.

2.3   Case Studies

Case studies are often used to focus on either narrow or broad subject areas. They tend to be more substantial than end-of-chapter exercises and can either be conducted as oral or written documentation of the learning objective. In business, they have been used in all types of subject areas. Also, assignments can be individually or collectively organized by the instructor.

2.4   Internships

Internships are field experiences where the student participates in real world activities related to the course matter under study. They are time consuming and require close coordination between the faculty and the organization sponsoring the internship [11]. There can be a high level of immersion on the part of the student. Often, a written report is required and some "sign-off" on the part of the sponsoring entity [12].

2.5   Experiential Exercises

As its name implies, this learning approach requires the student to perform an activity that is often a "hands on" in order to learn or master a subject area. Clearly, the degree of difficulty can range from a simple short-term exercise to a more complicated one involving a significant part of the term. A possible downside is that the person or group cannot be successful and lose interest along the way. This proves a challenge to the instructor and requires a degree of damage control [13].

2.6   Role Playing

In this approach to education, students are given background information via research or instructor-provided and they interact with each other according to specified norms of behavior. This approach is highly immersive and enables both the participants as well as the audience to comment on what is before them [14, 15].

2.7   Research Reports

In this traditional approach, a student or team of students, is assigned a topic and is expected to do independent research and present a report at the conclusion of the research. It is highly immersive for it to be successful.

2.8   Simulations/Games

These approaches are highly interactive and immersive by their very nature [16]. A situation is modeled either by the instructor and/or student and the participants are gamers that seek a particular objective as specified as the learning objective. In virtual world simulations there is often no specific general objective -- it is a world in which the participant can define his/her own objective in playing. Therefore, in the examples the authors present in this paper, specific educational subject areas are presented as well as how virtual world simulations can be used in them in order to achieve specific learning objectives [17].

3   Virtual World Simulations and Second Life

A virtual world is a massively multi-user online environment where users are represented by avatars. While some virtual worlds have rules and objectives, others, such as Second Life, are merely a container for the creations of their users [18].

In Second Life, 3D avatars own and develop land, build and create objects, trade items and currency, and interact with each other in any way they choose, thereby creating a simulation of the real world. While obviously no simulation is a perfect replica of its original, many of Second Life's discrepancies with the real world only enhance its usefulness as a tool in higher education to learn about countless aspects of the real world [19]. It should be noted that because Second Life is viewed as a relatively expensive open source virtual world platform some universities are exploring other virtual world platforms [20]. Following, we examine some of the characteristics of Second Life that lend themselves to being a learning environment.

3.1   Interactivity

A virtual world like Second Life is a prime example of social media. It is enhanced by, and in fact depends upon, the co-presence of and interaction between many users thus creating a community, or virtual society. For educational activities in Second

Life, this means that students can have an opportunity to be fully engaged and take an active role in the learning process [21]. Since Second Life is open to the public for anybody to become a member for free, it can be used as a laboratory to study phenomena that depend on human interaction in such fields as sociology, psychology and even consumer behavior.

3.2   Immersion

The virtual environment in which virtual world users interact is immersive. When they are "logged in" to the virtual environment users are embodied by avatars so that the interaction involves both visual and experiential aspects, appealing to various styles of learning [6]. Immersive interaction also allows users to share the same virtual surrounding so that they can actually experience the shared workspace rather than relying on the knowledge that others are co-present. Visual interaction and shared experiences strengthens the ability for users to feel connected to one another and to collaborate in a proactive manner.

3.3   Creativity

Second Life follows the trend of many Web 2.0 applications (such as YouTube, Wikipedia and Flickr) where content is generated by its users [19]. It is an online, digital world created by its residents. Areas of land, known as islands or sims, can be explored, built on, bought, and sold. The Second Life 3D world is also called the "grid" and at the backend is a collection of servers that run the simulators that make up Second Life's world. Users can build their own world by building and shaping their own place on the grid and adding communication to it [22]. Second Life has therefore become a place where among many other activities, users express and experiment with their creativity in the form of art, architecture, programming skills, etc. [21]. In turn, it is has the potential as an environment that is ripe for students of these arts and sciences.

3.4   Authenticity

While they originated as massively multi-user games where avatars, objects, and concepts are mere imitations of their real world counterparts, virtual worlds are becoming more and more inextricably intertwined with the real world [23]. Second Life consists of nearly a half million acres of virtual land either on Second Life's mainland or on private islands (http://secondlife.com/land/faq/). For all of that land, Second Life users have paid down-payments and continue to pay monthly fees for the privilege of customizing their virtual space to their liking. Much like the real world's real estate market, Second Life's real estate scene even comes complete with landlords and brokers. Most virtual worlds also include an in-world currency. Though the virtual currency, such as the Linden Dollar in Second Life, is invented by the creators of the virtual world, the value that people place on it gives it legitimacy in the real world [24]. In fact, virtual world economic activities have proven to closely mimic real world economies complete with in-world currency exchanges, wages, and statistics covering concepts such as labor supply, inflation, and foreign trade [25]. In a virtual world with such authentic [1] and true-to-life mechanisms that it actually overlaps with the real world [22], the stage is perfectly set to examine those mechanisms.

3.5   Fantasy

Still, even though the virtual world is sometimes very "real world"-like, its reality, ultimately, is fabricated and transcends the constraints of physical and other limitations that exist in the real world. The capacity to design and build any fictional scenario or modify any real one, in any part of the world, and then fly around it is an ideal setting in which to simulate objects of study to demonstrate, test, and acquire knowledge.

4   Application Areas of Virtual World Simulations in Higher Education

In this section we introduce several academic disciplines and how they could capitalize on the dimensions described above and employ virtual worlds as a component of the pedagogical experience [26-28].

4.1   Health Care

Health care is a very broad field that encompasses medical doctors, dentistry, psychology, psychiatry, nursing, therapy and nutrition. Therefore, the issue is where can experiential learning be an integral component of the education or training process? The first area is in the awareness category that transcends any disease and/or best practices of prevention. Virtual world sites have been created for AIDS education as well as for breast cancer awareness. Visitors to these sites can be exposed to videos and other literature is presented as well as the presentation of conferences on timely topics. Sex education is another area that can utilize virtual world simulations [28].

Another area where virtual worlds can be used is in diagnosis and protocol processes. In this application an avatar simulates the behavior of a patient while another avatar plays the role of the doctor. This behavior is monitored by medical educators who provide guidance and feedback to the doctors in training. "Scientists are developing ever more sophisticated versions of 'virtual patients' with the aim of testing medical devices and procedures that can't readily be assessed in real people" [29].

4.2   Sociology

One of the principal educational objectives of this discipline is to show how various cultures interact and how well "different" behaviors are received and assessed. For example, virtual world simulations can demonstrate in a highly interactive and immersive fashion what it is like for a minority person to experience life in a majority environment. This can be extended to people with alternative life styles who interact with "straight" populations. Students can role play either perspective and garner insights accordingly that would be more difficult to experience in a real world context. Extension to gender, age and physically challenged bias can also be explored in a very creative manner.

4.3   Language Education

Language education, at all levels, has utilized technology for a long time. Traditional language laboratories are a staple component in many language programs. No one questions the importance of practicing a language in order to learn the nuances of conversational speech. With virtual world simulations one is operating in a 24/7 global environment. At any time of the day the language student can be transported via their avatars to a varied set of "countries" and meet other people and interact with them in their host languages whether it be Spanish, Italian, French, Portuguese or any of a number of other languages. The student has the option of switching between a translator (built in) or to use an unedited level of interaction [30].

4.4   Visual Arts

This field of education relies on the understanding of theory as it is a prelude to the application of creative processes to the creation of art. One characteristic of the visual arts is the ability to express this creative design process on a canvas. Virtual world simulations provide such a canvas with extraordinary opportunity. Paintings, gardens, mountains, coastal areas can all be created with limitless potential. Both realistic and impressionistic experiences can be part of the experimental process of artistic creation.

4.5   Architecture

Education in this discipline involves sketches, building models and viewing the creations from multiple perspectives (3-D etc.). Technology has always been an integral part of this training. With the virtual world simulation platform students can design and build their creations. They can be viewed by other avatars from any angle since the avatars are not bounded by traditional laws of physics-they can fly! It should be noted that the creative mind of the architect can explore models that can be viewed by prospective clients who would have the capability to traverse the design by moving through it in life-size scale as they would be able to do in the real world. This would be virtually impossible to accomplish in other mediums.

Another creative implementation of virtual world simulations in this subject area is to fund design competitions according to a specified theme. An example was a Frank Lloyd Wright design competition that required architects to design and build examples in the style of this great architect.

4.6   Travel/Tourism

It is possible to create virtual world replicas of actual real world sites. These representations can then be used as part of a promotional campaign for potential visitors or even for educational purposes. With this medium, the visiting avatar can traverse the site and view examples of art (a museum), architecture (a building) or park or gardens. The site can be staffed with a tour guide who promotes the site, answers questions, and can offer discount coupons for visits to the actual location. Examples are the World Trade Center Memorial, Holocaust memorial, Versailles, the Vatican, miniature Netherlands and the Greek Theater in Taormina.

4.7   Music

A very popular component of virtual world simulation is the performing arts. At any time of the day (recall a 24/7 world), you can find live concerts being presented across a mind-boggling array of musical styles ranging from reggae, country, classical, rock and others. It should be noted that they are presented in many different languages. A side dimension is also the recording of these concerts so that budding directors and cinematographers can hone their craft by first using this medium. They can edit their work and possibly publish it on the Internet using a venue like YouTube. Real world artists supplement their careers by creating a following in the virtual world as well.

4.8   History

Most people associate the study of history with the memorization of facts from different eras. History and historical events can become alive for the student via immersive experiences in virtual world simulations that recreate an historical era. For example, an ancient Rome simulation can be created replete with avatars that dress and speak in the language of the time. Same for Victorian villages created with accurate designs, dress and the language of the period. Students can then "experience" role playing historical events and get far more involved than in the traditional educational process of lectures and readings.

4.9   Scripting Languages/Animation

This field of computer science can be approached in a very creative and exciting way using virtual world simulations as the platform. Each avatar in a virtual world is "embodied" by embedded scripts which control all avatar movements from the simple to the complex. These movements are done via scripting languages unique to the virtual world product. By utilizing a virtual world platform rather than traditional end-of-chapter exercise, the student is introduced to a richer application environment.

There is no limit to what a student can do via animation in a virtual world simulation. It is the perfect application environment. The animation can be targeted to any of the possible areas discussed in this paper. Animation can be done to any object-people, animals, cars, and airplanes – literally anything that is an "object" can be animated.

4.10 Forensics/Auditing

The education in this field is based on understanding critical theoretical concepts and then being able to apply them in a realistic context. This experiential component requires a degree of field experience that must be obtained in any number of forms. The cost of this experience can be prohibitive especially in the early stages of investigative training.

Examples of auditing can be the audit of a physical site. For example, a computer security audit of a data center can be conducted by students who must visit a site and perform a security review. They are tasked with looking for varying risk exposures and then documenting them via screen shots and then crafting a report on their findings. The cost of this valuable experience is minimal compared to the need to visit an actual site which in itself is a security risk. In fact, the nature of the physical site could be anything from a hospital, shipyard, airport, refinery, railroad station or national monument.

4.11 Military Science

An important aspect of military training is assessing a student's assessment and reaction to different combat situations. Virtual world simulations can be created that include hard assets such as tanks, barriers, buildings and troops organized according to the training exercise under study. The student via his/her avatar must traverse the targeted region, assess the situation and propose a course of action. The simulation can include vivid graphics that can be "tuned" to the desired training objectives.

4.12 Environmental Science

An important global concern affecting the environment is sustainability. Virtual world simulations can be developed that depict varied situations that can have adverse impact on the environment. Students would be charged to assess a particular situation as designed to reflect both good and bad examples of global behavior. For example, the location of toxic waste sources near sensitive areas can be portrayed in very creative and instructional ways that educate the students. Students would be challenged to "redesign" an area that is more environmentally friendly yet still achieves the specified objectives of a project.

4.13 Business

There are many potential educational possibilities for business in virtual world simulations. Virtual world simulations like Second Life utilize parcels in sizes of 64,000 square meters. These parcels can be subdivided by the owner/renter into two or more sub-parcels. These "sims", as they are called, can contain whatever the architect/designer wants them to be. This capability allows an entrepreneur to create a mall or a club or a housing development. Therefore, the first application we will discuss is *rental real estate business*. In order to accomplish this activity, one has to determine what type of development will be designed, how many homes to put on it, what rental to charge and what loan covenants will be stipulated. The key is to earn a profit so a developer must manage the cost and revenue side of the business. It is nearly impossible to do this in the classroom using real world models so a virtual world provides a perfect environment to explore the major facets of this business activity.

Another business application that can be modeled in virtual worlds is that of *hospitality management*. Students can be provided with valuable experience by being managers of entertainment clubs which provide live acts or DJ-provided dance music for virtual world clients. In this capacity, students must engage in booking new acts, managing any lottery games and serving as hosts or hostesses so that the clients are made to feel comfortable in attending the venue. If these customers are satisfied they can spread the word and encourage others to attend future events.

A third application in business education is *new product development*. Avatars engage in designing products throughout the virtual world environment. These products range from automobiles, houses, avatar shapes, hair, clothing, toys and other products that can be utilized by other business people. Here we will focus on designing clothes. This area is one of the most successful areas in virtual world simulations. Just as in the real world students of product management must understand what products are desired by consumers, create a product, market that product and continuously evolve their markets through cross-product marketing. Designers can specialize in men's clothing, women's clothing and children's clothing to name the obvious segments. Special tools are available such as pose stations, sculptys and other aids to enable the designer to create realistic designs that "flex" with the movements of the avatars.

A fourth application in business education is *advertising*. Virtually all segments of a virtual world environment are influenced by successful advertising. Land must be advertised. There are internal search engines which can be utilized by vendors to advertise their products or services. For a monthly fee, businesses can have their venues or products prominently positioned so that people searching can be informed of the existence of whatever it is they market. On the site of the actual sim there can be billboards, free-standing wall signs, animated neon signs or other forms of advertising. Several corporate sponsorships co-exist on virtual world sites as part of real world businesses seeking recognition of their brands. (See branding examples here.)

Promotions are another example of business activity in virtual worlds. Companies utilize contests, couponing, free gifts and demos to capture the consumer in virtual world simulations. It should be noted that both real world companies as well as virtual world businesses utilize this tool to advance their strategic agendas. This aspect is inherently interesting in its own right because real world businesses are exploring how they can utilize virtual world simulations to promote their products and services among a highly prized audience of gamers.

Leadership education is another area in which virtual world simulations can be a useful medium for education. Often, instructors organize their students in teams and assign them a project or a report to complete. Aside from the subject area under review, the object is to promote team-building and leadership skill development. Virtual worlds can be the venue to explore and develop these skills. The students can, via their avatars, participate in exercises that nurture these critical skills but in an entertaining and creative environment.

5   Summary and Conclusions

In this paper, virtual worlds have been posited and then demonstrated to have the potential to be a rich and robust tool for higher education in a wide range of disciplines. Already, many educators have tapped into the potential to create a learning environment that both appeals to many learning styles and includes the possibility for numerous teaching methodologies. Role playing, experiential exercises, research opportunities, case studies and simulation games all seem a natural fit to be uniquely applied in a virtual world. Keeping the caveats in mind (such as learning time required and technological issues [21]), we can likely begin to complement traditional passive pedagogical methods with new active and interactive ones, cultivating real educational possibilities in virtual worlds. Above all, it will be imperative to traverse all the hype for virtual worlds in higher education and focus on real contributions to the learning environment [31].

References

1. Warburton, S.: Second Life in Higher Education: Assessing the Potential for and the Barriers to Deploying Virtual Worlds in Learning and Teaching. British Journal of Educational Technology 40(3), 414–426 (2009)
2. Schroeder, R.: Defining Virtual Worlds and Virtual Environments. Journal of VirtualWorlds Research 1(1) (2008)
3. Bell, M.: Toward a Definition of 'Virtual Worlds'. Journal of VirtualWorlds Research 1(1) (2008)
4. Childress, M., Braswell, R.: Using Massively Multiplayer Online Role-Playing Games for Online Learning. Distance Education 27(2), 187–196 (2006)
5. Educational Uses of Second Life (2007),
   http://sleducation.wikispaces.com/educationaluses?f=print (cited June 26, 2007)
6. Middleton, A.J., Mather, R.: Machinima Interventions: Innovative Approaches to Immersive Virtual World Curriculum Integration. ALT-J Research in Learning Technology 16(3), 207–220 (2008)
7. Meyers, C., Jones, T.B.: Promoting Active Learning Strategies for the College Classroom. Jossey-Bass Inc., San Francisco (1993)
8. Meier, D.: The Accelerated Learning Handbook. McGraw-Hill, New York (2000)

9. Russell, L.: The Accelerated Learning Fieldbook. Jossey-Bass Pfeiffer, San Francisco (1999)

10. Kaufman, P., Mohan, J.: Video Use and Higher Education: Options for the Future. New York: Intelligent Television, New York University, New York, NY (2009), http://library.nyu.edu/about/ Video_Use_in_Higher_Education.pdf

11. Eyler, J.: Graduates' Assessment of the Impact of a Full-Time College Internship on Their Personal and Professional Lives. College Student Journal 29(2), 186–194 (1995)

12. Antelo, M.: Internships Have Value, Whether or Not Students Are Paid. Chronicle of Higher Education (April 24, 2011)

13. Cantor, J.A.: Experiential Learning in Higher Education: Linking Classroom and Community. ASHE-ERIC Higher Education Report No. 7. ERIC Clearinghouse on Higher Education, Graduate School of Education and Human Development, The George Washington University, Washington, DC (1995)

14. Heyward, P.: Emotional Engagement through Drama: Strategies to Assist Learning through Role-Play. International Journal of Teaching and Learning in Higher Education 22(4), 197–203 (2010)

15. Bender, T.: Role Playing in Online Education: A Teaching Tool to Enhance Student Engagement and Sustained Learning. Innovate (2007)

16. Damassa, D.A., Stiko, T.: Simulation Technologies in Higher Education: Uses, Trends, and Implications. Research Bulletin 3. EDUCAUSE Center for Applied Research, Boulder, CO (2010), http://www.educause.edu/ecar

17. Dalgamo, B., Lee, M.J.W., Carlson, L., Gregory, S.: 3d Immersive Virtual Worlds in Higher Education: An Australian and New Zealand Scoping Study. In: ASCILITE 2010, Sydney, Australia (2010)

18. Wankel, C., Kingsley, J.: Higher Education in Virtual Worlds Teaching and Learning in Second Life. Emerald Publishing Limited, London (2009)

19. Bowers, K.W., Ragas, M.W., Neely, J.C.: Assessing the Value of Virtual Worlds for Post-Secondary Instructors: A Survey of Innovators, Early Adopters and the Early Majority in Second Life. International Journal of Humanities and Social Sciences 3(1), 40–50 (2009)

20. Young, J.: After Frustrations in Second Life, Colleges Look to New Virtual Worlds. The Chronicle of Higher Education (January 14, 2010)

21. Baker, S.C., Wentz, R.K., Woods, M.M.: Using Virtual Worlds in Education. Teaching of Psychology 36, 59–64 (2009)

22. Altschuller, S., Moscato, D.R., Boekman, D.: Organizational Approaches to Global Brand Equity in 3-D Virtual Worlds: Second Life. In: 10th International Business and Economy Conference, Guadalajara, Mexico (2011)

23. Altschuller, S., Moscato, D.R.: Digital Ownership in 3-D Virtual Worlds: Are We Pushing the Limits of Propriety? In: 22nd Annual International Information Management Association (IIMA) Conference, New Orleans, LA (2011)

24. Castranova, E.: On Virtual Economies (2002), SSRN: http://ssrn.com/abstract=338500

25. Castranova, E.: Virtual Worlds: A First-Hand Account of Market and Society on the Cyberian Frontier. The Gruter Institute Working Papers on Law, Economics, and Evolutionary Biology 2(1) (2001)

26. Conklin, M.: 101 Uses of Second Life in the College Classroom. Version 2.0 (2007), doi: 10.1.1.133.9588.pdf

27. Harrison, D.: Real-Life Teaching in a Virtual World. Campus Technology (February 28, 2009)

28. Rapanotti, L., Nocha, S.M., Barroca, L., Boulos, M.N.K., Morse, D.R.: 3d Virtual Worlds in Higher Education. In: Olofsson, A.D., Lindberg, J.O. (eds.) Informed Design of Educational Technologies in Higher Education: Enhanced Learning and Teaching. IGI Global (2012)
29. Wang, S.S.: Scientists Find Safer Ways to Test Medical Procedures. Wall Street Journal (December 20, 2011)
30. Oaks, S.: Real Learning in a Virtual World. The International HETL Review 1, Article 3 (2011)
31. Zhang, J.X.: Second Life: Hype or Reality? Higher Education in the Virtual World. DE Oracle@UMUC (2007)

Evaluating the Impact of Virtual Reality on Liver Surgical Planning Procedures

Nashaat El-Khameesy[1] and Heba El-Wishy[2]

[1] Informatics Chair & Research and Development, Sadat Academy, Cairo-Egypt
wessasalsol@gmail.com
[2] Computers & Information System Dept., Sadat Academy, Cairo-Egypt

Abstract. In this paper, a proposed virtual reality (VR) based model is suggested to enhance the surgical planning procedures for liver patients. It's centered on the idea of developing a virtual reality environment (VE) based on the Virtual reality Surgical Planning System (VLSPS) where planning gets an easy and intuitive process in the clinical routine. The paper also presents a case study conducted on number of medical centers of liver in EGYPT in order to evaluate the applicability as well as the effectiveness of such techniques. The research has also conducted and analyzed a subjective questionnaire based on evaluating a Virtual Reality Surgical Planning System (VLSPS). Investigation analysis has been conducted to over 20 liver surgeons in the largest six liver centers in Egypt. The results of such investigations have proven quite a promising success to VR in liver surgery addressed by the effectiveness of the approach in training surgeons and shortening the surgical planning time. Enhancements have been agreed upon in a number of axioms including: (i) interactive qualitative analysis of already segmented liver related structures, liver segments, vasculature or respected liver tissue. Such gained enhancements in liver surgical planning procedures result in enabling surgeons doing more accurate quantitative analysis and distance related to liver structures ending in better decision making.

Keywords: Surgical Planning, Virtual Reality (VR), Virtual Environment (VE), Virtual Reality Surgical Planning System (VLSPS), Computerized Tomography Scanner (CT).

1 Introduction

While remarkable progress has been achieved in the field of surgical techniques only little efforts were spent on improving the pre-operative planning process. A successful surgical intervention is critically dependent on the surgical planning procedures, techniques and tools adopted. For instance, effective liver treatment require optimal accurate and detailed understanding of interior liver structure in order to guide decisions whether a patient should and/or can undergo surgery or not. Traditional surgical planning is mainly centered on volumetric information stored in a stack of intensity-based images gained via a computerized tomography scanner (CT).

K.J. Engemann, A.M. Gil Lafuente, and J.M. Merigó (Eds.): MS 2012, LNBIP 115, pp. 210–218, 2012.
© Springer-Verlag Berlin Heidelberg 2012

However, such techniques provide surgeons with just 2D images leaving surgeons to build their own 3D model of liver, tumor, and vasculature which is a challenging task for even experienced surgeons. Moreover, anatomical variability due to the lesion may occur making the situation more challenging. Consequently, some important information could be missed leading in non-optimal or even wrong treatment decisions.

On the other hand, 3D while might appear as a candidate solution to the aforementioned challenges, its visualization in such medical context is not even sufficient if presented on conventional displays. Moreover, surgical planning for liver resection requires a qualitative analysis of volumes and distances related to the liver structure. Examples include the relative size (volume) of the tumor to the overall liver tissues which is currently estimated on basis of the gained data collected via CT images. Currently, surgeons rely on their built mental model without having possible truthful simulation of the actual resection.

2   Potential Impact of Image Guided Systems to the Surgical Domain

The continuous progress in technology has made transition to clinical practice replacing traditional open surgical procedures with minimally invasive techniques. In contrast to open surgery, image guided procedures (IGP), physicians identify anatomical structures in images (segmentation) and mentally establish the spatial relationship between the imagery and the patient registration. To gain an insight into the major IGP technologies involved in surgical procedures, it becomes important to state first the three main phases of surgical planning procedures: (i) pre-operative planning, (ii) intra-operative plan execution and (iii) post-operative assessment. First, the key technologies involved in pre-operative planning are: (1) medical imaging including the correction of geometric and intensity distortions in images, (2) data visualization and manipulation of image and patients' data, (3) segmentation or classification of image data into anatomically meaningful structures and (4) registration: alignment of data into a single coordinate system. Second, the technologies related to the inter-operative plan execution include the aforementioned technologies in addition to tracking systems for localizing the spatial position and orientation of anatomy and tools and the human computer interaction (HCI). To sum up the key technologies involved are:

- Medical imaging and image processing
- Data visualization
- Segmentation
- Registration
- Tracking systems
- Human Computer Interaction (HCI)

3   The Proposed Virtual Liver Surgery Planning Model

It's our claim that VR presents a potential promising solution to enhance the tasks involved in surgical planning procedures as it efficiently counterpart the challenges and difficulties incurred by the CT data. The main idea of VLSPS is to support radiologists while preparing data sets of patient's liver and to provide surgeon with more insight and to navigate along 3D images. The addressed gain is to minimize the time needed to collect data during the pre-surgical planning stage and to automate the work done. While, a fully automated segmentation is not yet available due to the large variability of shape and gray level distribution of normal or diseased tissue, it's now possible to adopt a semi-automated approach.

3.1   Frame Work of the VLSP Model

The main idea of our proposed model is to employ the VLSPS in a hybrid user interface (e.g. both 2D and 3D) to enhance the user interaction and still take advantage of the gained information of the 2D data set which is to some extent a mature technique. The proposed system can be identified as three major subsystems: (1) medical image analysis, (2) interactive segmentation refinement and (3) resection planning. The following comments highlight some of the implied tasks in the adopted model as shown in figure (1):

- First part employs robust algorithms for segmentation of liver/tumor, vessel extraction and Voxel -based segment approximation
- Second part, aims at filtering, refining and verifying results gained from segmentation of first part via a hybrid interface.
- Third part provides the necessary components for a VR-based resection planning environment. It employs measurements tools, volumetric partitioning and general resection tools.
- Many VR hardware tools, in addition to high specification workstation of high resolution monitor, are utilized including shuttle glasses, head mount display (HMD), tracking pencil and transparent personal interaction pen (PIP).
- Actual surgical planning is completely gained via desktop by enabling surgeons gain 3D user interface at six levels of degree of freedom.

On the other hand, the image analysis part of the system aims at preparing raw data of the CT input in order to be used for the 3D visualization. Such objectives have to face a number of challenging tasks which can be summarized as: (1) difficulties of proper segmenting liver and tumor because in some cases local borders of the liver to neighboring structures (e.g. heart) are virtually not present in the image source, (2)

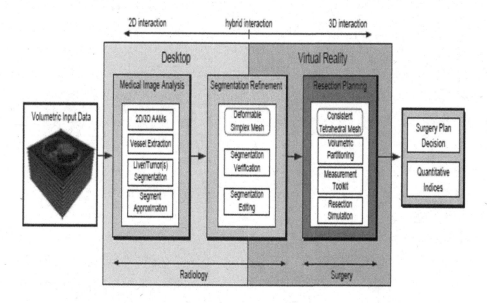

Fig. 1. Main subsystems of the Proposed Liver Surgical Planning System (VLPS)

challenges due to proper extraction of portal vein tree as good as possible, which is important for precise liver segment approximation and (3) challenges facing development of algorithms for generating a correct liver segment approximations based on a given portal labeling information. In order to counteract the preceding challenges, first task provides interaction with the segmentation refinement block of the system as fully automated approach is not viable yet due to the inhomogeneous structure of liver and tumor tissues. In the second task, interaction is limited to the specification of correct parameters for tweaking segmentation results. Liver partitioning requires surgical knowledge; therefore it's related to the resection planning block of the VLSPS. An interactive labeling technique of segment-feeding vessel branches is carried out, which can easily be done using direct 3D interactions. Segment classification of liver can't be visible in CT data but can be achieved in the model via segment approximation algorithms. Such algorithms can provide surgeons with automatically generated eight different liver segments labeled according to the labeled segment feeding vessel branches. Figure (2) shows an example for partitioning the liver into its main segments based on a labeled portal vein tree. Figure (2-a) shows the formal portal tree representation while Figure (2-b) shows the labeled surface based representation of the portal vein tree. Figures (2-c, d) show a front and a back view of the resulting liver segments displayed color-coded as surfaces.

3.2   VR-Based Segmentation Refinement Tools

The radiologist's task is to deliver correct segmentation results to surgeons before the actual resection planning takes place. The VLSPS therefore integrates different tools

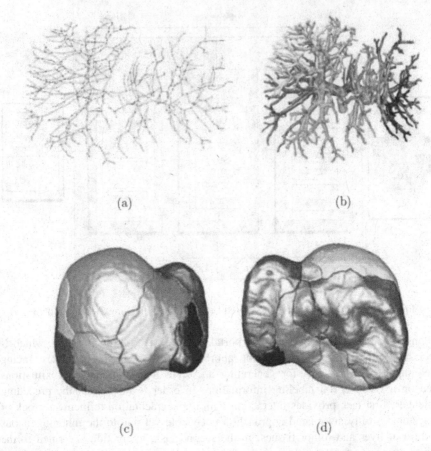

Fig. 2. *A liver Segment Approximation based on a labeled portal vein tree.* 2(a) Formal portal tree representation, 2(b) Labeled portal tree model, 2(c) Classified Liver Segments using nearest neighbor approximation (front view) and 2(d) Classified Liver Segments using nearest neighbor approximation (back view).

for segmentation verification and editing. The original CT data are projected as textures on the backside of the PIP which is a required information source during the whole refinement process. It is possible to make CT snapshots of arbitrarily oriented planes which can be positioned in space.

In order to enable interactive segmentation editing, a set of tools are embedded into the VR system for true 3D interaction. However, for specifying precise landmark points, 2D inputs may be useful and therefore, a hybrid user interface is currently developed which allows a combined 2D and 3D interaction. Figure (3) implies ideas behind the VR-based segmentation refinement using free-form deformation. A segmentation error can easily be corrected in a VR setup using the pencil for surface dragging and the PIP for showing additional context information.

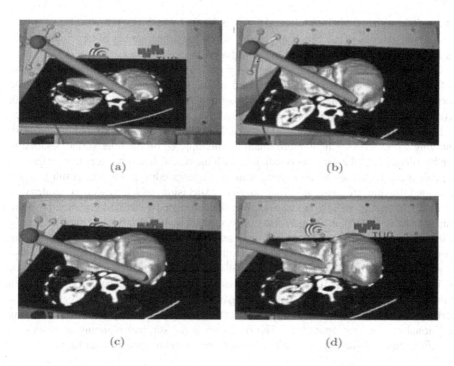

Fig. 3. Example of a VR-based Interactive Segmentation Refinement deforming the Liver Surface: (a) Detecting the region where segmentation failed. (b) Locating the best refinement position. (c) Starting the refinement process using context information on the PIP. (d) Observing the segmentation refinement result.

4   The Proposed Surgical Planning Workflow

Typically, the common practice for surgical planning starts after the radiologist assures correct segmentation of liver, tumor, and portal vein then surgeons start with the actual resection planning process. Meanwhile, in contrast to the hybrid user interface, the surgeons favor 3D interaction for planning, since they're highly 3D-oriented in their clinical routine. During an intervention, all movements with surgical devices take place in 3D, and the objects to interact with, are also 3D. **The suggested surgical planning workflow has been elaborated based on collaborative efforts with** surgeons; and such workflow is feasible for planning an intervention for patients suffering from HCC considered here as an empirical case study. The elaborated workflow is shown in Figure (4). It's assumed here, that each required object (i.e. liver, tumors, and vessel tree) has been already correctly segmented and approved by a radiologist. After radiological validation, each object is stored as an individual surface mesh (i.e. simplex mesh) and is then transformed into a triangular model. By using the mesh generation algorithms, a tetrahedral data model is generated including the liver boundary and all tumors. The vessel tree is treated separately, since its complexity would have negative effect on the overall performance of the tetrahedral mesh. A suitable number of elements, for tetrahedral meshes modeling a liver dataset, is about 100k. Since the vessel tree is very complex in geometry (e.g. 30k triangles are necessary), this number of target tetrahedral meshes cannot be reached without loss of information.

Once the model is generated, the surgeons can start their planning process by inspecting the data, especially the location of the tumor and the arrangement of the portal vessel tree. For malignant tumors, segmentation approximation is necessary in order to retrieve the information about affected liver segments. Therefore, the vessel tree is labeled according to anatomical knowledge about segment-feeding branches. By applying a preview (segment approximation only applied on the liver surface), labeling can be altered until the surgeon is satisfied with the result. In a next step, liver segments are calculated based on the underlying volumetric tetrahedral mesh. The result is again inspected by the surgeon, in order to find a decision which resection strategy is adequate. In case of an anatomical resection, the automated generated resection proposal can optionally be applied to perform collision detection with affected liver segments. The result of this proposal is verified and can be edited by adjusting the safety margin around the tumor. Additionally, measurement tools can be used for further quantitative analysis in an iterative process. If a typical resection is the only solution, one of the built-in resection tools is used allowing a classification of liver tissue into: resected and remaining regions and again, a validation using the oncological safety margin must be performed. Further adjustments of the resected region are also possible by applying additional partitioning operations. The outcome of the surgical planning consists of a resection plan and quantitative indices about resected and remaining liver tissue.

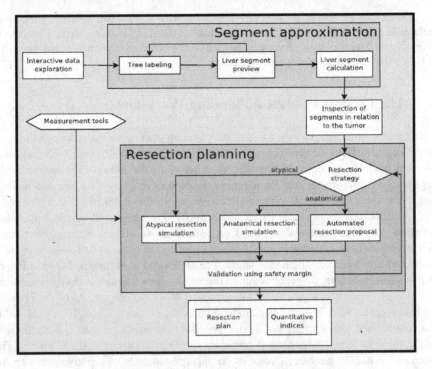

Fig. 4. The Proposed Surgical Planning Workflow

5   Results and Conclusions

This research has demonstrated the added value of collaborative research combining authors of both informatics interest and medical surgeons. Coupling their views enabled achieving applicable results in the medical domain of liver surgical planning. The proposed model has been highly accepted as a result of analyzing the response of over 20 surgeons in the largest six liver medical centers in Egypt. The research has highlighted and verified the potential advantages and the added value of using VR techniques and tools in the surgical planning domain in general and in liver surgical planning in particular. Results of adopting the proposed work flow of the VLPS in the surgical planning domain have shown quite an appreciation when demonstrated, tested and validated according to the considered surgeon set leading to a refined applicable liver surgical planning work flow. A significant gain has been proven from perspective of surgical planning task completion as it enabled trained surgeon to achieve a resection plan in less than 30 minutes including the calculation time for

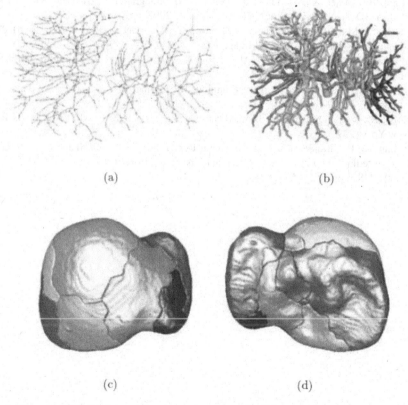

(a)   (b)

(c)   (d)

Fig. 5. The results gained from the proposed model provide surgeons with many valuable quantitative indices such as volume of liver segments, resected and remaining liver tissues. Consequently, the VLSPS would enable surgeons to elaborate more efficiently a better surgical planning strategy.

liver segments. Moreover, the adoption of such VR based models can be also adopted in enhancing clinical routines. However, cost as well as the need for more integrated VR tools still presents a challenge for spreading out such systems especially in small clinics and/or by liver specialists in their own offices.

References

1. Myers: Quantitative Research in Information Systems. Sage Publications (2009)
2. Bernard, R., Alexander, B., Beichel, R., Schmalstieg, D.: Liver Surgery Planning using Virtual Reality. IEEE Journal of Computer Graphics and Applications 26(6), 36–47 (2006)
3. Beichel, R.: Virtual Liver Surgery Planning: Segmentation of CT Data., PhD thesis, Graz University of Technology (2005)
4. Burdea, G., Coiffet, P.: Virtual Reality Technology, 2nd edn. Wiley-Interscience Pub. (2003)
5. Feiner, S.: Augmented Reality: A New way of Seeing. Journal of Scientific American Science and Technology (2002)
6. Preim, B., Tietjen, C., Spindler, W., Peitegen, H.: Integration of Measurement Tools in Medical 3D Visulaizations. In: IEEE Visualization 2002, pp. 21–28 (2002b)
7. Blackwell, M., Nikou, C., DiGioia, A., Kanade, T.: An Image Overlay System for Medical Visulaization. Journal of Medical Image Analysis 4(1), 67–72 (2000)
8. Maintz, J.B., Viergever, M.A.: A Survey of Medical Image Registration. Journal of Medical Image Analysis 2(1), 1–37 (1998)
9. Robb, R.A.: VR Assisted Surgery Planning. IEEE Mag. of Engineering Medicine and Biology 15(1), 60–69 (1996)
10. Robb, R.: Three Dimensional Biomedical Imaging: Principles and Practice. VCH Pub., New York (1994)
11. Delingette, H.: Simplex Meches: A General Representation for 3D Shape reconstruction. In: Proceedings the Int. Conf. on Computer Vision and Pattern Recognition (CVPR 1994), Seattle, USA, pp. 856–857 (1994)

A Model to Determine the Effect
of International Business Competences

Daniel Palacios-Marqués

Department of Business Administration, Universitat Politècnica de Valencia,
Camino Vera s/n, 46022 Valencia, Spain
dapamar@doe.upv.es

Abstract. This work aims to analyse the effect of the holistic view of business, continuous learning and the information technology infrastructure in the creation of international business competences. We also analyse whether the creation of this type of competences significantly affects firm performance. The methodology for hypotheses testing is structural equation modeling. This study verifies the mediator role of international business competences in a survey of 257 companies from the biotechnology and telecommunications industries. These findings suggest that managers should emphasise the creation of a holistic business view, promote continuous learning and improve the information technology infrastructure in order to develop international business competences, because these competences have a positive and significant impact on firm performance.

Keywords: international business competences, firm performance, holistic view of business, continuous learning, information technology infrastructure

1 Introduction

In a globalised context, firm internationalisation is a key strategic choice of growing interest in the fields of management and research. This article uses recent contributions from the knowledge-based approach to examine certain internal variables that determine a firm's internationalisation process. According to this approach the acquisition and use of relevant knowledge is the key to explaining a firm's competitive results. Very few studies, however, use a knowledge-based approach to examine a firm's entry into foreign markets [1,2,3].

Several authors [4,5] argue that the accumulation of market knowledge increases a firm's ability to coordinate its international activities and its willingness to commit resources to those activities [6]. In this regard internal knowledge transfer is very important at both individual and company level.

If the possession of different types of resources provides the base for internationalisation so that a company can spread beyond national borders, then firms should carry out an internal analysis to find out whether they have the strengths to access new markets [7]. Initially, a company can offset lack of size, scanty

K.J. Engemann, A.M. Gil Lafuente, and J.M. Merigó (Eds.): MS 2012, LNBIP 115, pp. 219–229, 2012.

international vocation or lack of own resources if it is able to develop other types of competences.

This paper is organised as follows. Section 2.1 reviews the effects of internationalisation from a resource-based approach presenting the motives that explain why some companies are better placed to internationalise than others. Section 2.2 develops the theoretical model, and posits the hypotheses. Section 3 includes all the methodological aspects of the study, explaining the operation of the theoretical constructs in the theoretical model, data collection and other matters concerning the research design. Section 4 shows the results of the hypothesis testing. The work ends with some final considerations in the light of the results of the empirical analysis and recommendations for management and future research.

2   Literature Review

2.1   Internationalisation from a Knowledge-Based Approach

In the resource-based approach, possession of resources [8,9] and strategic capabilities is key for firm internationalisation. This approach highlights the fact that firms need strategic resources to compete successfully in international markets and in particular, valuable knowledge that can provide a competitive advantage over local firms. If the company can make the most of this advantage from the home market, it will export; otherwise it must exploit the advantage in the target country.

When companies have resources and capabilities such as technology, reputation, brands, and so on, that provide them with a competitive advantage in their local market, they can consider the possibility of exploiting them abroad. A limitation of the resource-based approach and in particular the knowledge-based approach is that they are very static and do not take into account dynamic aspects of the creation and modification of capacities.

The Uppsala school [4], or internationalisation theory, recognises the importance of a key resource like information on foreign markets and on the internationalising process itself. Internationalisation theory maintains that the process takes place in successive stages as the firm gains experience in each stage. This theory assumes that only experience can affect management decisions to promote future international expansion; as a firm accumulates experience in foreign markets, its management of resources and transborder activities improves.

Consequently, there is a relationship between a firm's exposure to foreign market conditions and a) management of its own resources in order to develop transborder activities, or b) management of close competitors' resources [10]. Firms learn during the export process as they discover the characteristics of new markets and how to operate in them, thereby obtaining some of the information they need to make further progress abroad with less uncertainty.

Knowledge management ensures that knowledge becomes business competences which facilitate the task of successfully adapting to the market, and the process helps firms to commit new resources in markets other than the home market [11]. The ability to use existing knowledge enables firms to adapt with new innovations that are both radical and incremental, and to use resources efficiently, which requires greater involvement with the market.

2.2   Theoretical Model

The limitation with classical approaches to company organisation is that they consider organisations as closed systems. Furthermore, functionalist approaches divide firms into departments and units in pursuit of greater efficiency, often sacrificing the big picture. As firms grow, they tend to diversify their activity. A wide variety of different activities appear in firms which disperse in time and space and bring together actors from different places with different cognitive levels.

A firm should have a holistic view that aligns its corporate objectives with the goals that all its members have outlined so that they are committed to generating, sharing and socialising knowledge. Business structure affects knowledge transfer. A rigid vertical structure will generate barriers to communication and knowledge transfer. Furthermore, an excessive subdivision of work removes the firm's overview [12].

When small and medium-sized firms lack resources, other strategies are available such as establishing relations with other firms. These relations may be with shareholders, as happens when the company capital is open to external shareholders, or involve cooperation through strategic alliances. In these cases, trust is a key intangible asset. Transmitting a holistic overview is key for firms to develop lasting agreements [13].

One of the motivations for internationalisation is growth, either because firms want to enlarge the scope of their products or broaden their geographic reach. However, sometimes firms' collective knowledge disintegrates and hinders the internationalisation process [14]. The creation of collective, shared knowledge, through an efficient knowledge transfer system makes possession of specialised export management routines and knowledge fundamental.

On the basis of the above relations we posit the first hypothesis as follows:

H1: There is a positive and significant relationship between companies that develop a holistic view and the creation of international business competences.

The ability to learn relates to the ability to create, transfer and apply new knowledge, one of the main sources of competitive advantage [15,16]. Other studies such as the one [17] have already established a relationship between continuous learning systems and the development of competences. In fact, the above authors distinguish four dimensions for developing a continuous learning system: management commitment to continuous learning; a culture that promotes innovation and learning; the development of internal competences; learning-based business design.

Learning is essential to determine the degree of originality or the specific nature of a competence that fulfils the requirements of being valuable, rare, impossible to imitate perfectly and with no strategically equivalent substitutes (essential competence) as it is one of the reasons that make imitation difficult. The portfolio of business competences are closely related, so if the firm improves its learning-based distinctive competences, it will indirectly be improving other types of competences such as international business competences. Developing these abilities requires a process of transformation, using and combining standard or available resources in a business context alongside business routines to generate capabilities [18].

Similarly, the business environment must provide favourable conditions for integrating knowledge and applying it to create superior products and services. [19] also maintains that if people specialise in specific areas, efficiency in the acquisition of knowledge increases. Integrating knowledge in different departments makes it easier to apply and create competences that facilitate internationalisation [20,21].

One variable to consider in relation to learning is time delay. This variable takes into account the period of time that elapses between the development of knowledge flows and their effect on knowledge stocks. The magnitude of time delay varies in each organisation, influenced by the ease with which firms are able to integrate new knowledge as it arrives.

In order to produce positive variations in the knowledge stock, in addition to receiving positive flows, firms must integrate knowledge in an appropriate way. [22] states that at individual level and as a consequence of the appearance of cognitive problems, the acquisition and use of knowledge requires specialisation.

Therefore:

H2: There is a positive and significant relationship between continuous learning and the creation of international business competences.

Information and Communication Technologies (ICT) facilitate knowledge by capturing, storing and transmitting it [23]. The literature on information systems maintains that the introduction of ICT has no direct impact on business performance. In fact, there are intermediate variables that mediate the relationship between these two constructs. In this regard, [24] maintain that the effects on firm logistics are an intermediate variable which does have a direct impact on business performance. We consider that a firm must be able to transform ICT into business performance by identifying conversion effectiveness factors that mediate in the relationship between ICT and performance [25,26].

The function of production and process-oriented models describe the relationship between ICT investment and business performance from an input-output perspective that sometimes includes intermediate factors such as management decisions and business structure. A key variable for understanding internationalisation is knowledge management. [27] consider that information technologies and their abilities enable the creation of new knowledge and so contribute to different ways of managing knowledge: firstly, the technological infrastructure facilitates rapid compilation, storage and exchange of knowledge, thereby facilitating the process of knowledge creation [28]. Secondly, it enables integration of fragmented knowledge flows, eliminating the obstacles to communication between departments within a firm.

On the basis of these relationships we posit the third hypothesis:

H3: There is a positive and significant relationship between information technology infrastructure and the creation of international business competences.

The knowledge stock the company creates on the basis of its knowledge management processes increases its ability to adapt to market requirements. The internationalisation literature divides the knowledge stock into objective and experimental knowledge. By operating in the market the firm not only acquires market information but even manages to connect more closely with the market, making it difficult to use the firm's resources for other purposes. Lack of experience impacts negatively on the initial stages of business performance. The acquisition of external knowledge, including the compilation of customer and competitor data is particularly important when dealing with foreign markets.

Customers are a valuable source for developing or modifying products or services. When firms internalise knowledge acquired through relations with their customers, suppliers or competitors, individual learning creates new knowledge that increases the business knowledge base [29]. A company must commit resources to a foreign market, and spend a significant amount of time updating the competences it will use in subsequent international investments [30], which has a positive effect on business performance. Furthermore, dynamic markets demand dynamic responses and strategies and that require explicit commitment and a focus that takes market reality into account.

Therefore, we posit the final hypothesis:

H4: There is a positive and significant relationship between the creation of international business competences and business performance.

3   Methodology

3.1   Variables

There are five theoretical constructs in the theoretical section. The items for measuring the firm's holistic view and continuous learning come from [31] and the items for measuring information technology infrastructure come from [32]. The creation of international business competences comes from the work by [33] and business performance from the study by [34].

In the empirical study, we use two-stage structural equation modelling to validate the measurement scales and establish causal relations. This methodology enables representation of latent concepts of observed variables and study of causal relations with non experimental data, when the relations are linear. We use statistical software EQS 5.7b and the default Maximum Likelihood method offering consistent estimators for broad samples with continuous variables and multi-normal distributions [35].

3.2   Data Collection

The study sample are companies from the biotechnology and telecommunications industries in Spain. These companies have already begun to internalise and knowledge is a key competitive factor, there is a high degree of innovation and therefore they are appropriate for analysis of antecedent factors that improve the creation of international distinctive competences.

Email questionnaires were sent to the general managers of these companies because their holistic view enables them to provide reliable answers. We pre-tested the measurement instrument on 20 companies, 10 from the biotechnology industry and 10 from the telecommunications industry. The field work took place between February 2010 and July 2010. Elimination of incorrectly completed questionnaires gave 222 questionnaires, with a sample error of 5.7 and a 95% confidence interval.

4   Results

The following table validates the measurement scales for the five theoretical constructs in the theoretical model. We evaluate global fit using measures of absolute fit, incremental measurements and parsimony measurements. The statistical values reach the recommended thresholds.

Table 1. Fit Indices for the Initial Factorial Models

Scale	d.f.	Chi^2	p	BBNNFI	RCFI	GFI	RMR	NC
Company holistic vision	9	10.779	0.291	0.954	0.995	0.943	0.034	1.2
Continuous learning	2	2.732	0.255	0.986	0.998	0.975	0.027	1.37
IT technical infrastructure	2	1.794	0.407	0.940	1	0.982	0.040	0.89
International business competences	5	18.801	0.002	0.910	0.954	0.894	0.054	3.76
Firm performance	2	2.847	0.241	0.969	0.997	0.972	0,034	1.42

We assessed measurement scale reliability with composed reliability. Composed reliability takes into account the standardised loads and measurement errors for each item in the measurement scale. The minimum threshold for composed reliability is 0.7. Table 2 shows that the values are high and in all cases above the recommended threshold and therefore measurement scale reliability is high.

The content of the measurement scales is valid as they fulfil two conditions. Firstly, the dimensions and items are taken from the literature, that is, from prior theoretical arguments, scales and empirical studies. Secondly, the scales follow procedures accepted in the literature. We based all the measurement scales on previous works, so there is no new ad hoc scale. The references comply with all the

Table 2. Composed reliability value for each measurement scale

Measurement scale	Composed reliability
Company holistic vision	0.89
Continuous learning	0.84
IT technical infrastructure	0.82
International business competences	0.92
Firm performance	0.91

sociometric properties required from measurement scales and the statistical tool enables us to validate the hypotheses while also re-testing the measurement scales.

4.1   Empirical Testing of the Hypotheses

Relationship between holistic business view and international business competences (H1)

The model shows the following fit. The model is over-identified, the chi-square statistic is statistically significant, the GFI index (0.904) is above 0.99 and RMR (0.024) is around 0 which indicates a good fit in absolute terms. BBNFI (0.950) and RCFI (0.991) exceed the minimum threshold levels. The parsimonious fit measure (NC=1.6) is between 1 and 2. Structural equation reliability is $R^2=0.938$. Therefore we can accept the first hypothesis with a structural relation coefficient of 0.89.

Relationship between continuous learning and international business competences (H2)

In this case the fit indexes are also positive. Absolute fit measures (GFI=0.923; RMSEA=0.022; chi-square value=10.584), incremental fit measures (BBNFI=0.956) and the parsimonious fit measure (NC=1.32) are all above the threshold levels Structural model reliability is high ($R^2=0.948$). The coefficient for the equation is 0.92 which indicates that the effect of continuous learning on the creation of international business competences is high and therefore the hypothesis is verified.

Relationship between IT infrastructure and international business competences (H3)

As in the two previous cases, the statistics obtain statistically significant values and the model is also over-identified. Absolute fit measures (GFI=0.960; RMSEA=0.010; chi-square value=11.851), incremental fit measures (BBNFI=0.983) and the parsimonious fit measure (NC=1.48) are statistically significant.

Structural model reliability is 0.941. The structural equation coefficient is 0.79 and is the weakest of the three antecedent variables of international business competences. However, the hypothesis is fulfilled and it is possible to infer a direct and significant relationship between IT infrastructure and the creation of international business competences.

Relationship between international business competences and firm performance (H4)

As for the previous hypotheses, we divide fit measurements into absolute fit (GFI=0.937; RMSEA=0.030; chi-square value=25.761), incremental fit measures (BBNFI=0.951) and the parsimonious fit measure (NC=1.29) The values are good and corroborate the hypotheses. Structural model reliability is 0.943 and the coefficient of the equation is 0.89.

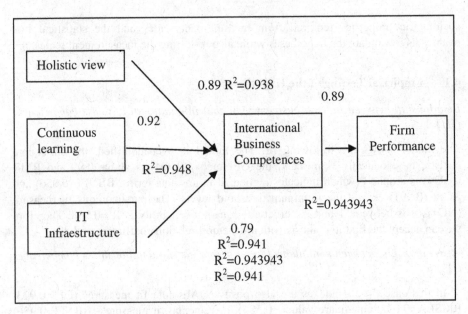

Fig. 1. Hypotheses coefficients and structural equations reliability

5   Conclusions and Managerial Implications

In this study we examined the impact of three antecedent variables on the creation of international business competences, in particular the holistic business view, the promotion of continuous learning and improved information technology infrastructure. The empirical study shows a positive and significant relationship in the three hypotheses. The coefficients in the respective structural equations indicate that continuous learning is the most influential variable in the creation of international business competences (0.92) while information technology infrastructure (0.79) has the least impact although it is still positive and significant. Furthermore, the relationship between international business competences and firm performance show a positive significant relationship with a coefficient in the structural equation of 0.89.

One aspect to consider is that businesses develop specific capabilities and create competences to respond to changes in the business environment. Competences depend on each firm's individual strategies and its sector of activity. It is particularly interesting to develop international business competences based on joint exploitation of various resources in which the organisation is particularly skilled and which give it a competitive advantage. An appropriate ICT infrastructure enables firms to standardise processes and accumulate knowledge of foreign markets more efficiently.

Companies must propose a model of business learning that contributes to the development of essential capabilities in a holistic approach. Thus, if the company improves its distinctive internationalisation competences, it will indirectly be improving other types of competences such as, for example distinctive knowledge and innovation competences. Developing this type of capacities requires a transformation process, using and combining standard or available resources in an business context, together with business routines to generate capacities.

The empirical study in firms from the biotechnology and telecommunications industries, two knowledge-intensive sectors where intellectual capital has a key role and a clear export vocation. However, according to the theoretical model, we can generalise these relations to other industries that intend to internationalise.

Future research must examine other antecedent factors that positively affect the creation of international business competences to further understanding of the way managers can create a culture that develops knowledge of foreign markets and promotes internationalisation. Longitudinal studies would provide further information on how those competences evolve over time.

References

1. Li, T., Cavusgil, S.T.: Decomposing the effects of market knowledge competence in new product export: a dimensionality analysis. European Journal of Marketing 34(1), 57–79 (2000)
2. Wang, G., Olsen, J.E.: Knowledge, performance, and exporter satisfaction: An exploratory study. Journal of Global Marketing 15(3/4), 39–64 (2002)
3. Morgan, N.A., Zou, S., Vordhies, D.W., Katsikeas, C.S.: Experiential an Informational Knowledge, Architectural Marketing Capabilities, and the Adaptive Performance of Export Ventures: A Cross-National Study. Decisions Sciences 34(2), 287–321 (2003)
4. Johanson, J., Vahlne, J.E.: The internationalization process of the firm: A model of knowledge development and increasing forein market commitments. Journal of International Business Studies 8(1), 23–32 (1977)
5. Blomstermo, A., Eriksson, K., Lindstrand, A., Sharma, D.: The perceived usefulness of network experiential knowledge in the internationalizing firm. Journal of International Management 10(3), 355–373 (2004)
6. Hadjikhani, A.: A note on the criticisms against the internationalization process model. Management International Review 37(2), 43–66 (1997)
7. Yu, J., Gilbert, B.A., Oviatt, B.J.: Effects of alliances, time and network cohesion on the initiation of foreign sales by new ventures. Strategic Management Journal 32, 424–446 (2011)

8. Kogut, B., Zander, U.: Knowledge of the Firm, Combinative Capabilities and Replication of Technology. Organization Science 3(3), 383–397 (1992)
9. Hitt, M.A., Hoskisson, R.E., Kim, H.: International diversification: effects on innovation and firm performance in product-diversified firms. Academy of Management Journal 40(4), 767–798 (1997)
10. Delios, A., Henisz, W.: Political Hazards and the Sequence of Entry by Japanese Firms. Journal of International Business Studies 34(3), 227–241 (2003)
11. Cumbers, A., Mackinnon, D., Chapman, K.: Innovation, collaboration, and learning in regional clusters: a study of SMEs in the Aberdeen oil complex. Environment and Planning 35(9), 1689–1706 (2003)
12. Magnier-Watanabe, R., Senno, D.: Shaping knowledge management: organization and national culture. Journal of Knowledge Management 14(2), 214–227 (2010)
13. Chesbrough, H.W.: Business model innovation: opportunities and barriers. Long Range Planning 43, 354–363 (2010)
14. Añón, D., Driffield, N.: Exporting and innovation performance: analysis of the Annual Small Business Survey in the UK. International Small Business Journal 29(1), 4–24 (2011)
15. Fiol, M.C., Lyles, M.A.: Organizational Learning. Academy of Management Review 10(4), 803–813 (1985)
16. Grant, R.M.: Towards a Knowledge-Based Theory of the Firm. Strategic Management Journal 17, 109–122 (1996)
17. Palacios, D., Garrigos, F.J., Gil, I.: The relationship between organizational learning and firm performance in Spanish hotel firms. Industrial Economy 375, 77–87 (2010)
18. Khara, N., Dogra, B.: Examination of export constraints affecting the export performance of the Indian sports goods industry. European Journal of International Management 3(3), 382–392 (2009)
19. Demsetz, H.: The Theory of the Firm Revisited. In: Williamson, O.E., Winter, S. (eds.) The Nature of the Firm: Origins, Evolution, and Development, pp. 159–178. Oxford University Press, New York (1991)
20. Bruneel, J., Yli-Renko, H., Clarysse, B.: Learning from experience and learning from others: how congenital and interorganizational learning substitute for experiential learning in young firm internationalization. Strategic Entrepreneurship Journal 4(2), 164–182 (2010)
21. Simsek, Z., Heavey, C.: The mediating role of knowledge-based capital for corporate entrepreneurship effects on performance: a study of small to medium-sized firms. Strategic Management Journal 5, 81–100 (2011)
22. Simon, H.A.: Organizations and Markets. Journal of Economic Perspectives 5, 25–44 (1991)
23. Gururajan, V., Fink, D.: Attitudes towards knowledge transfer in an environment to perform. Journal of Knowledge Management 14(6), 828–840 (2010)
24. Byrd, T.A., Pitts, J.P., Adrian, A.M., Davidson, N.W.: Examination of a Path Model Relating Information Technology Infrastructure with Firm Performance. Journal of Business Logistics 29(2), 161–187 (2008)
25. Soh, C., Markus, M.L.: How IT creates business value: a process theory synthesis. In: Ariav, G., et al. (eds.) Proceedings of the 16th International Conference on Information Systems, Amsterdam, Netherlands, December 10-13 (1995)
26. Morgan-Thomas, A.: Online activities and export performance of the smaller firm: a capability perspective. European Journal of International Management 3(3), 266–285 (2009)

27. Lee, H., Choi, B.: Knowledge enablers, processes and organizational performance: An integrated view and empirical examination. Journal of Management Information Systems 20(1), 179–228 (2003)
28. Nassimbeni, G.: Technology, innovation capacity, and the export attitude of small manufacturing firms: a logit/probit model. Research Policy 30, 245–262 (2001)
29. Majocchi, A., Bacchiocchi, E., Mayrhofer, U.: Firm size, business experience and export intensity in SMEs: A longitudinal approach to complex relationships. International Business Review 14, 719–738 (2005)
30. Harris, R., Li, Q.C.: Exporting, R&D and absorptive capacity in UK establishments. Oxford Economic Papers 61, 74–103 (2009)
31. Palacios, D., Garrigos, F.: A measurement scale for knowledge management in the biotechnology and telecommunications industries. International Journal of Technology Management 31(3/4), 358–374 (2005)
32. Byrd, T.A., Turner, D.E.: Measuring the Flexibility of Information Technology Infrastructure: Exploratory Analysis of a Construct. Journal of Management Information Systems 17(1), 167–208 (2000)
33. Knight, G., Kim, D.: International business competence and the contemporary firm. Journal of International Business Studies (40), 255–273 (2009)
34. Nakata, C., Zhu, Z., Kraimer, M.L.: The Complex Contribution of Information Technology Capability to Business Performance. Journal of Managerial Issues 20(4), 485–506 (2008)
35. Bollen, K.A.: Structural equations with latent variables. John Wiley & Son, New York (1989)

Ant Colony Optimization for Solving the Vehicle Routing Problem with Delivery Preferences

Herminia I. Calvete[1], Carmen Galé[2], and María-José Oliveros[3]

[1] Dpto. de Métodos Estadísticos, IUMA, Universidad de Zaragoza, Pedro Cerbuna 12, 50009 Zaragoza, Spain
[2] Dpto. de Métodos Estadísticos, IUMA, Universidad de Zaragoza, María de Luna 3, 50018 Zaragoza, Spain
[3] Dpto. de Ingeniería de Diseño y Fabricación, Universidad de Zaragoza, María de Luna 3, 50018 Zaragoza, Spain
{herminia,cgale,mjoliver}@unizar.es

Abstract. This paper addresses a variant of the vehicle routing problem with time windows in which each customer has two time windows associated. A hard time window indicates the period of time in which the delivery has to take place and a soft time window which lays out the preferences of the customer. In these problems, for the distribution company is important not only minimizing total travel time but satisfying customer preferences. Assuming that vehicles are allowed to wait, we must decide the order in which customers are served and the delivery start time at every customer. Both decisions determine the feasibility of routes and the satisfaction of customers. For solving this problem an Ant Colony System is developed. To deal with delivery preferences, it is assigned to every customer artificial beacons to indicate delivery start time. Drops of pheromone are put on the beacons to guide ants when building routes.

Keywords: Ant colony optimization, Vehicle routing problem, Hard and soft time windows.

1   Introduction

The vehicle routing problem (VRP) consists of designing a set of routes for serving a number of geographically dispersed customers from a central depot. The objective is to minimize total distance or total travel time. In this problem customer demand is fixed and known in advance, vehicles are assumed to be identical and cannot be overloaded, all routes start and end at the depot and each customer is visited exactly once by a single vehicle. Laporte ([1]) gives a survey of main techniques proposed to solve the VRP in the fifty years elapsed since the first paper introducing the VRP was published until 2009.

Throughout these years, a lot of variants of the VRP have been studied which try to capture more precisely different characteristics of the real situation modeled. Comprehensive reviews of these problems are presented in Cordeau et al. ([2, 3]); Toth and Vigo ([4]). These papers also give an overview of main algorithms developed in the literature. The VRP and its variants have proven to be NP-hard.

K.J. Engemann, A.M. Gil Lafuente, and J.M. Merigó (Eds.): MS 2012, LNBIP 115, pp. 230–239, 2012.

In fact, these problems can be solved to optimality within reasonable computational effort only for relatively small instances. Although some researchers have focused on exact methods to solve the VRP, most techniques described in the literature are heuristics or metaheuristics which aim to provide cuasi-optimal solutions in acceptable computational times.

The VRP with time windows (VRPTW) is a variant of the VRP in which there are side constraints specifying the periods of the day in which deliveries should take place. Each customer has a time interval or time window associated which indicates the earliest and the latest allowable times within which the service should begin. In these problems, main concern is about minimizing the number of vehicles and then minimizing the total distance traveled. Bräysy and Gendreau ([5, 6]) give a survey on heuristics and metaheuristics for this problem. There are also a number of papers in the literature which deal with the VRP with soft time windows (VRPSTW) in which the time window can be violated at a penalty. These papers propose slightly relaxing the time window aiming to further reduce the number of vehicles involved or the total distance. Most papers consider soft time windows as relaxed time windows, but they do not specifically focus on the problem of satisfying customer preferences about the delivery period. A multiobjective point of view that bears in mind the preferences of customers is taken by Min ([7]). The customer preferred vehicle arrival time is incorporated into the modeling process and a goal approach is proposed to solve the problem. The objectives are to minimize the total travel time and the deviations from preferred due time. Only small-sized problems can be solved due to the large number of constraints and binary integer variables required. Also a multiobjective approach is taken by Calvete et al. ([8]), who propose a goal programming model with four goals: minimize total operational cost, satisfy time window preferences of customers, and avoid underutilization of vehicles capacity and labor. In order to solve the model they develop an exact method based on enumerating feasible routes, which is adequate for solving medium-sized problems.

This paper addresses a variant of the VRPTW in which each customer has two time windows associated. There is a hard time window which indicates the period of time in which the delivery has to take place and a soft time window which lays out the preferences of the customer regarding the period within he would like to be served. In these problems, it is important for the distribution company not only minimizing total travel time but satisfying customer preferences, that is to say, minimizing total deviation from preferences. The weight assigned to each goal is decided by the company taking into account their relative importance. Assuming that vehicles are allowed to wait at customer locations, what distinguishes this problem is that we must decide the order in which customers are served and the delivery start time at every customer. Both decisions determine the feasibility of routes and the satisfaction of customers.

For medium-sized problems, this model could be solved by developing a similar exact approach to that proposed in Calvete et al. ([8]). In general, to solve large problems requires heuristics or metaheuristics methods. The contribution of this paper is to develop an Ant Colony Optimization (ACO) approach for dealing with this VRP with hard and soft time windows (VRPHSTW) that allows us to handle large problems. ACO is one of the most powerful metaheuristics developed for solving complex combinatorial problems [9, 10]. It is inspired by real behavior of ants when looking for food. Ants are able to communicate information about food sources by laying a chemical pheromone trail on the ground which guides other ants. First ACO algorithms where proposed in the early 1990s for solving the traveling salesman problem. Since then, several variants and extensions have been developed and a variety of optimization models have been solved by using ACO algorithms. In particular, Bullnheimer et al. ([11, 12]) and Bell and McMullen ([13]) propose ACO algorithms for solving the VRP. Gambardella et al. ([14]) deal with the VRPTW and propose a multiple ant colony system for solving a lexicographic optimization problem in which first objective is to minimize the number of vehicles involved and second objective is to minimize traveled distance.

In this paper we develop an ACO algorithm for solving the VRPHSTW, whose objective is to minimize a weighted sum of the total travel time and the total deviation of customer preferences. The general framework of the algorithm with respect to pheromone updating and selection of the next customer to be visited is that of an ACS algorithm [9]. Moreover, it allows dealing with the problem of determining the customer delivery start times, which is a distinctive feature of the VRPHSTW. For this purpose, when an ant builds a solution, customers are divided in two groups, those whose preferences are satisfied and those which not. This classification is marked with beacons which guide the ants. Pheromone drops are put in these beacons and successively updated in order to reflect the quality of solutions obtained.

Moreover, several strategies of local search are also implemented which improve the quality of solutions obtained. The remainder of the paper is organized as follows. Section 2 describes the problem and Section 3 develops the algorithm. In Section 4 the computational performance of the algorithm is analyzed. Finally, conclusions are presented in Section 5.

2   Setting the Problem

Let $G = (N, A)$ be a directed network where $N = \{0, 1, \cdots, n\}$ is the set of nodes and $A = \{(i,j): i, j \in N\}$ is the set of directed arcs. Node 0 represents the location of a depot at which a fleet of identical vehicles is based. Nodes 1 to n represent customer locations. Each arc $(i,j) \in A$ refers to a direct connection from i to j and has an associated non-negative travel time t_{ij}.

We assume that every route starts at the depot, visits a number of customers and ends at the depot. Each route is assigned to a vehicle with fixed capacity U, that cannot be overloaded. At customer i a fixed quantity of goods q_i has to be delivered. This delivery cannot be divide up amongst vehicles, i.e. every customer is served by a single vehicle. Therefore, it is assumed that $q_i \leq U$ for all $i = 1, \cdots, n$. Otherwise, the corresponding customer could not be served.

Customer i requires a service time s_i and has two time windows. Let $\left[e_i^{sg}, l_i^{fg}\right]$ be the service time window guaranteed by the distribution company. That is to say, the company guarantees that the delivery to the customer i will not start before time e_i^{sg} (earliest start guaranteed) and will not finish after time l_i^{fg} (latest finish guaranteed). Moreover, the customer i has a preferred delivery time window $\left[e_i^{sp}, l_i^{fp}\right]$ that represents the period of the day within which he would like to be served. He would not like the delivery to start before time e_i^{sp} and to finish after time l_i^{fp}. Needless to say, $s_i \le l_i^{fp} - e_i^{sp}$. Time windows verify that $e_i^{sg} \le e_i^{sp} \le l_i^{fp} \le l_i^{fg}$.

Vehicles are allowed to wait to begin the service when they arrive too early at a customer location. If the vehicle arrives before the earliest time allowable e_i^{sg}, it should wait until a delivery start time in the interval $[e_i^{sg}, l_i^{fg} - s_i]$. The precise time in which is convenient to begin the delivery will be determined by the model. Once the service starts, it is carried out until completion and the vehicle leaves the customer as soon as the delivery has been completed. The model would be similar if it was not possible to wait at customer locations. In this case, the service would start as soon as the vehicle arrived at the customer location and the model should guarantee that the vehicle arrives within the interval $[e_i^{sg}, l_i^{fg} - s_i]$.

The depot has a hard time window $[e_0^s, l_0^f]$, where e_0^s refers to the earliest start time of every route and l_0^f refers to the latest return time of every vehicle to the depot. Moreover, due to labor regulations, we assume that there are limits to the total daily time that a driver can be driving and can be working. Let these limits be t_{max}^d and t_{max}^w, respectively. Notice that t_{max}^w establishes a limit on the duration of each route.

The VRPHSTW consists of determining a set of routes that minimizes a weighted sum of the total travel time and the total deviation from time window preferences of customers. A route involves information about the customers which are visited and the order in which they are visited. Moreover, taking into account the existence of hard and soft time windows, the time at which every customer delivery starts must be specified.

The metaheuristic method developed in the next Section builds routes in order to provide a feasible solution of the VRPHSTW. Therefore, next we pay attention to the constraints relative to vehicle capacity, labor regulations and time windows, as well as to the objective function value of a feasible solution expressed in terms of routes.

Let R be a feasible solution, i.e. R is a set of routes verifying above constraints. Let $r = \{0, i_1, i_2, i_3, \cdots, i_k, 0\}$ be a route in R which sequentially visits customers $i_j, j = 1, \cdots, k$. Since the vehicle cannot be overloaded:

$$q_{i_1} + q_{i_2} + \cdots + q_{i_k} \le U \tag{1}$$

Taking into account driving hours and working hours regulations:

$$t_{0i_1} + t_{i_1 i_2} + \cdots + t_{i_k 0} \le t_{max}^d \tag{2}$$
$$t_{0i_1} + s_{i_1} + t_{i_1 i_2} + s_{i_2} + \cdots + s_{i_k} + t_{i_k 0} \le t_{max}^w$$

Regarding time windows, let x_{i_j} be the delivery start time of customer i_j. Then, the delivery end time is $x_{i_j} + s_{i_j}$ and the hard time window constraints (3) have to be hold:

$$e_i^{sg} \le x_{i_j} \quad \text{and} \quad x_{i_j} + s_{i_j} \le l_{i_j}^{fg} \tag{3}$$

Note that this is equivalent to require that the delivery begins within the interval $[e_{i_j}^{sg}, l_{i_j}^{fg} - s_{i_j}]$.

The total travel time of the feasible solution R is

$$\sum_{r \in R} t_r \tag{4}$$

where $t_r = t_{0i_1} + t_{i_1 i_2} + \cdots + t_{i_k 0}$.

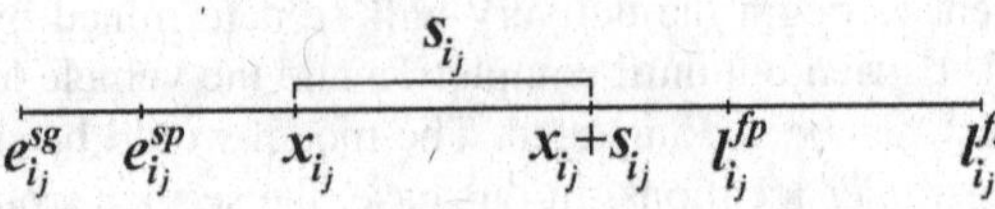

Fig. 1. Time windows of customer i_j. Customer i_j is satisfied.

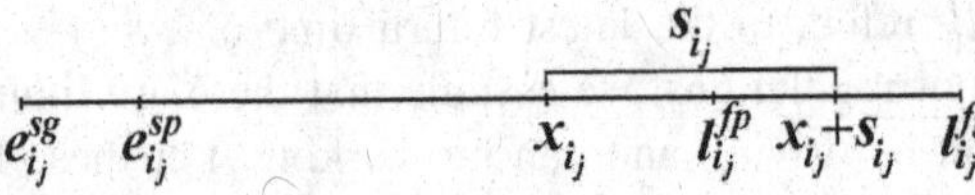

Fig. 2. Time windows of customer i_j. Deviation of preferences: $x_{i_j} + s_{i_j} - l_{i_j}^{fp}$.

In order to compute the total deviation from time window preferences of customers, we realize that the customer i_j would like to be served in his preferred time windows, that is to say, he would like $e_{i_j}^{sp} \leq x_{i_j} \leq l_{i_j}^{fp} - s_{i_j}$ to be hold. Fig. 1 and Fig. 2. display the satisfaction of the customer depending on the delivery start time and the duration of service. In Fig. 1 the service is completed during the preferred time window and so the customer is satisfied. In Fig. 2 the delivery finishes after the preferred time. We measure the customer dissatisfaction in terms of the deviation $x_{i_j} + s_{i_j} - l_{i_j}^{fp}$. A similar expression is obtained if the delivery begins earlier than is preferred, i.e. $x_{i_j} \leq e_{i_j}^{sp}$. In this case, the dissatisfaction is $e_{i_j}^{sp} - x_{i_j}$. Therefore, the total deviation from time window preferences of the feasible solution R is :

$$\sum_{r \in R} d_r \tag{5}$$

where $d_r = \sum_{j=1}^{k} max \left\{ 0, e_{i_j}^{sp} - x_{i_j} \right\} + \sum_{j=1}^{k} max \left\{ 0, x_{i_j} + s_{i_j} - l_{i_j}^{fp} \right\}$.

Bearing in mind that the decision maker can assign a different importance to each of above objectives and even to every customer, we define the objective function value of the feasible solution R as:

$$F_R = W \sum_{r \in R} t_r + \sum_{r \in R} \sum_{j=1}^{k} \widetilde{W}_{i_j} max \left\{ 0, e_{i_j}^{sp} - x_{i_j} \right\} + \sum_{r \in R} \sum_{j=1}^{k} \widehat{W}_{i_j} max \left\{ 0, x_{i_j} + s_{i_j} - l_{i_j}^{fp} \right\} \tag{6}$$

where W, $\widetilde{W}_{i_j}$ and $\widehat{W}_{i_j}$ are different weights assigned to the travel time and to the deviation of preferred time window in every customer. As a consequence, the VRPHSTW consists of selecting a feasible set of routes R so that minimizes F_R.

3   HSACS: Solving the VRPHSTW Using an Ant Colony System

The algorithm HSACS developed in this paper to solve the VRPHSTW builds feasible solutions using artificial ants. Every ant builds a feasible solution, i.e. a set of routes visiting all customers and satisfying the constraints. For this purpose, each ant leaves the depot and constructs a route by successively selecting a customer which has not yet been visited. This selection process continues as long as constraints (1)-(3) are verified. When it is not possible to keep on adding customers to the route, the artificial ant returns to the depot and starts a new route. This process continues until every customer has been visited. When constructing the route, the ant lays a pheromone trail. The pheromone trail together with an heuristic value which represents a priori information on the problem are used to select next movement of ants based on a stochastic rule. After a prescribed number of ants have constructed feasible solutions, their objective function value is compared. Then, the best one is selected and used to guide other ants by depositing a quantity of pheromone in its arcs which depends on the quality of the feasible solutions obtained. The customer selection process and the guide process are main characteristics of ACO. On the other hand, in the VRPHSTW it is also necessary to make a decision about the delivery start time, since it affects the feasibility of the route and the quality of the solution in terms of customer satisfaction/dissatisfaction. A change of the delivery start time at a customer impacts in succeeding customers that can change their status from accessible to non-accessible due to constraints on time windows and labor regulations. Moreover, the delivery start time has also influence in the objective function as it acts on the deviation of the preferred time window.

The literature on VRPTW using ACO algorithms pays no particular attention to the delivery start time. The service just starts as soon as possible. That is, if the arrival happens within the allowed time window, the delivery starts immediately the vehicle arrives at the customer location; otherwise, the vehicle waits until the earliest start. The underlying idea in this process is to allow for more flexibility in the reachable customers from a customer. Note that it is nonsense to wait at a customer location if the delivery can be made. This leads to finish the service later and so increases the possibility of some customers not being reachable from the incumbent customer.

The distinctive aspect of the VRPHSTW considered in this paper is the existence of soft and hard time windows and the importance of satisfying customer preferences, which ask for a decision on the precise time in which the delivery should start. Selecting, as before, the earliest possible start can lead to very bad solutions due to customer dissatisfaction. On the contrary, selecting the beginning of the delivery according to the earliest preferred start can lead to solutions very costly in terms of travel time. Note that, when constructing a route, the ant has a myopic behavior and is unable to anticipate the consequences of its selection in terms of future reachable customers.

Let i be the incumbent customer and j be the next customer to be visited. Bearing in mind that x_i is the delivery start time at customer i and s_i is its service time, the vehicle arrives at customer j at time $x_i + s_i + t_{ij}$. We look for the time at which the delivery of customer j must start.

If $x_i + s_i + t_{ij} \geq e_j^{sp}$ then the delivery should start immediately, i.e. $x_j = x_i + s_i + t_{ij}$. This conforms with previous remarks on the importance of an early start in order to get more flexibility for the selection of the next customer of the route. The service at customer j will finish at time $x_j + s_j = x_i + s_i + t_{ij} + s_j$. Hence, for the customer j to be reachable from i, the hard time window constraint (3) has to be verified, i.e. $x_i + s_i + t_{ij} + s_j \leq l_j^{fg}$.

If $x_i + s_i + t_{ij} < e_j^{sp}$, the delivery at customer j should start within the interval $\left[e_j^{sg}, e_j^{sp}\right]$. This guarantees the hard time window constraints (3) and do not delay the beginning of service unnecessarily. Moreover, in this case it is unpractical for drivers to be mindful of waiting at every customer location up to a particular time in that interval. Hence, we propose to classify customers in categories in accordance with the satisfaction of their preferences. In the implementation of the algorithm we have considered two categories. Type 1 matches customers whose preferences are satisfied. If j is a type 1 customer, the delivery will start at time $x_j = e_j^{sp}$. Type 2 collects the remaining customers and their delivery start as soon as possible. If j is a type 2 customer, the delivery will start at time $x_j = \max\{x_i + s_i + t_{ij}, e_j^{sg}\}$.

For the purpose of classifying customers, several routines can be envisaged. For instance, this classification can be made a priori, taking into account the importance of customers. Another possibility is to assign every customer a probability of being satisfied. Then, while building a feasible solution, this probability is used to decide the time at which the customer delivery must start.

Above rules are naive in the sense that they do not get feedback of the quality of solutions obtained during the iterations of the algorithm. Hence, in this paper we propose a method to classify customers that keeps in mind the philosophy of ants when using pheromone trails to guide other ants. The idea is to put in every customer two "artificial beacons", each associated to a type, with some pheromone drops. When the ant is building a route, it chooses one beacon, and so a delivery start time, using these pheromone drops. Let τ_i^1 and τ_i^2 be the pheromone associated to types 1 and 2, respectively. In order to select the type assigned to customer i a number v is generated from a uniform variable in the interval $(0,1)$. If $v \leq \tau_i^1/(\tau_i^1 + \tau_i^2)$ then the type 1 is selected; otherwise the type 2 is chosen. At the end of each iteration, the information on the best-so-far solution is transmitted to artificial beacons in order to proceed to global update of current pheromone drops.

In general, in ACS algorithms the heuristic information of a link (i, j) is defined as the inverse of the travel time between the two customer locations t_{ij}. In this paper, we have also evaluated the use of another heuristic information referred, in a loose way, to the existence of time windows and the urgency of serving the customer j. This heuristic value is defined as the inverse of the difference $e_j^{sg} - t_{ij}$.

4   Computational Study

The algorithm has been tested on problems derived of standard problem sets R1, C1, RC1, R2, C2, RC2, found in http://web.cba.neu.edu/~msolomon/problems.htm. We have randomly selected two problems in each group. In order to introduce the soft time window we have followed the ideas of Balakrishnan [15]. Hence, from the original hard time windows we derive the preferred time windows i.e. e_i^{sp} is the ready time and l_i^{fp} is the due time plus the service time. Then, hard time windows are defined by allowing a deviation on the left and on the right, expressed as a percentage of the maximum route time allowed, which is equal to the due date of the depot minus the ready time of the depot. Hence,

$$e_i^{sg} = \max\{0, e_i^{sp} - 0.05(l_0^f - e_0^s)\}$$

and

$$l_i^{fg} = \min \{l_i^{fp} + 0.05(l_0^f - e_0^s), l_0^f - e_0^s\}$$

In the experiment we have studied the influence of two factors in the performance of the algorithm. The first factor is the method of classifying customers to determine the delivery start time. We have defined two levels of the factor, codified as level 1 and level 2. In the level 1, every customer is assigned, a priori, a fixed probability 0.5 of being classified as type 1 or type 2. The level 2 means that artificial beacons are used as explained in section 3. The second factor refers to the heuristic information associated to time windows. We have defined two levels of the factor, meaning that this heuristic information is either considered or not. For each problem, 5 runs of 10 minutes have been done. Moreover local search and resetting procedures have been applied to obtain better solutions.

Problems of sets R1, C1 and RC1 seem to be less influenced by both factors. These insights are confirmed by the statistical analysis. Only the main effects of factor 1 are significant for problems R112 (p =0.016) and RC102 (p = 0.081). Factor 2 is not significant in any of problems R107, R112, C103, C108, RC102 and RC108.

With regard to the remaining problems, the statistical analysis confirms that the factor 1 is significant for problems R205 (p=0.008), R210 (p = 0), C203 (p = 0.089) and C204 (p = 0009). For problem RC201 both factors are significant (p = 0.013 and p= 0.031). For problem RC207 only factor 2 is significant (p = 0.099). Remember that the original problem sets R1, C1 and RC1 have a short scheduling horizon and allow only a few customers per route. In contrast, the sets R2, C2 and RC2 have a long scheduling horizon permitting many customers to be serviced by the same vehicle.

Generally speaking, although the characteristics of the problem are very influential, we can conclude that, when there are differences, the level 2 of the factor 1 provides better results. Factor 2 does not interestingly influence the results of the algorithm.

5   Conclusions

In this paper we have developed an Ant Colony Optimization algorithm to deal with a variant of the VRP in which each customer has two time windows associated. The hard time window reflects the period of time in which the delivery must take place. The soft time window expresses the preferences of the customer regarding the period within he would like to be served. In these problems the decision maker is interested in minimizing the total travel time as well as the total deviation from preferences. The objective function is formulated as a weighted combination of both goals, in which weights reflect their relative importance. A main feature of the VRPHSTW is that, in addition to the sequence of customers, the routes have to include information about the delivery start time at every customer.

In order to solve this problem, an Ant Colony Optimization algorithm is developed. The algorithm includes the general characteristics of an Ant Colony System. But, for the purpose of dealing with the decision on the delivery start time, artificial beacons are associated to every customer which indicate a different time of beginning service. Drops of pheromone are put on the beacons which guide ants when building routes. Global updating of beacon pheromone drops is made at the same time as global updating of pheromone trails. Moreover, the algorithm includes the possibility of using two heuristic values. The classical one associated to travel time and a new one associated to time windows.

A computational study has been carried out to analyze the influence of two factors: the method of selecting the delivery start time and the use of the new heuristic value. Twelve problems were randomly selected from benchmark problem sets R1, C1, RC1, R2, C2 and RC2 and properly modified to take into account customer preferences, i.e. hard and soft time windows. From the study, we conclude that the most influential factor is the method of selecting the delivery start time: The use of beacons provide better results. Except for problems RC201 and RC207, the use of the new heuristic information is not significant.

Acknowledgments. This research work has been funded by the Spanish Ministry of Education and Science under grant MTM2010-17559 and by Ibercaja under grant UZ2011-CIE-01.

References

1. Laporte, G.: Fifty years of vehicle routing. Transportation Science 43(4), 408–416 (2009)
2. Cordeau, J.F., Gendreau, M., Laporte, G., Potvin, J.Y., Semet, F.: A guide to vehicle routing heuristics. J. of the Operational Research Society 53(5), 512–522 (2002)
3. Cordeau, J.F., Laporte, G., Savelsbergh, M.W.P., Vigo, D.: Vehicle routing. In: Barnhart, C., Laporte, G. (eds.) Handbook in Operations Research and Management Science, vol. 14, ch. 6, pp. 367–428. Elsevier (2007)
4. Toth, P., Vigo, D.: The vehicle routing problem. SIAM Monographs on Discrete Mathematics and Applications, Philadelphia, USA (2002)
5. Bräysy, O., Gendreau, M.: Vehicle routing problem with time windows, part I: Route construction and local search algorithms. Transportation Science 39(1), 104–118 (2005)
6. Bräysy, O., Gendreau, M.: Vehicle routing problem with time windows, part II: Metaheuristics. Transportation Science 39(1), 119–138 (2005)
7. Min, H.: A multiobjective vehicle routing problem with soft time windows: The case of a public library distribution system. Socio- Economic Planning Science 25(3), 179–188 (1991)

8. Calvete, H.I., Galé, C., Oliveros, M.J., Sánchez-Valverde, B.: A goal programming approach to vehicle routing problems with soft time windows. European J. of Operational Research 177(3), 1720–1733 (2007)

9. Dorigo, M., Stützle, T.: Ant Colony Optimization. MIT Press, Cambrigde (2004)

10. Dorigo, M., Stützle, T.: Ant colony optimization: Overview and recent advances. In: Gendreau, M., Potvin, J.Y. (eds.) Handbook of Metaheuristics, 2nd edn., pp. 227–263. Springer (2010)

11. Bullnheimer, B., Hartl, R.F., Strauss, C.: Applying the ant system to the vehicle routing problem. In: Voß, S., Martello, S., Osman, I.H., Roucairol, C. (eds.) Meta-Heuristics: Advances and Trends in Local Search Paradigms for Optimization, pp. 109–120. Kluwer, Boston (1998)

12. Bullnheimer, B., Hartl, R.F., Strauss, C.: An improved ant system algorithm for the vehicle routing problem. Annals of Operations Research 89, 319–328 (1999)

13. Bell, J.E., McMullen, P.R.: Ant colony optimization techniques for the vehicle routing problem. Advanced Engineering Informatics 18, 41–48 (2004)

14. Gambardella, L., Taillard, E.D., Agazzi, G.: Macs-vrptw: A multiple ant colony system for vehicle routing problems with time windows. In: Corne, F.G.D., Dorigo, M., Glover, F. (eds.) New Ideas in Optimization, pp. 63–76. McGraw Hill, London (1999)

15. Balakrishnan, N.: Simple heuristics for the vehicle routing problem with soft time windows. J. of the Operational Research Society 44(3), 279–287 (1993)

Metaheuristic for the Multiple Level Warehouse Layout Problem

Caline El Khoury and Jean-Paul Arnaout

Industrial and Mechanical Engineering Department, Lebanese American University,
Byblos, Lebanon
jparnaout@lau.edu.lb

Abstract. In this paper, we introduce an ant colony optimization (ACO) algorithm for solving the NP-hard Multiple Level Warehouse Layout Problem (MLWLP). The problem description consists of a warehouse made up of several levels, each divided into a known number of cells, and different product types need to be allocated while respecting the capacity constraints. There is one I/O port located at different horizontal distances from each cell, and the objective is to minimize the total horizontal and vertical costs. The ACO comprises two stages, and its performance is evaluated by comparing its solutions to that of Branch and Bound (B&B) and Genetic Algorithm (GA). Experimental results show that ACO attains optimal solutions for small problems, and superior solutions to B&B and GA for larger problems.

Keywords: Multiple Level Warehouse Layout Problem, Ant Colony Optimization, Genetic Algorithm, Branch and Bound.

1   Introduction

Efficient warehouse design and planning contributes in decreasing material handling costs, promises a smoother flow of products within the facility, and enhances the number of products to store; thus, finding optimal solution for allocating products within a given space is critical. Unfortunately, this is a very time consuming especially when dealing with systems whose boundaries may greatly add or deduct from the objective function value. As the number of the given parameters grows in terms of set size, heuristic algorithms become the most practical tool to solve such problems.

In The Multiple Level Warehouse Problem addressed in this paper, J types of items are to be assigned to K cells on L levels to minimize the total vertical and horizontal transportation costs. The warehouse has only one I/O port with sufficient capacity to transport items vertically from ground floor to every other level at all times. The capacity A of each cell is the same on all levels, at which we assume the same number of cells. A monthly demand Q_j and inventory requirement S_j are associated with every type along with a vertical unit transportation cost (i.e. the cost to move one unit of the item between the ground and other level) and horizontal unit transportation cost (i.e. the cost to move one unit of the item one meter in a horizontal distance). Each item must be assigned to exactly one cell, and a cell may store more than one item type. No fractional distribution of products is allowed.

K.J. Engemann, A.M. Gil Lafuente, and J.M. Merigó (Eds.): MS 2012, LNBIP 115, pp. 240–248, 2012.

The above problem along with its mathematical formulation are adopted from [7]. Developing an efficient metaheuristic adds great values to such problem as exact solutions are not possible for medium and large size problems.

The rest of this paper is organized as follows. In Section 2, related studies and proposed algorithms for the MLWLP are described briefly. In Section 3, the mathematical formulation of the problem at hand is given along with a description of the parameters involved. A description of ACO approach and its application to the problem at hand are respectively presented in Sections 4 and 5. Experiments and results are highlighted in Section 6. Finally, we conclude with the significant outcomes this study and future research in Section 7.

2   Literature Review

Research related to MLWLP can be summarized as follows. Johnson [1] and Buffa et al. [5] adressed the multiple-level layout problem by introducing the CRAFT algorithm which is a part of the combining path relinking algorithm. Kusiak and Heragu [6] used the CRAFT algorithm along with branch and bound and cutting plane algorithms by first determining the cost of the initial layout, then evaluating all possible location exchanges between pairs of facilities which either are adjacent to each other or are of the same area. The location exchange that results in the greatest estimated cost reduction is made. This procedure continues until there is no location exchange resulting in a lesser cost than that of the current layout. Foulds [3] and Scriabin and Vergin [8] noted that CRAFT can handle only forty facilities and does not perform well when the facilities are of unequal areas. A more general approach of the combining path relinking algorithm was introduced by [4] suggesting that selecting an unattractive move to generate the path at each step tends to produce high quality improving final moves.

Some researchers developed algorithms using a two-stage heuristic approach. Lai et al. [12] introduced an algorithm to solve the problem and pointed out its NP-hardness. A natural decomposition of the problem enables a simple optimal solution method and a simulated annealing method to be used iteratively to solve the problem. The computational results seem to indicate that the proposed solution method is extremely effective in finding high quality solutions, and efficient in solving large size problems.

In the study of [2], the aim was to design a multiple-level warehouse shelf configuration which minimizes the annual carrying costs. Since the proposed mathematical model was shown to be NP-hard, a particle swarm optimization algorithm (PSO) as a novel heuristic was developed for determining the optimal layout. Zhang and Lai [11] introduced a TABU approach for the same problem with adjacent constraints. An integer programming model is proposed to formulate the problem, which is NP hard. Along with a cube-per-order index policy based heuristic, the standard tabu search (TS), greedy TS, and dynamic neighborhood based TS are presented to solve the problem. The computational results show that the proposed approache can reduce the transportation cost significantly.

Yang and Feng [9] implemente a fuzzy layout optimization algorithm to explain an approach for the MLWLP and to deal with a multi-level warehouse layout problem under fuzzy environment, in which different types of items need to be placed in a multi-level warehouse and the monthly demand of each item type and horizontal distance traveled by clamp track are treated as fuzzy variables. In order to minimize the total transportation cost, chance-constrained programming model is designed for the problem based on the credibility measure and then tabu search algorithm based on the fuzzy simulation is designed to solve the model. Some mathematical properties of the model are also discussed when the fuzzy variables are interval fuzzy numbers or trapezoidal fuzzy numbers. Finally, a numerical example is presented to show the efficiency of the algorithm.

Savic et al. [7] introduced a genetic algorithm for the problem addressed in this paper. The authors noted that the GA is useful in solving robust optimization problems where genetic operators are used in order to solve this kind of problems such as selection, crossover and mutation leading to an optimal solution in a reasonable computational time. The GA implementation was coded in C programming language and good results were reported. An integer linear programming model was also developed by the authors.

To the best of the authors' knowledge, there does not exist published research that developed Ant Colony Optimization algorithm for the MLWLP. Having said this, in this study, a newly designed Ant Colony Optimization (ACO) algorithm is introduced for the problem and the results are compared to those obtained using GA and B&B by [7].

3   Mathematical Formulation

The following variables represent the problem parameters:

$j \in \{1, 2, ..., J\}$ item types
$l \in \{1, 2, ..., L\}$ levels
$k \in \{1, 2, ..., K_l\}$ available cells for the l-th level
  Q_j monthly demand of the item type j
  S_j inventory requirement of the item type j
  C_j^h horizontal unit transportation cost of the item type j
  C_{jl}^v vertical unit transportation cost of the item type j to level l
  A storage capacity of a cell (same for all cells)
   D_{lk} horizontal distance from the cell k on the level l to the I/O port
  or elevator

The mathematical formulation for MLWLP is summarized below. We recall that this formulation has been adopted from [7].

$$\text{Min } \sum_{j=1}^{J} \sum_{l=1}^{L} \sum_{k=1}^{K} \quad Q_j(D_{lk} * C_j^h + C_{jl}^v) * x_{jlk} \tag{1}$$

S.T:

$$\sum_{l=1}^{L} \sum_{k=1}^{Kl} \; x_{jlk} = 1, \text{ for } j=1,2,....,J \tag{2}$$

$$\sum_{j=1}^{J} \; S_j x_{jlk} \leq A, \text{ for } l=1, 2... \text{ L } / \text{ k= 1, 2... K} \tag{3}$$

$$x_{jlk} \in \{0,1\}, \text{ for all } j, l, k. \tag{4}$$

Constraints (1) represent the objective function where X is the random variable that carries in its value which item type to store to which level and in what cell. As you can notice, both the horizontal and vertical costs affect the value in addition to the monthly demand Q_j. Constraints (2) guarantee the assignment of an item type to just one level and one cell in that level. Constraints (3) guarantee that the residual space of cell K is not violated, i.e. the inventory required by the item does not exceed the available space. Constraints (4) are the binary constraints: 1 if item j is assigned to level l and cell k; 0 otherwise.

4 Ant Colony Optimization (ACO)

In the fields of computer science and operations research, the ant colony optimization algorithm is a probabilistic tool that solves a problem by converging to the better path in a given network. The root of the ACO is traced back to Marco Dorigo in the 1990's and inspired by the natural behavior of ants in finding the shortest path that links the food to their nest. This meta-heuristic optimization approach branches to solve numerical problems with different constraints. Initially, the movement of ants is characterized by equal probabilities to take any of the given paths and of which we find one that is shorter than the other. On the latter, one can find more pheromone trails than the rest of the paths and if other ants find such a path, they are likely to diverge from the randomness to follow the trail until they find food. With time, the attractive appeal of the pheromones evaporates in longer paths and becomes more concentrated in the good path attracting more ants. Evaporation is the factor that ensures that the optimality at the end is not local but rather a global one.

To solve using ACO, the problem can be seen as a connected graph with nodes and edges. Initially, all edges have a certain equal amount of pheromone. Specific parameters and terms should be set. First, we need to define the probability of moving from node i to node j: $P^k_{ij} = \dfrac{\left(\tau_{ij}^{\alpha} * \mu_{ij}^{\beta}\right)}{\sum_{l \in \omega}\left(\tau_{ij}^{\alpha} * \mu_{ij}^{\beta}\right)}$ where ω is the set of nodes, μ represents the visibility, and τ the amount of pheromone deposited. To explain the approach, we use the Traveling Salesman Problem (TSP) as an application. It is based on the usage of the greedy heuristic of moving to the next closest unvisited city, and the visibility (μ) will be the reciprocal of the distance between the current city (i) and another unvisited one (j), i.e, $\mu_{ij} = \dfrac{1}{d_{ij}}$. Two important parameters to direct the search are α and β, which are the exponents in the probability function that determine the importance of the pheromone amount over the visibility amount. After all ants construct their tours, pheromone will be updated locally and globally according to the quality of the constructed tour solutions and evaporation rate. The first term accounts for the pheromone reduction (local update) due to evaporation, and the second term (global update) represents the additional deposit of pheromone in the links(arcs) according to the produced tour length when traveled by all ants.

$$\tau_{ij} \leftarrow (1\text{-}\rho)\,\tau_{ij} + \sum_{K=1}^{k} \Delta\,\tau^K_{ij}\,,\ \forall(i,j)$$

$$\Delta\,\tau^K_{ij} = \begin{cases} \frac{1}{l^k} & \text{if arc } (i,j)\text{is used by ant } k \\ 0 & \text{Otherwise} \end{cases}$$

where l^k represents the length of the tour taken by ant k

5   ACO Application to MLWLP

We solve the MLWLP in two stages: *Assigning* items to levels then *assigning* items to cells in each level. Both include sorting of items according to a certain criteria to be viewed. The two stages will be based on the pheromone amounts to be defined and adjusted according to the minimum total cost of each iteration represented by $Q_j(D_{lk} * C_j^h + C_{jl}^v) * X_{jlk}$. In stage one, we start by sorting from largest to smallest Q_j. The first visibility amount is defined as $\mu_{jl} = \frac{1}{C_{jl}}$ for every L where C_{jl} refers to the vertical cost of item j at level l. Before every item j assignment to a level l, we need to check if the inventory space required by item j does not exceed the residual space of the current level (i.e., $S_j <=$ Residual Space L_i (K_l*A)). Since the same item type needs to be stored in the same space and cannot be split between two or more cells, we maintain a matrix that keeps track of the cell residual leaving the assignment to cells to the next stage. The two previous steps ensure that by the end of stage 1, the total sum of inventory of items assigned to each level respect the capacity limit of levels, and indirectly checks for a feasible assignment to cells in order not to fall in the trap of fractional distribution of item types to cells. The first pheromone amount is τ_{jl} which is constant for all at the beginning of the algorithm. The probability to allocate item *j* to level *l* is calculated as \

$$\pi_{l,jl} = \frac{(\mu_{jl})^{\alpha} * (\tau_{jl})^{\beta}}{\sum_j (\mu_{jl})^{\alpha} * (\tau_{jl})^{\beta}} \qquad \text{for every L}$$

The results of stage one are represented with a 2-D matrix (LxJ). If the item *j* is to be assigned to level *l*, then the entry [l][j] will be set to true. Otherwise [l][j] is set to false.

Table 1. Result matrix from stage one

	Item 1	Item 2	Item 3	Item 4	Item 5
Level 1	True	True	True	False	False
Level 2	False	False	False	True	True

In stage two, several observations should be taken into account. Currently, we need to fit the cells on the chosen level, such that the closest cell is saturated before its neighboring cells of larger distances.

Step1: Arrange cells according to their horizontal distance in an ascending manner. The criteria for step one is min $\{d_l^k\}$. The dependency of the distance on the cell rather than on the item shapes our objective at the second stage to fill the nearest cells from the I/O.

Step2: Check whether the residual of the current cell fits the min inventory found between items to be assigned. If the condition is met, then the second visibility amount is defined as $\mu_{kj} = C_j^h * Q_j * S_j$ for each j. The first two terms follow the reasoning that the bigger the demand the better it is to keep its relevant item closer. S_j serves here to increase the chance of items with big inventory to be assigned closer. If the condition is not met, the second visibility amount is set to 0 ($\mu_{kj}=0$) if the item j is already assigned or if the inventory of item j is greater than the remaining residual cell capacity (i.e. $Res_K A < S_j$).

The second stage's pheromone amount is τ_{jk} which is constant for all and the probability to allocate item j to cell k for each level l is calculated as

$$\pi_{II,jk} = \frac{(\mu_{jk})^\alpha * (\tau_{jk})^\beta}{\Sigma_J (\mu_{jk})^\alpha * (\tau_{jk})^\beta} \quad \text{for every J}$$

The results of stage 2 are also stored in a 2-D matrix (JxK), and the same rationale is applied. For instance, consider five item types, two levels, and two cells with capacity A at each level. An example of Stage 2 output is described in Table 2.

Table 2. Result matrix from stage two

	Cell 1	Cell 2
Item 1	True	False
Item 2	False	True
Item 3	False	True
Item 4	False	True
Item 5	True	False

Combining both the results of stage 1 and 2, we get the distribution of items between and within levels.

After all ants finish their paths' traversals, the pheromone amounts in each edge are reduced locally by evaporation. The path that results in the best objective function value i.e. the minimum transportation costs has the pheromone amounts increased on it globally.

$$\tau_{jl} \leftarrow (1-\rho)*\tau_{jl}^I + \varphi.\Delta\,\tau_{jl}^{I,Best}$$
$$\tau_{jk} \leftarrow (1-\rho)*\tau_{jk}^I + \varphi.\Delta\,\tau_{jk}^{II,Best}$$

Where:

$$\Delta\,\tau_{jl}^{I,Best} = \begin{cases} \dfrac{1}{Total\ Cost} & if\ arc\ (j,l)is\ used\ by\ best\ ant \\ 0 & Otherwise \end{cases}$$

$$\Delta\,\tau_{jk}^{II,Best} = \begin{cases} \dfrac{1}{Total\ Cost} & if\ arc\ (j,k)is\ used\ by\ best\ ant \\ 0 & Otherwise \end{cases}$$

Note that the Total Cost is defined as $Q_j(D_{lk} * C_j^h + C_{jl}^v) * X_{jlk}$

6   Computational Tests

The integer linear programming formulation for our MLWLP (B&B) presented in Section 3 was modeled on Lingo 11 from Lindo Systems in order to provide optimal solutions for small size problem instances. These instances were available each with several combinations of items, levels, and number of cells. Data was obtained from [7] for benchmarking. Results show that with small numerical instances, B&B solved for global optimal in less than a second. Larger instances are the main issue where B&B was run on some instances for more than 48 hrs and still was not able to converge to a solution, reconfirming the need for Metaheuristics.

The proposed ant colony optimization was written in Java and complied with Netbeans7.0.1. The ACO parameters' values used for the preliminary test conducted in this paper are as follows:

Number of Ants = 60 ; Iterations = 15000 ; $\rho = 0.001$; $\varphi = 0.034$; $\tau_1 = 0.5$; $\tau_2 = 5.76$; $\alpha = 2$; $\beta = 2$.

The first observation was the optimality results that ACO attained for small instances (varying between 5 to 35 different item types). The smaller data gives optimal solution with minimal computational time; the same result was observed in *GA* and *B&B*. The noticeable achievement is in the larger problems. where ACO did much better for large number of items with a saving in the cost between 50000 (200 items, 3 levels) and 2000000 (400 items, 3 levels) relatively to GA. For the rest, ACO did either close enough or better than the genetic algorithm. One instance gave a better cost in the GA, with only one instance where the latter performed slightly better.

Table 3 summarizes the values of the objective function relevant to some small and large instances.

Table 3. Total Cost Values for each algorithm

Item Number	Level	ACO	GA	B&B
10	2	21062.3	21062.3	21062.3*
10	3	22324.2	22324.2	22324.2*
15	3	28124.8	28124.8	28124.8*
15	4	29878.8	29878.8	29878.8*
20	3	35682.8	35682.8	35682.8*
20	4	35470.8	35470.8	35470.8*
150	2	2153074	4324473	N/A**
150	3	1485267	2330456	N/A**
200	2	2934149	6363821	N/A**
200	3	2788524	3253969	N/A**
200	4	1499194	3471625	N/A**
250	2	733543.3	12477844	N/A**
350	2	17012609	17019324	N/A**
350	4	4183995	9987552	N/A**
400	3	17465532	20004111	N/A**

*   Global Optimal Solution attained by all algorithms

**   None of the problem instances reached a feasible solution within one hour limit

7   Conclusions and Future Research

In this paper, we have introduced a two stage ant colony optimization ACO for the MLWLP with the objective of minimizing the total vertical and horizontal costs. The algorithm was compared to the results obtained by a genetic Algorithm GA, as well as optimal solutions generated using B&B. The computational tests that were carried out proved the superiority of the ACO especially in large instances.

As an extension to this research, it would be interesting to decide on the ACO parameters using Design of Experiments in order to generate superior solutions. Enhancement of computational time is another target to be studied as well. Finally, it would be interesting to compare the results obtained by ACO to algorithms that are more advanced than Genetic Algorithm.

Acknowledgments. The authors would like to thank Ms. Melanie El-Kaddissi and Mr. Rami Otayek for their contributions in the literature review and the mathematical formulation's modeling on Lingo respectively.

References

1. Buffa, E.S., Armour, G.C., Vollmann, T.E.: Allocating Facilities with Craft. Harvard Business Review 42, 136–158 (1964)
2. Doga, B., Onut, S., Tuzkaya, U.R.: A particle swarm optimization algorithm for the multiple-level warehouse layout design problem. Computers and Industrial Engineering 54(4), 783–799 (2008)
3. Foulds, L.R.: Techniques for facilities layout: Deciding which pairs of activities should be adjacent. Management Science 29(2), 1414–1426 (1983)
4. Glover, F.: Scatter Search and Path Relinking. In: Corne, D., Dorigo, M., Glover, F. (eds.) New Ideas in Optimisation. Wiley (1999)
5. Johnson, R.V.: SPACECRAFT for multi-floor layout planning. Management Science 30(5), 648–649 (1982)
6. Kusiak, A., Heragu, S.S.: The Facility Layout Problem. European Journal of Operation Research 29(3), 229–251 (1987)
7. Savic, A., Dragan, M., Filipovic, V., Stanimirovic, Z.: A genetic algorithm for solving multiple warehouse layout problem. Kragujevac Journal of Mathematics 35(1), 119–138 (2011)
8. Scriabin, M., Vergin, R.C.: Computer and visual methods for plant layout-A rejoinder. Management Science 23(1), 105 (1976)
9. Yang, L., Feng, Y.: Fuzzy multi-level warehouse layout problem: New model and algorithm. Journal of Systems Science and Systems Engineering 15(4), 493–503 (2006)
10. Zhang, G.Q., Lai, K.K.: A Combining path relinking and genetic algorithms for the multiple-level warehouse layout problem. European Journal of Operational Research 169(2), 413–425 (2006)
11. Zhang, G.Q., Lai, K.K.: Tabu search approach for multi-level warehouse layout problem with adjacent constraints. Engineering Optimization 42(6), 775–790 (2010)
12. Zhang, G.Q., Xue, J., Lai, K.K.: Layout design for a paper reel warehouse: A two-stage heuristic approach. International Journal of Production Research 40(3), 731–744 (2002)
13. Zhang, G.Q., Xue, J., Lai, K.K.: Layout design for a paper reel warehouse: A two-stage heuristic approach. International Journal of Production Economics 75(3), 231–243 (2002)

Study of the Reaction of Dissolution
of Sphalerite Using the Model of Pitzer-Variation
in the Activity Coefficient of Zinc Sulfate

Abdelhakim Begar[1], Mohamed-El-Amine Djeghlal[2], and Abdelhafidh Saada[3]

[1] Department of Mechanical Engineering, University of Biskra,Biskra BP145, Algeria
begarabdelhakim@yahoo.fr
[2] LSGM Laboratory, Department of Metallurgy, National Polytechnic School, 10 Avenue
Pasteur B.P182, El-Harrach, Algiers, Algeria
[3] Department of Mining Engineering, National Polytechnic School, 10 Avenue Pasteur B.P182,
El-Harrach, Algiers, Algeria

Abstract. The present study concerns the dissolution process of sphalerite in synthetic aqueous solution of sulfuric acid in the absence of oxygen, which allows zinc sulfate to be obtained from a sphalerite. The reaction product of the reaction solution in the absence of oxygen is determined using the Pitzer model used to calculate the various activity coefficients. As the leaching experiments of the present study covered the temperature range from 25° C to 200° C, it is necessary to consider the expressions giving the coefficients activity of zinc sulfate and sulfuric acid as a function of temperature.

Keywords: sphalerite, Pitzer, sulfuric acid, activity coefficients, leaching.

1   Introduction

Environmental restrictions imposed on sulfide smelters have stimulated the development of alternative methods in particular hydrometallurgical processes that avoid the production of SO_2, a pollutant. In recent decades, attention has been paid to the leaching of Sphalerite in sulfuric acid medium [5-9].

Electrolytes are important in many applications that typically occur in the areas of leaching of minerals in the areas of corrosion, the effects of the fight against water pollution, the food processing andoil fields [3.10 to 14]. There are many models to represent the thermodynamic properties of aqueous solutions of electrolytes.

Pitzer and al. [1,2,4] have developed a model for calculating the properties of electrolytes from an improved analysis of the Debye-Huckel model and semi-numerical models. This model links the intermolecular forces and the distribution of ions to the osmotic pressure and reflects the influence of short-range forces in binary interactions. The equations obtained are similar to those of Guggenheim.

The terms of the activity coefficient and osmotic coefficient are deduced from the equation of the Gibbs energy molar excess.

K.J. Engemann, A.M. Gil Lafuente, and J.M. Merigó (Eds.): MS 2012, LNBIP 115, pp. 249–258, 2012.

The ultimate goal of our work is the optimization of operating parameters of leaching and leaching to allow the processing of ores low in zinc and prevent pollution from sulfur dioxide.

2   Thermodynamic Model

In water, sulfuric acid is dissociated according to equilibrium:

$$H_2SO_4 \Leftrightarrow H^+ + HSO_4^- \quad K_1$$
$$HSO_4^- \Leftrightarrow H^+ + SO_4^{-2} \quad K_2$$

The dissociation constants at 25 ° C of these reactions are calculated from the values of standard Gibbs energies of the species, that is to say:

$$K_1 = 80{,}72 \qquad\qquad K_2 = 0{,}0125$$

The high value of the equilibrium constant K1 translated into strong shift in equilibrium towards the formation of HSO_4^- and H^+. We admit later that the first dissociation of sulfuric acid is complete. Only the second dissociation equilibrium will be considered.

Pitzer and al. [1] give a value substantially different for K_2, given the uncertainty associated with various experimental methods: $K_2 = 0.0105$. This value will be used in subsequent calculations.

The law of mass action applied to the dissolution equilibrium of sulfuric acid is then:

$$K_2 = \frac{a_{H^+} \cdot a_{SO_4^{2-}}}{a_{HSO_4^-}} = \frac{\gamma_{H^+} \cdot \gamma_{SO_4^{2-}}}{\gamma_{HSO_4^-}} \cdot \frac{m_{H^+} \cdot m_2}{m_1} \tag{1}$$

In the absence of oxygen, the reaction of dissolution of Sphalerite in aqueous sulfuric acid is written as:

$$ZnS + 2H^+ + SO_4^{-2} \Leftrightarrow Zn^{+2} + SO_4^{-2} + H_2S$$

The expression of Pitzer to calculate the activity coefficients of ions H^+, SO_4^{-2} in the presence of HSO_4^- and Zn^{+2}. In this case, the coefficients of B'_{Zn_1} and C_{Zn_1} can be neglected at species interactions Zn^{+2} and SO_4^- as B'_{Zn_1} represents the derivative with respect to the ionic strength of a term $B_{Zn_1}^{(0)}$ which is independent of I and C_{Zn_1} is an adjustable parameter that depends on interactions triples. These parameters are important only for high concentrations (usually greater than 2 mol/kg) which give:

$$\ln(\gamma_H^2 \gamma_{SO_4}) = 6f^\gamma + 4m_2(B_{H_2}^{(0)} + (m_H + 2m_{Zn})C_{H_2})$$
$$+ 4m_1(B_{H_1}) + 2m_{Zn}(B_{Zn_2} + (m_H + 2m_{Zn})C_{Zn_2})$$
$$+ 2m_H(B_{H_2}^{(0)} + (m_H + 2m_{Zn})C_{H_2}) + 6m_1 m_H B_{H_1}' \tag{2}$$
$$+ 4m_2 m_H C_{H_2} + 6m_2 m_{Zn} B_{Zn_2}' + 4m_2 m_{Zn} C_{Zn_2}$$

Applying the expression of the activity coefficient for a mixture of electrolytes in ion pair H^+, HSO_4^- in the presence of ions SO_4^{-2} and Zn^{+2}, we obtain the following equation:

$$\ln(\gamma_H \gamma_{HSO_4}) = 2f^\gamma + 2m_1 B_{H_1} + 2m_2(B_{H_2} + (m_H + 2m_{Zn})C_{H_2})$$
$$+ 2m_H B_{H_1} + 2m_1 m_H B_{H_1}' + 2m_2 m_H C_{H_2} \tag{3}$$
$$+ 2m_2 m_{Zn}(B_{Zn_2}' + C_{Zn_2})$$

As the difference between (2) and (3), we obtain the term $\ln(\dfrac{\gamma_H^2 \gamma_{SO_4}}{\gamma_{HSO_4}})$ which is used to calculate the molarity of H+ ions by applying eq.(1). The calculation is done by the numerical method using as an initial value of molality m_H slightly higher than the initial molality of sulfuric acid.

$$\ln(\dfrac{\gamma_H^2 \gamma_{SO_4}}{\gamma_{HSO_4}}) = 4f^\gamma + 2m_2(B_{H_2}^{(0)} + (m_H + 2m_{Zn})C_{H_2}) + 2m_1(B_{H_1})$$

$$+ 2m_{Zn}(B_{Zn_2} + (m_H + 2m_{Zn})C_{Zn_2}) + 2m_H(B_{H_2}^{(0)} - B_{H_2}^{(1)}) \tag{4}$$
$$+ (m_H + 2m_{Zn})C_{H_2} + 4m_1 m_H B_{H_1}' + 2m_2 m_H C_{H_2}$$
$$+ 4m_2 m_{Zn} B_{Zn_2}' + 2m_2 m_{Zn} C_{Zn_2}$$

For all these expressions the terms $f^\gamma, A^\gamma, B_{H_1}', B_{H_1}^{(0)}, B_{H_2}^{(0)}, B_{H_1}^{(1)}, C_{H_2}$ are given by the following formulas:

$$f^\gamma = A^\gamma(\dfrac{I^{\frac{1}{2}}}{1 + 1.2I^{\frac{1}{2}}} + \dfrac{2}{1.2}\ln(1 + 1.2I^{\frac{1}{2}})) \tag{5}$$

with

$$A^\gamma = 0,0000043T^2 + 0,002709T + 0,583022 \tag{6}$$

Pitzer and al. [4] gives the equations $B_{H_1}^{(0)}, B_{H_2}^{(0)}, B_{H_1}^{(1)}, C_{H_2}$ and K_2 as a function of temperature:

$$B_{H1}^{(0)} = 0,05584 + \frac{46,040}{T}$$

$$B_{H_1}^{(1)} = -0,65758 + \frac{336,514}{T}$$

$$B_{H_2}^{(0)} = -0,32806 + \frac{98,607}{T} \qquad (7)$$

$$C_{H_2} = 0,25333 - \frac{63,124}{T}$$

$$k_2 = e^{-14,0321 + \frac{2825,2}{T}}$$

B_{Zn_2} is a function expressing the binary interaction between Zn^{+2} ions and sulfate ions $SO4^{-2}$, its expression is given by:

$$B_{Zn_2} = B_{Zn_2}^{(0)} + \frac{2B_{Zn_2}^{(1)}}{\alpha_2^2 I}(1-(1+\alpha_1 I^{\frac{1}{2}})\exp(-\alpha_1 I^{\frac{1}{2}}))$$

$$+ \frac{2B_{Zn_2}^{(2)}}{\alpha_2^2 I}(1-(1+\alpha_2 I^{\frac{1}{2}})\exp(-\alpha_2 I^{\frac{1}{2}})) \qquad (8)$$

B'_{Zn_2} is the derivative of B_{Zn_2} compared to ionic strength, as:

$$B'_{Zn_2} = \frac{2B_{Zn_2}^{(1)}}{\alpha_1^2 I^2}(-1+(1+\alpha_1 I^{\frac{1}{2}}+\frac{1}{2}\alpha_1^2 I)\exp(-\alpha_1 I^{\frac{1}{2}}))$$

$$+ \frac{2B_{Zn_2}^{(2)}}{\alpha_2^2 I^2}(-1+(1+\alpha_2 I^{\frac{1}{2}}+\frac{1}{2}\alpha_2^2 I)\exp(-\alpha_2 I^{\frac{1}{2}})) \qquad (9)$$

In the case of the electrolyte 2-2 (ZnSO4), it was necessary to introduce an additional factor $B^{(2)}$ compared to the analogous expressions valid for electrolytes 1-1 and 1-2. $B_{Zn_2}^{(0)}, B_{Zn_2}^{(1)}, B_{Zn_2}^{(2)}$ are parameters dependent on temperature.

Pitzer and Mayorga [2] give the values of these parameters at 25°C :

$$B^{(0)}_{Zn_2} = 0,1949, B^{(1)}_{Zn_2} = 2,883, B^{(2)}_{Zn_2} = 32,81$$

$$\frac{\partial B^{(0)}_{Zn_2}}{\partial T} = -3,68.10^{-3}$$

$$\frac{\partial B^{(1)}_{Zn_2}}{\partial T} = 2,33.10^{-2} \tag{10}$$

$$\frac{\partial B^{(2)}_{Zn_2}}{\partial T} = -3,33.10^{-1}$$

The term C_{Zn_2} corresponding to the ternary interactions here is equal to:

$$C_{Zn_2} = \frac{3}{2} C^{\delta}_{Zn_2}$$

The general expression giving the Pitzer activity coefficient for a mixture of electrolytes can be applied to zinc sulfate in the presence of sulfuric acid. In this case we obtain:

$$\begin{aligned}
\ln[(\gamma_{Zn^{+2}})(\ln \gamma_{SO_4})] = {} & 8 f^{\gamma} + 2m_2[B_{Zn2} + (m_H + 2m_{Zn})C_{Zn2}] \\
& + 2m_{Zn}[B_{Zn2} + (m_H + 2m_{Zn})C_{Zn2}] \\
& + 2m_H[B_{H2} + (m_H + 2m_{Zn})C_{H2}] \\
& + 2m_2 m_{Zn}[4B'_{Zn2} + 2C_{Zn2}] \\
& + 2m_2 m_H[2C_{H2}] + 2m_1 m_H[4B'_{H1}]
\end{aligned} \tag{11}$$

with $B_{Zn2} = B^{\gamma}_{Zn2} - B^{\varphi}_{Zn2}$ et $B'_{Zn2} = \dfrac{(2B'_{Zn2} - B^{\gamma}_{Zn2})}{I}$

and C_{Zn2} expressed according to C^{φ}_{Zn2} or C^{γ}_{Zn2} :

$$C_{Zn2} = \frac{C^{\varphi}_{Zn2}}{4} = \frac{C^{\gamma}_{Zn2}}{6} \tag{12}$$

The activity coefficient using the zinc sulfate, will be expressed as below by expanding the terms f^γ, B^γ et C^γ :

$$\ln \gamma_{ZnSO_4} = -4A^\varphi\left(\frac{I^{\frac{1}{2}}}{I+bI^{\frac{1}{2}}} + \frac{2}{b}\ln(I+bI^{\frac{1}{2}})\right) + \frac{3}{2}m^2 C^\varphi_{ZnSO_4}$$

$$+ m \times \left\{ \begin{array}{l} B^{(0)}_{ZnSO_4} + 2\dfrac{B^{(1)}_{ZnSO_4}}{\alpha_1^2 I}[1-(1+\alpha_1 I^{\frac{1}{2}})\exp(-\alpha_1 I^{\frac{1}{2}})] \\[2ex] + 2\dfrac{B^{(2)}_{ZnSO_4}}{\alpha_2^2 I}[(1+\alpha_2 I^{\frac{1}{2}})\exp(-\alpha_2 I^{\frac{1}{2}})] \end{array} \right\} \qquad (13)$$

As the leaching experiments of this study cover the temperature range from 25°C to 200°C, it was necessary to supplement the expressions giving the coefficients of activity of zinc sulfate at 25°C, introducing the dependence with respect to temperature.

In this expression the coefficients $B^{(0)}_{ZnSO_4}, B^{(1)}_{ZnSO_4}, B^{(2)}_{ZnSO_4}$, $C^\varphi_{ZnSO_4}$ and A^φ are temperature dependent. The partial derivative of the activity coefficient of zinc sulfate in relation to temperature is:

$$\frac{\partial \ln \gamma_{ZnSO_4}}{\partial T} = -4\left(\frac{\partial A^\varphi}{\partial T}\right)_p \left(\frac{I^{\frac{1}{2}}}{(1+bI^{\frac{1}{2}})} + \frac{2}{b}\ln(1+bI^{\frac{1}{2}})\right) + m\left\{2\left(\frac{\partial B^{(0)}_{ZnSO_4}}{\partial T}\right)\right.$$

$$+ \left(\frac{2}{\alpha_1^2 I}\right)\left(\frac{\partial B^{(1)}_{ZnSO_4}}{\partial T}\right)[1-(1+\alpha_1 I^{\frac{1}{2}})\exp(-\alpha_1 I^{\frac{1}{2}})] + \left(\frac{2}{\alpha_2^2 I}\right)\left(\frac{\partial B^{(2)}_{ZnSO_4}}{\partial T}\right) \qquad (14)$$

$$\left. [1-(1+\alpha_2 I^{\frac{1}{2}})\exp(-\alpha_2 I^{\frac{1}{2}})]\right\} + \frac{3}{2}m^2\left(\frac{\partial C^\varphi_{ZnSO_4}}{\partial T}\right)$$

Solving this equation requires knowledge of the partial derivatives of A^φ with respect to temperature:

$$\frac{A_H}{RT} = 4I\frac{\partial A^\varphi}{\partial T} \qquad (15)$$

Bradley and Pitzer [4] provide the values of the apparent molar enthalpy reduced $\dfrac{A_H}{RT}$ for temperatures ranging from 0 to 350 °C.

For the convenience of numerical calculation, the apparent molar enthalpy is expressed as a function of temperature and pressure equal to the saturation vapor pressure in the form:

$$\frac{A_H}{RT} = 11{,}0679053 - 0{,}145798089\ T + 6{,}95581035\ T^2$$

$$- 1{,}41494867 \cdot 10^{-6} T^3 + 1012268758 \cdot 10^{-9} T^4$$

(16)

3   Optimization of Model Parameters

The Pitzer model was applied to calculate the activity coefficients and the average H^+ ion molality in aqueous solution of sulfuric acid we have developed for this purpose as a machine code in Fortran that allowing the calculation of the H^+ ion molality in aqueous solution of sulfuric acid and the activity coefficients.

4   Results and Discussion

The curves in Figure1 which gives the activity coefficient of sulfuric acid molality for between 0.01 and 5 mol.kg^{-1} at 25°C, and molality of 0.01, 0.05 and 0.1 mol.kg^{-1} in zinc sulfate, can say that the presence of zinc sulfate affects the average ionic activity coefficient $\gamma_{H_2SO_4}$ in the lower bound, especially at low stoichiometric molality of sulfuric acid. This molality of zinc sulfide is of the order of magnitude of those we find in the solution from the dissolution of zinc sulfide in sulfuric acid medium. The attack of 5 g of zinc sulfide in 400ml of sulfuric acid solution corresponds to a 0.128 M solution of zinc sulfate to total dissolution.

The curves corresponding to temperatures from 50°C to 200°C show that the presence of zinc sulfate resulted in lower average activity coefficients of sulfuric acid, regardless of temperature. This influence becomes neglected at low molality zinc sulfate and sulfuric acid molality important.

The change in the activity coefficient of sulfuric acid depending on the temperature data for molality of zinc sulfate and sulfuric acid is comparable to those obtained in the absence of zinc sulfate. It may be noted that the coefficient $\gamma^{\pm}_{H_2SO_4}$ of a solution without zinc sulfate at 190 ° C is greater than the coefficient $\gamma^{\pm}_{H_2SO_4}$ of the solution containing zinc sulfate.

The values of activity coefficients of zinc sulfate at room temperature are slightly higher between 10^{-3} and 10^{-2} mol/kg the values of 50 °C. At a given temperature, the activity coefficient using the zinc sulfate passes through a minimum which is located in a zinc sulfate molality between 1mol/kg at 25 ° C and 0.01 mol / kg at 200°C.

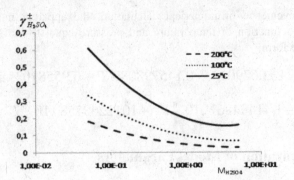

Fig. 1. Activity coefficient of sulfuric acid in the presence of zinc sulfate as a function of molality of sulfuric acid at 25°C, 100°C and 200°C

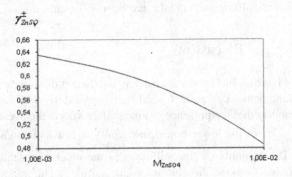

Fig. 2. Stoichiometric activity coefficient of zinc sulfate in the presence of sulfuric acid as a function of molality in zinc sulfate at 25 ° C

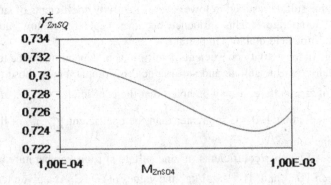

Fig. 3. Stoichiometric activity coefficient of zinc sulfate in the presence of sulfuric acid as a function of molality in zinc sulfate at 100 ° C

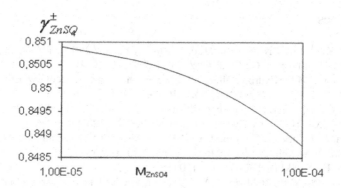

Fig. 4. Stoichiometric activity coefficient of zinc sulfate in the presence of sulfuric acid as a function of molality in zinc sulfate at 200 ° C

5   Conclusions

The values of activity coefficients of zinc sulfate in the presence of sulfuric acid, show that at a given temperature, the presence of sulfuric acid lowers the activity coefficient of zinc sulfate. It decreases as the molality increases in zinc sulphate which results in a weak sulfuric acid molality.

The study area is divided into three distinct parts: 25° C and 100°C, 110°C and 150°C and 160°C and 200°C. These areas appear to be related to the structure of sulfur, the melting temperature of crystalline sulfur of between 110 ° C and 120°C while the temperature of 160°C corresponds to the transition $S\lambda \rightarrow Sp$

Nomenclature

m_i	Molality of the species considered
A^φ	Debye-Huckel parameter
I	ionic strength
T	temperature
R	gas constant
f^γ	term describing the effect of electrostatic forces
K	equilibrium constant
Bx	is a function expressing the binary interaction
Cx	is a function expressing the triple interactions
B'x	represents the derivative with respect to the ionic strength
b	adjustable parameter that was optimized and set equal to 1.2 $kg^{1/2}$. $mol^{-1/2}$ for all temperatures
$\alpha_1 = 2$	for most electrolytes
$\alpha_2 = 12$	for electrolytes 2-2
γ	Activity coefficient of the body considered

References

1. Pitzer, K.S., Roy, R.N., Silvester, L.F.: J. Am. Chem. Soc. 99(15), 4930–4936 (1977)
2. Pitzer, K.S., Mayorga, G.: J. Sol. Chem. 3(7), 539–546 (1974)
3. Wadsworth, M.E.: Hydrometallurgical processes in rate processes of extractive metallurgy, pp. 133–197. Plenum Press, New York (1979)
4. Bradley, D.J., Pitzer, K.S.: J. Phys. Chem. 83(12), 1599–1603 (1979)
5. Peng, P., Xie, H., Lu, L.: Hydrometallurgy 80, 265–271 (2005)
6. Pecina, T., Franco, T., Castillo, P., Orrantia, E.: Minerals Engineering, 23–30 (2008)
7. Forward, F.A., Veltman, H.: J. Metals 11, 836–840 (1959)
8. Parker, E.G.: CIM Bull. 74(5), 145–150 (1961)
9. Demopoulos, G.P., Baldwin, S.A.: Stoichiometric and kinetic aspects on the pressure leaching of zinc concentrates. In: Mishra, Brajendra (eds.) Proceedings of Sessions and Symposia held at the TMS Annual Meeting, EPD Congress 1999, San Diego, February 28-March 4, pp. 567–583 (1999)
10. Wang, P., Anderko, A., Springer, R.D., Young, R.D.: J. Molecular Liquids 125, 37–44 (2006)
11. Lu, X.-M., Xu, W.-G., Chang, X.-H., Lu, D.-Z., Yang, J.-Z.: J. Chem. Thermodynamics 36, 253–257 (2004)
12. Marion, G.M.: Geochimica and Cosmochimica Acta 14(15), 2499–2516 (2002)
13. Pitzer, K.S.: Activity coefficients in electrolyte solutions, 2nd edn. CRC, USA (1991)
14. Zemaitis, J.F., Clark, D.M., Rafal, M., Scrivner, N.C.: Handbook of aqueous electrolyte thermodynamics. Design Institute for Physical Property Data-AIChE, N.Y., USA (1986)

Application of Geometric Explicit Runge–Kutta Methods to Pharmacokinetic Models

Moses A. Akanbi[1] and Kailash C. Patidar[2]

[1] Department of Mathematics, Lagos State University, P.M.B. 0001 LASU Post Office
Lagos, Nigeria
akanbima@gmail.com
[2] Department of Mathematics & Applied Mathematics, University of the Western Cape Private
Bag X17, Bellville 7535, Cape Town, South Africa
kpatidar@uwc.ac.za

Abstract. In an earlier work, the authors proposed a class of geometric explicit Runge– Kutta methods for solving one-dimensional first order Initial Value Problems (IVPs). In this work, some members of this class of schemes which were found to be more accurate are applied to systems of first order ordinary differential equations (ODEs). We present the development of these selected schemes and also study their basic properties vis-a-vis systems of ODEs. We then apply this approach to solve some mathematical models arising in Pharmacokinetics.

Keywords: Pharmacokinetic model, Geometric Mean, Explicit Rung–Kutta Method, Stability, Convergence, Absolute stability, Initial Value Problems.

1 Introduction

Many problems in engineering and science can be formulated in terms of differential equations. A large part of the motivation for building the early computers came from the need to compute ballistic trajectories accurately and quickly. Numerical methods are truly a crucial part of solving differential equations which cannot be neglected. Today, computer are used extensively to solve the equations of ballistic-missile and artificial-satellite theory, as well as those of electrical networks, bending of beams, stability of aircraft, vibration theory, population biology and especially Pharmacokinetics [15,17].

The mathematical models of physical phenomena often leads to one or a set of nth order differential equations having the form

$$y' = f(x, y) \quad , \quad y(x_0) = \eta \tag{1}$$

(where the 1st , 2nd , 3rd and 4th order partial derivatives of f with respect to y for purpose of derivation of the schemes are respectively given as f_y, f_{yy}, f_{yyy} and f_{yyyy}). Yet, the number of instances where an exact solution can be found by analytical means is very limited. Apparently, only a small class of differential equations possesses analytic solution $y(x)$ expressible in terms of known

K.J. Engemann, A.M. Gil Lafuente, and J.M. Merigó (Eds.): MS 2012, LNBIP 115, pp. 259–269, 2012.

tabulated transcendental functions that satisfy the differential equation as well as the initial conditions. Even when the analytic solutions to certain differential equations are available, their numerical evaluation may be quite difficult. Thus, we need to develop effective and efficient numerical methods for advancing the solution of (1). Since the advent of widespread digital computing in the 1960s, a great many theoretical and practical developments have been made in this area, and new ideas continue to emerge [10].

There are generally three families of methods for solving (1) numerically. These are Taylor series method, Linear Multistep Methods (LMM) and Runge–Kutta (RK) methods. The three families can all be viewed as generalizations of the Euler method which is the simplest and most analyzed numerical scheme for IVPs. Runge-Kutta methods have the high-order local truncation error of the Taylor methods while eliminating the need to compute and evaluate the derivatives of $f(x, y(x))$.

Runge [21], first proposed the extension to Euler method by computing the first derivative, f several times per step in 1895. Further contributions were made by [11], and by [13]. The latter completely characterized the set of RK methods of order 4, and proposed the first methods of order 5. Special RK methods for second-order differential equations were proposed by [16], who also contributed to the development of methods for first-order equations. It was not until the work of [12], that sixth-order methods were introduced. Since the advent of digital computers, a lot of effort has been made to improve the efficiency of ERK schemes. In recent times, the formulation of ERK methods was extended by applying geometric approach to the internal functions evaluation by many authors [3–9, 14, 18–20, 22–26, 28–34].

In this work, we shall discuss the theory of geometric Explicit Runge-Kutta (GERK) methods for the solution of systems of IVPs (1). We also apply this scheme to Biological systems using Pharmacokinetic models as a case study. The pharmacokinetics of a drug in the body is a complex process, governed by a variety of factors, including:

- the properties of the drug molecule,
- the blood flows to and the volumes of various tissues,
- the permeability of various membranes,
- tissue composition and
- the affinity of tissues for the administered compound [15].

The ability to characterize and predict pharmacokinetics is of utmost importance both to academia and pharmaceutical industry.

The rest of the paper is organized as follows, Section 2 presents a brief overview of ERK schemes, and Section 3 deals with the summary of derivation of a class of GERK methods. In Section 4 we implement the class of GERK methods on Pharmacokinetic model. Finally, Section 5 deals with concluding remarks and plans for future research.

2   A Brief Overview of Explicit Runge–Kutta Methods

Although, a general overview of ERK algorithm has been presented in an earlier work
[1], but for purpose of clarity we present a brief overview below. As it is well known,
one-step method is of the form

$$y_{n+1} - y_n \;=\; \Phi(x_n, y_n; h) \tag{2}$$

The Taylor's algorithm of order p is obtained from (2) by setting

$$\Phi(x_n, y_n; h) \;=\; \Phi_T(x_n, y_n; h)$$

$$= \sum_{r=0}^{\infty} \frac{h^{r+1}}{(r+1)!} \left(h\frac{\partial}{\partial x} + h\frac{\partial}{\partial y} \right)^r f(x, y) \tag{3}$$

and whenever f does not depend on x explicitly, we have the incremental function

$$\Phi_T(y_n; h) \;=\; \sum_{r=0}^{\infty} \frac{h^{r+1}}{(r+1)!} \left(\frac{\partial}{\partial y} \right)^r f(y)$$

$$= hf + \frac{h^2}{2} ff_y + \frac{h^3}{6}\left(ff_y^2 + f^2 f_{yy}\right) + \frac{h^4}{24}\left(ff_y^3 + 4f^2 f_y f_{yy} + f^3 f_{yyy}\right) \tag{4}$$

$$+ \frac{h^5}{120}\left(ff_y^4 + 11f^2 f_y^2 f_{yy} + 4f^3 f_{yy}^2 + 7f^3 f_y f_{yyy} + f^4 f_{yyyy}\right)$$

$$+ O(h^5).$$

An s-stage ERK method is of the form

$$y_{n+1} - y_n \;=\; \Phi_{RK}(x_n, y_n; h),$$

$$\text{where } \Phi_{RK}(x_n, y_n; h) \;=\; \sum_{r=1}^{s} b_r K_r$$

$$K_1 \;=\; hf(x, y)$$

$$K_r \;=\; hf\left(x + c_r h, y + \sum_{u=1}^{r-1} a_{ru} K_r \right) , r = 2, 3, ..., s \tag{5}$$

$$c_r \;=\; \sum_{u=1}^{r-1} a_{ru} , r = 2, 3, ..., s.$$

The basis of derivation of ERK schemes is to equate the coefficients of the incremental
functions $\Phi_T(x_n, y_n; h)$ and $\Phi_{RK}(x_n, y_n; h)$ to $O(h^p)$ for a pth order method.

3   Geometric Explicit Runge–Kutta (GERK) Method and Its Derivation

The computation of the first derivative, f several times per step, in the methods
discussed above is viewed as averages. In particular, arithmetic averages have been
used. In this work, we extend the use of geometric averages to the internal stages
evaluation as follows

$$y_{n+1} - y_n = \Phi_G(x_n, y_n; h) \tag{6}$$

where

$$
\begin{aligned}
\Phi_G(x_n, y_n; h) &= b_1\sqrt{K_1 K_2} + b_2\sqrt{K_2 K_3} + b_3\sqrt{K_3 K_4}, \\
K_1 &= hf(y_n) \\
K_2 &= hf(y_n + a_{21}K_1) \\
K_3 &= hf(y_n + a_{31}K_1 + a_{32}K_2). \\
K_4 &= hf(y_n + a_{41}K_1 + a_{42}K_2 + a_{43}K_3)
\end{aligned}
\tag{7}
$$

The expansion of (7) gives

$$
K_2 = hf + a_{21}h^2 ff_y + \frac{1}{2}a_{21}^2 h^3 f^2 f_{yy} + \frac{1}{6}f^3 h^4 a_{21}^3 f_{yyy} + O(h^5)
$$

$$
\begin{aligned}
K_3 =\ & hf + (a_{31} + a_{32})h^2 ff_y + \left(a_{21}a_{32}ff_y^2 + \frac{1}{2}(a_{31} + a_{32})^2 f^2 f_{yy}\right)h^3 \\
& + h^4\left(f^2\left(\frac{1}{2}a_{21}^2 a_{32} + a_{21}a_{31}a_{32} + a_{21}a_{32}^2\right)f_y f_{yy}\right. \\
& \left. + f^3\left(\frac{a_{31}^3}{6} + \frac{1}{2}a_{31}^2 a_{32} + \frac{1}{2}a_{31}a_{32}^2 + \frac{a_{32}^3}{6}\right)f_{yyy}\right) + O(h^5)
\end{aligned}
$$

$$
\begin{aligned}
K_4 =\ & fh + fh^2(a_{41} + a_{42} + a_{43})f_y + h^3\left(f(a_{21}a_{42} + a_{31}a_{43} + a_{32}a_{43})f_y^2\right. \\
& + f^2\left(\frac{a_{41}^2}{2} + a_{41}a_{42} + \frac{a_{42}^2}{2} + a_{41}a_{43} + a_{42}a_{43} + \frac{a_{43}^2}{2}\right)f_{yy}\right) \\
& + h^4\left(fa_{21}a_{32}a_{43}f_y^3 + f^2\left(\frac{1}{2}a_{21}^2 a_{42} + a_{21}a_{41}a_{42} + a_{21}a_{42}^2 + \frac{1}{2}a_{31}^2 a_{43} + a_{31}a_{32}a_{43} + \frac{1}{2}a_{32}^2 a_{43}\right.\right. \\
& \left.+ a_{31}a_{41}a_{43} + a_{32}a_{41}a_{43} + a_{21}a_{42}a_{43} + a_{31}a_{42}a_{43} + a_{32}a_{42}a_{43} + a_{31}a_{43}^2 + a_{32}a_{43}^2\right)f_y f_{yy} \\
& + f^3\left(\frac{a_{41}^3}{6} + \frac{1}{2}a_{41}^2 a_{42} + \frac{1}{2}a_{41}a_{42}^2 + \frac{a_{42}^3}{6} + \frac{1}{2}a_{41}^2 a_{43} + a_{41}a_{42}a_{43} + \frac{1}{2}a_{42}^2 a_{43}\right. \\
& \left.\left. + \frac{1}{2}a_{41}a_{43}^2 + \frac{1}{2}a_{42}a_{43}^2 + \frac{a_{43}^3}{6}\right)f_{yyy}\right) + O(h^5)
\end{aligned}
$$

such that

$$
K_1 K_2 = h^2 f^2 + h^3 a_{21}f^2 f_y + \frac{1}{2}h^4 a_{21}^2 f^3 f_{yy} + \frac{1}{6}h^5 a_{21}^3 f^4 f_{yyy} + O(h^6)
$$

$$
\begin{aligned}
K_2 K_3 =\ & h^2 f^2 + h^3(a_{21} + a_{31} + a_{32})f^2 f_y + h^4\left((a_{21}a_{31} + 2a_{21}a_{32})f^2 f_y^2 + \frac{1}{2}\left(a_{21}^2 + (a_{31} + a_{32})^2\right)f^3 f_{yy}\right) \\
& + h^5\left(a_{21}^2 a_{32}f^2 f_y^3 + \frac{1}{2}a_{21}\left(a_{21}(a_{31} + 2a_{32}) + (a_{31} + a_{32})(a_{31} + 3a_{32})\right)f^3 f_y f_{yy}\right. \\
& \left. + \frac{1}{6}\left(a_{21}^3 + (q_{31} + a_{32})^3\right)f^4 f_{yyy}\right) + O(h^6)
\end{aligned}
$$

$$
\begin{aligned}
K_3K_4 \;=\;& h^2 f^2 + h^3 \left(a_{31} + a_{32} + a_{41} + a_{42} + a_{43}\right) f^2 f_y \\
&+ h^4 \left(\left(a_{21}\left(a_{32} + a_{42}\right) + a_{31}\left(a_{41} + a_{42} + 2a_{43}\right) + a_{32}\left(a_{41} + a_{42} + 2a_{43}\right)\right) f^2 f_y^2 \right. \\
&\left. + f^3 \left(\frac{a_{31}^2}{2} + a_{31}a_{32} + \frac{a_{32}^2}{2} + \frac{a_{41}^2}{2} + \frac{a_{42}^2}{2} + a_{42}a_{43} + \frac{a_{43}^2}{2} + a_{41}\left(a_{42} + a_{43}\right) \right) f_{yy} \right) \\
&+ h^5 \left(f^2 \left(a_{31}^2 a_{43} + 2a_{31}a_{32}a_{43} + a_{32}^2 a_{43} + a_{21}\left(a_{31}a_{42} + a_{32}\left(a_{41} + 2a_{42} + 2a_{43}\right)\right) \right) f_y^3 \right. \\
&+ f^3 \left(a_{21}^2 \left(\frac{a_{32}}{2} + \frac{a_{42}}{2} \right) + a_{31}^2 \left(\frac{a_{41}}{2} + \frac{a_{42}}{2} + a_{43} \right) + a_{32}^2 \left(\frac{a_{41}}{2} + \frac{a_{42}}{2} + a_{43} \right) \right. \\
&+ a_{21}\left(a_{31}a_{32} + a_{32}^2 + a_{41}a_{42} + a_{42}^2 + a_{42}a_{43} \right) + a_{32}\left(\frac{a_{41}^2}{2} + \frac{a_{42}^2}{2} + 2a_{42}a_{43} + \frac{3a_{43}^2}{2} + a_{41}\left(a_{42} + 2a_{43}\right) \right) \\
&\left. + a_{31}\left(\frac{a_{41}^2}{2} + \frac{a_{42}^2}{2} + 2a_{42}a_{43} + \frac{3a_{43}^2}{2} + a_{41}\left(a_{42} + 2a_{43}\right) + a_{32}\left(a_{41} + a_{42} + 2a_{43}\right) \right) \right) f_y f_{yy} \\
&+ f^4 \left(\frac{a_{31}^3}{6} + \frac{1}{2}a_{31}^2 a_{32} + \frac{1}{2}a_{31}a_{32}^2 + \frac{a_{32}^3}{6} + \frac{a_{41}^3}{6} + \frac{a_{42}^3}{6} + a_{41}^2 \left(\frac{a_{42}}{2} + \frac{a_{43}}{2} \right) + \frac{1}{2}a_{42}^2 a_{43} + \frac{1}{2}a_{42}a_{43}^2 \right. \\
&\left. \left. + \frac{a_{43}^3}{6} + a_{41}\left(\frac{a_{42}^2}{2} + a_{42}a_{43} + \frac{a_{43}^2}{2} \right) \right) f_{yyy} \right) + O(h^6)
\end{aligned}
$$

Using

$$
(1+x)^n = 1 + \frac{n}{1!}x + \frac{n(n-1)}{2!}x^2 + \frac{n(n-1)(n-2)}{3!}x^3 + \cdots
$$

we have

$$
\begin{aligned}
\sqrt{K_1K_2} \;=\;& hf + \frac{1}{2}h^2 a_{21} f f_y - \frac{1}{8}a_{21}^2 h^3 \left(f f_y^2 - 2f^2 f_{yy} \right) + \\
&+ h^4 \left(\frac{1}{16}f a_{21}^3 f_y^3 - \frac{1}{8}f^2 a_{21}^3 f_y f_{yy} + \frac{1}{12}f^3 a_{21}^3 f_{yyy} \right) + O(h^5) \tag{8}
\end{aligned}
$$

$$
\begin{aligned}
\sqrt{K_2K_3} \;=\;& fh + \frac{1}{2}fh^2 \left(a_{21} + a_{31} + a_{32}\right) f_y \\
&+ h^3 \left(\frac{1}{8}f \left(-a_{21}^2 - \left(a_{31} + a_{32}\right)^2 + 2a_{21}\left(a_{31} + 3a_{32}\right) \right) f_y^2 + \frac{1}{4}f^2 \left(a_{21}^2 + \left(a_{31} + a_{32}\right)^2 \right) f_{yy} \right) \\
&+ h^4 \left(\frac{1}{16}f \left(a_{21} - a_{31} - a_{32}\right)\left(a_{21}^2 + 4a_{21}a_{32} - \left(a_{31} + a_{32}\right)^2 \right) f_y^3 \right. \\
&+ \frac{1}{8}f^2 \left(-a_{21}^3 - \left(a_{31} + a_{32}\right)^3 + a_{21}^2\left(a_{31} + 3a_{32}\right) + a_{21}\left(a_{31} + a_{32}\right)\left(a_{31} + 5a_{32}\right) \right) f_y f_{yy} \\
&\left. + \frac{1}{12}f^3 \left(a_{21}^3 + \left(a_{31} + a_{32}\right)^3 \right) f_{yyy} \right) + O(h^5) \tag{9}
\end{aligned}
$$

$$
\begin{aligned}
\sqrt{K_3K_4} \;=\;& fh + \frac{1}{2}fh^2 \left(a_{31} + a_{32} + a_{41} + a_{42} + a_{43}\right) f_y + h^3 \left(\frac{1}{8}f \left(-\left(a_{31} + a_{32} - a_{41} - a_{42}\right)^2 + 4a_{21}\left(a_{32} + a_{42}\right) \right. \right. \\
&\left. + 2\left(3a_{31} + 3a_{32} - a_{41} - a_{42}\right)a_{43} - a_{43}^2 \right) f_y^2 + \frac{1}{4}f^2 \left(\left(a_{31} + a_{32}\right)^2 + \left(a_{41} + a_{42} + a_{43}\right)^2 \right) f_{yy} \Big) \\
&+ h^4 \left(\frac{1}{16}f \left(\left(a_{31} + a_{32} - a_{41} - a_{42}\right)\left(\left(a_{31} + a_{32}\right)^2 + 4a_{21}\left(-a_{32} + a_{42}\right) - \left(a_{41} + a_{42}\right)^2 \right) \right. \right. \\
&\left. + \left(4a_{21}\left(3a_{32} - a_{42}\right) + 3\left(a_{31} + a_{32} - a_{41} - a_{42}\right)^2\right)a_{43} + \left(-5a_{31} - 5a_{32} + 3\left(a_{41} + a_{42}\right)\right)a_{43}^2 + a_{43}^3 \right) f_y^3 \\
&+ \frac{1}{8}f^2 \left(-a_{31}^3 + 2a_{21}^2\left(a_{32} + a_{42}\right) - \left(-a_{32} + a_{41} + a_{42}\right)^2\left(a_{32} + a_{41} + a_{42}\right) + 3\left(a_{32}^2 + 2a_{32}\left(a_{41} + a_{42}\right) \right. \right. \\
&- \left(a_{41} + a_{42}\right)^2\right)a_{43} + \left(5a_{32} - 3\left(a_{41} + a_{42}\right)\right)a_{43}^2 - a_{43}^3 + a_{31}^2\left(-3a_{32} + a_{41} + a_{42} + 3a_{43}\right) \\
&+ 4a_{21}\left(a_{32}^2 + a_{42}\left(a_{41} + a_{42} + a_{43}\right)\right) + a_{31}\left(4a_{21}a_{32} - 3a_{32}^2 + 2a_{32}\left(a_{41} + a_{42} + 3a_{43}\right) \right. \\
&\left. \left. + \left(a_{41} + a_{42} + a_{43}\right)\left(a_{41} + a_{42} + 5a_{43}\right)\right) \right) f_y f_{yy} + \frac{1}{12}f^3 \left(\left(a_{31} + a_{32}\right)^3 + \left(a_{41} + a_{42} + a_{43}\right)^3 \right) f_{yyy} \right) \\
&+ O(h^5) \tag{10}
\end{aligned}
$$

Using (9) – (11) in (7) gives

$$\Phi_G(y_n; h) = b_1\sqrt{K_1K_2} + b_2\sqrt{K_2K_3} + b_3\sqrt{K_3K_4}$$

$$= fh(b_1 + b_2 + b_3) + \frac{1}{2}fh^2\left((a_{31} + a_{32})b_2 + a_{21}(b_1 + b_2) + (a_{31} + a_{32} + a_{41} + a_{42} + a_{43})b_3\right)f_y$$

$$+ \frac{h^3}{8}\left((-(a_{31} + a_{32})^2 b_2 - a_{21}^2(b_1 + b_2) - ((a_{31} + a_{32} - a_{41} - a_{42})^2 - 2(3a_{31} + 3a_{32} - a_{41} - a_{42})a_{43} + a_{43}^2)b_3\right.$$

$$+ 2a_{21}((a_{31} + 3a_{32})b_2 + 2(a_{32} + a_{42})b_3))ff_y^2 + 2\left((a_{31} + a_{32})^2 b_2 + a_{21}^2(b_1 + b_2) + ((a_{31} + a_{32})^2\right.$$

$$\left. + (a_{41} + a_{42} + a_{43})^2)b_3\right)f^2 f_{yy}) + h^4\left(\frac{1}{16}f\left(-a_{21}^2(a_{31} - 3a_{32})b_2 + (a_{31} + a_{32})^3 b_2 + a_{21}^3(b_1 + b_2)\right.\right.$$

$$+ ((a_{31} + a_{32} - a_{41} - a_{42})^2(a_{31} + a_{32} + a_{41} + a_{42}) + 3(a_{31} + a_{32} - a_{41} - a_{42})^2 a_{43}$$

$$+ (-5a_{31} - 5a_{32} + 3(a_{41} + a_{42}))a_{43}^2 + a_{43}^3)b_3 + a_{21}(-(a_{31} + a_{32})(a_{31} + 5a_{32})b_2$$

$$- 4((a_{32} - a_{42})(a_{31} + a_{32} - a_{41} - a_{42}) + (-3a_{32} + a_{42})a_{43})b_3))f_y^3 + \frac{1}{8}f^2\left(-(a_{31} + a_{32})^3 b_2 - a_{21}^3(b_1 + b_2)\right.$$

$$+ (-(a_{31} + a_{32} - a_{41} - a_{42})^2(a_{31} + a_{32} + a_{41} + a_{42}) + 3(a_{31}^2 + a_{32}^2 + 2a_{32}(a_{41} + a_{42}) - (a_{41} + a_{42})^2$$

$$+ 2a_{31}(a_{32} + a_{41} + a_{42}))a_{43} + (5a_{31} + 5a_{32} - 3(a_{41} + a_{42}))a_{43}^2 - a_{43}^3)b_3 + a_{21}^2((a_{31} + 3a_{32})b_2$$

$$+ 2(a_{32} + a_{42})b_3) + a_{21}((a_{31} + a_{32})(a_{31} + 5a_{32})b_2 + 4(a_{32}(a_{31} + a_{32}) + a_{42}(a_{41} + a_{42} + a_{43}))b_3))f_y f_{yy}$$

$$\left.\left. + \frac{1}{12}f^3\left((a_{31} + a_{32})^3 b_2 + a_{21}^3(b_1 + b_2) + ((a_{31} + a_{32})^3 + (a_{41} + a_{42} + a_{43})^3)b_3\right)f_{yyy}\right) + O(h^5)\right. \tag{11}$$

Comparing the coefficients of h in (12) with the Taylor's expansion (4) yields the following system of non-linear equations:

$$b_1 + b_2 + b_3 = 1$$

$$(a_{31} + a_{32})b_2 + a_{21}(b_1 + b_2) + (a_{31} + a_{32} + a_{41} + a_{42} + a_{43})b_3 = 1$$

$$-(a_{31} + a_{32})^2 b_2 - a_{21}^2(b_1 + b_2) - ((a_{31} + a_{32} - a_{41} - a_{42})^2$$

$$+ 2(-3a_{31} - 3a_{32} + a_{41} + a_{42})a_{43} + a_{43}^2)b_3 + 2a_{21}((a_{31} + 3a_{32})b_2 + 2(a_{32} + a_{42})b_3) = \frac{4}{3}$$

$$(a_{31} + a_{32})^2 b_2 + a_{21}^2(b_1 + b_2) + ((a_{31} + a_{32})^2 + (a_{41} + a_{42} + a_{43})^2)b_3 = \frac{2}{3}$$

$$-a_{21}^2(a_{31} - 3a_{32})b_2 + (a_{31} + a_{32})^3 b_2 + a_{21}^3(b_1 + b_2)$$

$$+ ((a_{31} + a_{32} - a_{41} - a_{42})^2(a_{31} + a_{32} + a_{41} + a_{42}) + 3(a_{31} + a_{32} - a_{41} - a_{42})^2 a_{43} + (-5a_{31} - 5a_{32} + 3(a_{41} + a_{42}))a_{43}^2 + a_{43}^3)b_3$$

$$+ a_{21}(-(a_{31} + a_{32})(a_{31} + 5a_{32})b_2 - 4((a_{32} - a_{42})(a_{31} + a_{32} - a_{41} - a_{42}) + (-3a_{32} + a_{42})a_{43})b_3) = \frac{2}{3}$$

$$-(a_{31} + a_{32} - a_{41} - a_{42})^2(a_{31} + a_{32} + a_{41} + a_{42}) + 3(a_{31}^2 + a_{32}^2 + 2a_{32}(a_{41} + a_{42})$$

$$- (a_{41} + a_{42})^2 + 2a_{31}(a_{32} + a_{41} + a_{42}))a_{43} + (5a_{31} + 5a_{32} - 3(a_{41} + a_{42}))a_{43}^2 - a_{43}^3)b_3 + a_{21}^2((a_{31} + 3a_{32})b_2$$

$$+ 2(a_{32} + a_{42})b_3) + a_{21}((a_{31} + a_{32})(a_{31} + 5a_{32})b_2 + 4(a_{32}(a_{31} + a_{32}) + a_{42}(a_{41} + a_{42} + a_{43}))b_3) = \frac{4}{3}$$

These were solved to obtain 4–stage GERK schemes of order 4 stated below

$$y_{n+1} = \tfrac{1}{4}\sqrt{K_1 K_2} + \tfrac{1}{2}\sqrt{K_2 K_3} + \tfrac{1}{4}\sqrt{K_3 K_4}$$
$$K_1 = hf(y_n)$$

$$K_2 = hf\left(y_n + \tfrac{1}{3}K_1\right)$$
$$K_3 = hf\left(y_n + \tfrac{13-\sqrt{793}}{36}K_1 + \tfrac{11+\sqrt{793}}{36}K_2\right)$$
$$K_4 = hf\left(y_n + \tfrac{-22+\sqrt{793}}{6}K_1 + \tfrac{71-3\sqrt{793}}{12}K_2 + \tfrac{-15+\sqrt{793}}{12}K_3\right)$$

$\left.\right\}$ - GM-1

$$y_{n+1} = \tfrac{1}{4}\sqrt{K_1 K_2} + \tfrac{1}{2}\sqrt{K_2 K_3} + \tfrac{1}{4}\sqrt{K_3 K_4}$$
$$K_1 = hf(y_n)$$
$$K_2 = hf\left(y_n + \tfrac{1}{3}K_1\right)$$
$$K_3 = hf\left(y_n + \tfrac{13+\sqrt{793}}{36}K_1 + \tfrac{11-\sqrt{793}}{36}K_2\right)$$
$$K_4 = hf\left(y_n + \tfrac{-22-\sqrt{793}}{6}K_1 + \tfrac{71+3\sqrt{793}}{12}K_2 + \tfrac{-15-\sqrt{793}}{12}K_3\right)$$

$\left.\right\}$ -GM-2

4   Application of GERK to Pharmacokinetic Models

Pharmacokinetic and dynamic population models provide the means to store past experience with the behavior of drugs, and to apply it to the care of future patients. One of the most important outcomes in pharmacokinetic modeling is the ability to predict drug levels and/or dynamic behaviors of drug entities in the body. Another is the deduction of mechanistic insight into what events have happened. Sometimes the question is not how realistic the models are but what is their consistency and adequacy in making predictions [17]. Among all the means for the delivery of therapeutic drugs to the bloodstream, oral ingestion/gastrointestinal absorption is by far the most popular [27].

In 1996, Yeargers [35], constructed a model for ingestion of drug into the body. The drug is taken orally and is delivered to the gastrointestinal (GI) tract. The drug then moves into the blood stream, without delay, at a rate proportional to its concentration in the GI tract and independent of its concentration in the blood. The drug is metabolized and cleared from the blood at a rate proportional to its concentration there. The model is based on the two compartments, GI tract and blood [2, 27].

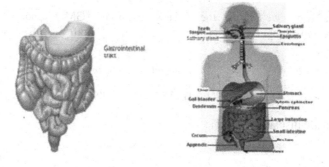

Fig. 1. Gastrointestinal tract and the Circulatory System

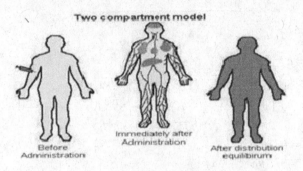

Fig. 2a. Two Compartments model

Compartmental diagram of drug concentration in the GI tract and blood

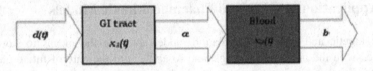

Fig. 2b. Two Compartments model

Figures 1 and 2 show the GI tract and the circulatory system as well as the two compartment model. Denoting

- the concentration of the drug in the GI tract by $x(t)$,
- the concentration in the blood by $y(t)$,
- and the drug dosage by $d(t)$,

the process can be modeled as a system of linear, non-homogeneous differential equations:

$$\left.\begin{array}{l} \frac{dx}{dt} = -a\,x(t) + d(t) \\[2mm] \frac{dy}{dt} = a\,x(t) - b\,y(t) \end{array}\right\} \tag{12}$$

In this section, we apply the GERK methods (GM-1 and GM-2) to the pharmacokinetic model (13). The absolute error for both of these components are calculated as

$$x^e = |x^{exact} - x^{numerical}|$$
$$y^e = |y^{exact} - y^{numerical}|$$

The results are displayed in Table 1.

Table 1. The absolute values of error of $x(t)$ and $y(t)$ for Pharmacokinetic model (13) using the GERK approach with $a = \ln 4, b = (\ln 2)/5, h = 0.001, 0.005, 0.025, 0.125$

h	t	xExact	yExact	GM-1		GM-2	
				x^e	y^e	x^e	y^e
0.001	1.00E-01	9.34E-02	6.59E-03	6.28E-01	7.21E+00	6.28E-01	7.21E+00
	2.00E-01	1.75E-01	2.51E-02	5.47E-01	7.19E+00	5.47E-01	7.19E+00
	3.00E-01	2.45E-01	5.38E-02	4.76E-01	7.16E+00	4.76E-01	7.16E+00
	4.00E-01	3.07E-01	9.12E-02	4.14E-01	7.12E+00	4.14E-01	7.12E+00
	5.00E-01	3.61E-01	1.36E-01	3.61E-01	7.08E+00	3.61E-01	7.08E+00
	6.00E-01	4.07E-01	1.87E-01	3.14E-01	7.03E+00	3.14E-01	7.03E+00
	7.00E-01	4.48E-01	2.43E-01	2.73E-01	6.97E+00	2.73E-01	6.97E+00
	8.00E-01	4.83E-01	3.04E-01	2.38E-01	6.91E+00	2.38E-01	6.91E+00
	9.00E-01	5.14E-01	3.69E-01	2.07E-01	6.84E+00	2.07E-01	6.84E+00
	1.00E+00	5.41E-01	4.36E-01	1.80E-01	6.78E+00	1.80E-01	6.78E+00
0.005	1.00E-01	9.34E-02	6.59E-03		6.71E+00		6.71E+00
	2.00E-01	1.75E-01	2.51E-02	5.47E-01	7.16E+00	5.47E-01	7.16E+00
	3.00E-01	2.45E-01	5.38E-02	4.76E-01	7.16E+00	4.76E-01	7.16E+00
	4.00E-01	3.07E-01	9.12E-02	4.14E-01	7.12E+00	4.14E-01	7.12E+00
	5.00E-01	3.61E-01	1.36E-01	3.61E-01	7.08E+00	3.61E-01	7.08E+00
	6.00E-01	4.07E-01	1.87E-01	3.14E-01	7.03E+00	3.14E-01	7.03E+00
	7.00E-01	4.48E-01	2.43E-01	2.73E-01	6.97E+00	2.73E-01	6.97E+00
	8.00E-01	4.83E-01	3.04E-01	2.38E-01	6.91E+00	2.38E-01	6.91E+00
	9.00E-01	5.14E-01	3.69E-01	2.07E-01	6.84E+00	2.07E-01	6.84E+00
	1.00E+00	5.41E-01	4.36E-01	1.80E-01	6.78E+00	1.80E-01	6.78E+00
0.02	1.00E-01	9.34E-02	6.59E-03		2.61E+00		2.61E+00
	3.00E-01	2.45E-01	5.38E-02		5.64E+00		5.64E+00
	4.00E-01	3.07E-01	9.12E-02		6.25E+00		6.25E+00
	5.00E-01	3.61E-01	1.36E-01		6.58E+00		6.58E+00
	6.00E-01	4.07E-01	1.87E-01		6.74E+00		6.74E+00
	7.00E-01	4.48E-01	2.43E-01		6.80E+00		6.80E+00
	8.00E-01	4.83E-01	3.04E-01	2.38E-01	6.81E+00	2.38E-01	6.81E+00
	9.00E-01	5.14E-01	3.69E-01	2.07E-01	6.79E+00	2.07E-01	6.79E+00
	1.00E+00	5.41E-01	4.36E-01	1.80E-01	6.75E+00	1.80E-01	6.75E+00
0.125	5.00E-01	3.61E-01	1.36E-01		2.48E+00		2.48E+00
	1.00E+00	5.41E-01	4.36E-01		4.13E+00		4.13E+00

5 Concluding Remarks and Future Research

GERK method is found to be superior as compared to other methods. We intend to compare them with other adaptive RK methods (those developed by group of Jeff Cash). We are also extending them to solve the delay models arising in Pharmacokinetics.

References

1. Akanbi, M.A.: On 3–stage Geometric Explicit Runge–Kutta Method for Singular Autonomous Initial Value Problems in Ordinary Differential Equations. Computing 92, 243–263 (2011)
2. Allen, L.J.S.: An Introduction To Mathematical Biology. Pearson Education, Inc., Upper Saddle River (2007)
3. Evans, D.J., Sanugi, B.B.: A new fourth order Runge-Kutta method for inilia/value problems. In: Fatunla, S.O. (ed.) Computat. Math. II. Boole Press (1986)
4. Evans, D.J., Sanugi, B.B.: A new 4th order Runge-Kutta formula for y = Ay with stepwise control. Camp. Math. Applic. 15, 991–995 (1988)
5. Evans, D.J.: New Runge-Kutta methods for initial value problems. Appl. Math. Letters 2, 25–28 (1989)
6. Evans, D.J., Yaacob, N.B.: A New Fourth Order Runge-Kutta Formula Based On Harmonic Mean. Department of Computer Studies. Loughborough University of Technology, Loughborough (1993)
7. Evans, D.J., Yaacob, N.B.: A Fourth order Runge-Kutta Method Based on the Heronian Mean Formula. Intern. J. Comput. Math. 58, 103–115 (1995)
8. Evans, D.J., Yaacob, N.B.: A new Runge Kutta RK (4,4) method. Intern. J. Comput. Math. 58, 169–187 (1995)
9. Evans, D.J., Yaacob, N.B.: A Fourth Order Runge-Kurla Method Based On The Heronian Mean. Intern. J. Computer Math. 59, 12 (1995)
10. Griffiths, D.F., Higham, D.J.: Numerical Methods for Ordinary Differential Equations. Springer Undergraduate Mathematics Series. Springer, London (2010), doi:10.1007/978-0-85729-148-6
11. Heun, K.: Neue Methoden zur approximativen Integration der Differential-gleichungen einer unabhangigen Veranderlichen. Z. Math. Phys. 45, 23–38 (1900)
12. Hǔta, A.: Une amelioration de la methode de Runge-Kutta-Nyström pour la resolution numerique des equations differentielles du premier ordre. Ada Fac. Nat. Univ. Comenian. Math. 1, 201–224 (1956)
13. Kutta, W.: Beitrag zur Naherungs-weissen Integration tolaken Differential-gleichungen. Z. Maths Phys. 46, 435–453 (1901)
14. Murugesan, K., Paul Dhayabaran, D.P., Amirtharaj, E.C.H., Evans, D.J.: A fourth order Embedded Runge-Kutta RKCeM(4,4) Method based on Arithmetic and Centroidal Means with Error Control. International J. Comput. Math. 79(2), 247–269 (2002)
15. Nestorov, I.: Whole body Pharmacokinetic Models. Clinical Pharmacokinetic 42(10) (2003)
16. Nyström, E.J.: Uber die numerische Integration von Differentialgleichungen. Ada Soc. Sci. Fennicae 50(13), 55 (1925)
17. Pang, K.S., Weiss, M., Macheras, P.: Advanced Pharmacokinetic Models Based on Organ Clearance. Circulatory, and Fractal Concepts The AAPS Journal 9(2), Article 30, E268–E283 (2007)
18. Ponalagusamy, R., Senthilkumar, S.: A comparison of RK-fourth orders of variety of means on multilayer raster CNN simulation. Trends in Applied Science and Research 3(3), 242–252 (2008)
19. Ponalagusamy, R., Senthilkumar, S.: A New Fourth Order Embedded RKAHeM(4,4) Method with Error Control on Multilayer Raster Cellular Neural Network. Signal Image and Video Processing (2008) (accepted in press)

20. Razali, N., Ahmad, R.: New Fifth-Order Runge-Kutta Methods for Solving Ordinary Differential Equation. Proceeding of Seminar on Engineering Mathematics 2, 155–162 (2008)
21. Runge, C.: Uber die numerische Auflosung von differntialglechungen. Math. Ann. 46, 167–178 (1895)
22. Sanugi, B.B., New Numerical Strategies for Initial Value Type Ordinary Differential Equations, Ph.D. Thesis, Loughborough University of Technology, U.K. (1986)
23. Sanugi, B.B.: Ph.D. Thesis Loughborough University of Technology (1986)
24. Sanugi, B.B., Evans, D.J.: A New Fourth Order Runge-Kutta method based on Harmonic Mean. Comput. Stud. Rep., LUT (June 1993)
25. Sanugi, B.B., Evans, D.J.: A new fourth order Runge-Kutta formulae based on harmonic mean. Intern. J. Comput. Math. 50, 113–118 (1994)
26. Sanugi, B.B., Yaacob, N.B.: A new fifth order Runge-Kurra method for Initial Value type problems in ODEs. Intern. J. Comput. Math. 59, 187–207 (1995)
27. Shonkwiler, R.W., Herod, J.: Mathematical Biology: An Introduction with Maple and Matlab, 2nd edn. Undergraduate Texts in Mathematics. Springer, Heidelberg (2009)
28. Wazwaz, A.M.: A modified third order Runge-Kutta method. Appl. Math. Letter 3, 123–125 (1990)
29. Wazwaz, A.M.: A Comparison of Modified Runge-Kutta formulas based on Variety of means. Intern. J. Comput. Math. 50, 105–112 (1994)
30. Yaacob, N.B., Evans, D.J.: A new fourth order Runge-Kutta method based on the Root Mean Square formula, Computer Studies, Report, 862, Louborough University of Technology, U.K (December 1993)
31. Yaacob, N.B., Sanugi, B.B.: A New Fifth-Order Five-Stage Explicit HaM-RK5(5) Method For Solving Initial Value Problems in ODEs. Laporan Teknik. Jab. Mat., UTM (1995)
32. Yaacob, N.B., Sanugi, B.B.: A New Fourth-Order Embedded Method Based on the Harmonic mean. Mathematika, Jilid, hml. 1–6 (1998)
33. Yaacob, N.B., Evans, D.J.: New Runge-Kutta Starters for Multi-step Methods. Intern. J. Comput. Math. 71, 99–104 (1999)
34. Yaacob, N.B., Evans, D.J.: A fourth order Runge-Kutta RK(4,4) method with Error Control. Intern. J. Comput. Math. 71, 383–411 (1999)
35. Yeargers, E.K., Shonkwiler, R.W., Herod, J.V.: An Introduction to the Mathematics of Biology, Birkhau

Frequency Tuning Algorithm for Loudspeaker Driven Thermoacoustic Refrigerator Optimization

Yuriy P. Kondratenko[1,2], Oleksiy Korobko[2], and Oleksiy V. Kozlov[2]

[1] Petro Mohyla Black Sea State University, 10 68th Desantnykiv st., 54003,
Mykolaiv, Ukraine
y_kondrat2002@yahoo.com
[2] Admiral Makarov National University of Shipbuilding, 9 Geroyiv Stalingrada ave., 54025
Mykolaiv, Ukraine
oleksii.korobko@nuos.edu.ua, kozlov_ov@ukr.net

Abstract. Article describes the experimental method of acoustic pressure amplitude-frequency response measurement in resonator of thermoacoustic refrigerator with electromechanical generator of acoustic oscillations. Based on the results obtained the nonlinear mathematical model of thermoacoustic refrigerator with electromechanical generator is synthesized using the statistical methods of construction and evaluation of pairwise dependencies. The resonance frequency identification algorithm for thermoacoustic system parametric optimization is developed using the obtained mathematical model. Experimental verification of the proposed resonance frequency identification algorithm is considered.

Keywords: thermoacoustic device, amplitude-frequency response, resonance frequency, parametric optimization.

1   Introduction

Heat engines that produce mechanical or electrical energy by heat energy conversion are one of the most common energy sources in modern industry. Basically, thermal machines can be divided into two groups: direct action machines (heat engines) and reverse action machines (heat pumps) [15]. The most common are mechanical thermal machines in which the mutual energy conversions are based on the use of special mechanical devices [15]. Among them there are piston mechanisms (internal combustion engines, steam engines, Stirling engines) and rotation devices (gas and steam turbines).

Thermoacoustic devices (TAD) [9,13] are the newest type of alternative heat engines, whose work is based on the use of acoustic energy as "the moving" mechanism. The use of acoustic pulsations as a carrier of energy significantly simplifies the design of these machines compared to traditional. Fig. 1 shows the scheme of thermoacoustic refrigerator (TAR), which consists of sound wave electromechanical generator, hollow resonator filled with gas medium, heat exchange surfaces (T_C, T_H) for heat adding and subtraction and a special heat exchange surface (the stack), which is the main catalyst of acoustic and thermal energies mutual conversion.

K.J. Engemann, A.M. Gil Lafuente, and J.M. Merigó (Eds.): MS 2012, LNBIP 115, pp. 270–279, 2012.

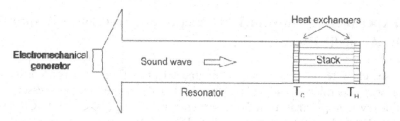

Fig. 1. Structure of the thermoacoustic refrigerator with electromechanical generator of sound waves

Absence of moving parts [14] in such heat engines [7] increases their reliability and also reduces energy losses in the mechanical connections. However, for high power thermoacoustic systems operations the acoustic pressure level of 150-180 dB must be maintained in resonator, which leads to the appearance of parasitic nonlinear effects [16]. Acoustic pressure magnitude in the thermoacoustic devices resonator is one of the determining factors of their efficiency. It is known [5] that in standing wave devices maximum level of sound pressure can be reached only on resonance frequency of sound wave in resonator. Therefore, the problem of the resonance frequency exact value determination is important.

Traditionally, the sound wave is considered as a set of pressure and velocity (displacement) oscillations of the working environment [12,15], whose values are determined by both structural features of thermoacoustic plant and external influences. The frequency of a sound signal, which is formed by electromechanical generator, significantly affects the amplitude of acoustic pressure and particle displacement oscillations in the cavity, and is an important factor in design and functioning of TAD.

In general, the resonant frequency of the acoustic pressure oscillations (1) for the empty resonator TAD can be calculated as [5]:

$$f = \frac{a}{\lambda} = \frac{\sqrt{\gamma k T / m}}{4L},\tag{1}$$

where f – sound wave frequency; a – speed of sound; λ – sound wave length; k – Boltzmann constant; γ – the adiabatic coefficient of gases; T – absolute temperature; m – molecular mass of gas; L – resonator length.

It should be noted that the sound frequency calculated according to (1) is only approximate value [3,5], because its calculation does not consider the specific design features of TAD and external environment basic physical parameters. Therefore, conduction of experimental measurements and processing of the results obtained for specific types of TAD is one of the favorable ways for resonance frequency of sound vibrations parametric identification.

The aim of this paper is a synthesis of algorithm for parametric optimization of thermoacoustic refrigerator with electromechanical generator based on experimental studies of the acoustic pressure amplitude-frequency response in the resonator of TAR.

2   Experimental Study of TAR Amplitude-Frequency Response by Acoustic Pressure

Given research were held on created experimental plant (Fig. 2) which consist of: standing wave thermoacoustic refrigerator [13] and remote microprocessor data acquisition system [3] that includes a programmable logic controller ICP DAS uPAC7186 EX-SM, extension module ICP DAS I7018P to collect signals from thermocouples, microcontroller Atmel Xmega A3 to measure the output signals of the acoustic pressure sensors Freescale MPXV7007DP (170 dB SPL , $\Delta P = \pm 7$ KPa), a personal computer, loudspeaker for sound waves generation and an acoustic amplifier MMF LV103.

Developed software allows to generate the acoustic signals with different shapes at the output of the electromechanical generator and to store data received from the sensors in real time.

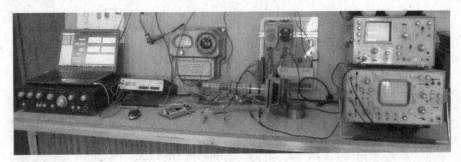

Fig. 2. Standing wave thermoacoustic refrigerator and remote data acquisition system

2.1   The Real-Time Amplitude-Frequency Response Determination Method

Described microprocessor system (Fig. 2) allows the measurement of thermoacoustic system amplitude-frequency characteristics (AFC) in the automatic mode. In this case the electromechanical generator consistently forms a set of sinusoidal acoustic signals (2) with a constant amplitude $P_{in0}(\omega)$ and variable frequency $\omega = 2\pi f$.

$$P_{in}(t) = P_{in0}(\omega)\sin \omega t, \qquad (2)$$

where $P_{in0}(\omega)$ – peak value of input signal, $\omega = 2\pi f$ – frequency of the harmonic signal.

The collected data on the acoustic pressure magnitude (3) for each sinusoidal signal is stored in the memory of microprocessor data acquisition system for further processing.

$$P_{out}(t) = P_{out0}(\omega)\sin\left[\omega t + \phi_0(\omega)\right], \qquad (3)$$

where $P_{out0}(\omega)$ – peak value of the output signal, $\phi_0(\omega)$ – phase shift between output and input harmonic signals.

Three samples of each signal are measured and then the value of their average maximum is stored as a peak value of pressure at the given frequency. AFC ordinates (4) for each input signal are calculated as the ratio of output signal $P_{out0}(\omega)$ to the input signal $P_{in0}(\omega)$ amplitudes at constant frequency:

$$A(\omega) = \frac{P_{out0}(\omega)}{P_{in0}(\omega)},\tag{4}$$

where $A(\omega)$ – value of the amplitude-frequency characteristics.

After processing all the data obtained the amplitude-frequency response representing arrays (5), (6) are formed:

$$A(\omega) = \left\{\begin{array}{l} 35.27, 35.3, 35.27, 35.3, 35.24, 35.89, 36.53, 36.82, 37.18, \\ 37.39, 37.56, 37.59, 37.85, 37.98, 37.88, 37.92, 38.14, 38.43, \\ 39.5, 41.36, 41.49, 43.14, 44.85, 46.2, 47.39, 45.52, 45.3, \\ 41.53, 40.2, 40.17, 40.91, 42.4, 43.59, 39.79, 37.53, 36.92, \\ 36.27, 35.85, 35.79, 35.79, 36.21, 37.02, 37.21, 36.95, 36.43, \\ 36.14, 35.92, 35.69, 35.56, 35.53, 35.37 \end{array}\right\}\tag{5}$$

$$f = \frac{\omega}{2\pi} = \left\{\begin{array}{l} 0.01, 0.1, 0.5, 0.8, 1, 2, 5, 8, 10, 20, 30, 40, 50, 60, 70, 80, 90, \\ 100, 120, 150, 180, 200, 210, 215, 220, 230, 250, 280, 300, 320, \\ 350, 380, 400, 420, 450, 480, 500, 550, 600, 650, 700, 750, 800, \\ 820, 850, 880, 900, 920, 950, 980, 1000 \end{array}\right\}\tag{6}$$

2.2   Synthesis of TAR Acoustic Pressure Nonlinear Mathematical Models

Synthesis of mathematical models is implemented on the basis of experimental data (5), (6) of the sound pressure $P_{out0}(\omega)$ in thermoacoustic refrigerator resonator.

It should be noted that the required mathematical model should convey the exact shape of AFC and be relatively simple to implement on peripheral devices of remote data acquisition system (Fig. 2). Therefore, as approximating functions were chosen the Fourier series (7), the number of Gaussian functions (8) and ninth-order polynomial (9):

$$p_{a1}(f) = \frac{a_0}{2} + \sum_{n=1}^{8} \left(a_n \sin(nwf) + b_n \cos(nwf)\right),\tag{7}$$

$$p_{a2}(f) = a_1 e^{-\left(\frac{f-b_1}{c_1}\right)^2} + a_2 e^{-\left(\frac{f-b_2}{c_2}\right)^2} + \ldots + a_8 e^{-\left(\frac{f-b_8}{c_8}\right)^2}, \tag{8}$$

$$p_{a3}(f) = a_1 f^9 + a_2 f^8 + \ldots + a_9 f + a_{10}, \tag{9}$$

where $p_{aj}(f), j = \overline{1,3}$ – functional dependence of the acoustic pressure on the frequency; a_n, b_n, c_n, w – factors of the mathematical model; n – models order.

The statistical methods [6] of construction and evaluation of pairwise dependencies are used for the mathematical models synthesis. In particular, the task of regression analysis by least squares method is used to build the regression line from known k number of sample points $P_{out0}(\omega)$, so that the sum Z of squared deviations Δ_i^2 of these points along the axis of the performed regression line would remain minimal.

The task of the least squares method can be expressed analytically by the following expression [4,6]:

$$Z = \sum_{i=1}^{k} \Delta_i^2 \to \min, \tag{10}$$

where Δ_i – deviation along the vertical axis of the analytic function $p_{aj}(f)$ value from the experimental value $P_{out0}(\omega)$.

For the physical model it can be written as:

$$Z_j = \sum_{i=1}^{k} \left[P_{out0i} - p_{aj}(f_i) \right]^2 \to \min, i = \overline{1,k}, j = 1,2,3. \tag{11}$$

The basis of condition (11) allows the regression line with a certain probability predict the values of functions $p_{aj}(f)$ on the interval $f \in [f_1; f_n]$ that are missing in table $P_{out0}(\omega)$ of results obtained.

As a result of calculations the following values of the coefficients for selected mathematical models were obtained:

- Fourier series (7)

$$a \in \begin{cases} -5.776 \cdot 10^{12}, 9.312 \cdot 10^{12}, -4.686 \cdot 10^{12}, 1.187 \cdot 10^{12}, \\ 7.557 \cdot 10^{10}, -1.497 \cdot 10^{11}, 4.079 \cdot 10^{10}, -3.809 \cdot 10^{9} \end{cases};$$

$$b \in \begin{cases} 4.018 \cdot 10^{12}, -4.969 \cdot 10^{12}, 3.264 \cdot 10^{12}, -1.291 \cdot 10^{12}, \\ 2.979 \cdot 10^{11}, -3.423 \cdot 10^{10}, 1.142 \cdot 10^{9} \end{cases}; \tag{12}$$

$$w = 0.0002992;$$

- A number of Gaussian functions (8)

$$a \in \{5.529, 4.961, 6.239, 1.983, 0.1906, 35.32, -0.03014, -1.413\} ;$$

$$b \in \{228.6, 394.8, 241.3, 784.6, 1294, 2693, 2200, 0.4843\} ; \qquad (13)$$

$$c \in \{33.77, 25.39, 214.5, 120.6, 250.9, 24230, 22.22, 2.106\} ;$$

- Polynomial (9)

$$a \in \left\{ \begin{array}{l} -6.902 \cdot 10^{-27}, 8.766 \cdot 10^{-23}, -4.573 \cdot 10^{-19}, 1.257 \cdot 10^{-15}, -1.924 \cdot 10^{-12}, \\ 1.56 \cdot 10^{-9}, -5.004 \cdot 10^{-7}, -7.833 \cdot 10^{-5}, 0.06299, 35.36 \end{array} \right\} \qquad (14)$$

Fig. 3 shows the graphs of mathematical models obtained (in comparison with the experimental sample) for research frequency band $f \in [0;1000]$ Hz.

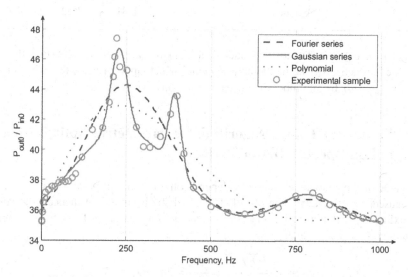

Fig. 3. Comparison of TAR's AFC mathematical models obtained

2.3   Analysis of the Synthesized Mathematical Models Adequacy

A necessary requirement for the implemented mathematical model usage is its adequacy. Adequacy is a reproduction of all important object properties in synthesized model with the completeness required for the purposes of research. Typically, the adequacy of the model is determined based on statistically estimated divergence between the output values of real object $P_{out0}(\omega)$ and implemented model $p_{aj}(f)$ for the same values of input frequency.

The adequacy analysis of the obtained mathematical models (12) – (14) is performed using the methods of mathematical statistics hypotheses evaluation [4,6], such as: sum of squared errors (SSE), coefficient of determination (R^2), root mean square error (RMSE), Fisher's F-test.

The calculation results of TAR's AFC mathematical models adequacy are shown in Table 1.

Table 1. Comparison of TAR's AFC mathematical models adequacy

	Approximating function	SSE	R^2	RMSE	F-test
$j = 1$	Fourier series	74.9	0.89	1.24	7.16
$j = 2$	A number of functions Gauss	7.69	0.98	0.43	69.27
$j = 3$	Polynomial	127.25	0.81	1.52	4.42

Based on calculated statistical data (Table 1) it can be concluded that the best outcome is reached by the mathematical model which approximates the experimental sample $P_{out0}(\omega)$ by a number of Gaussian functions (8), (13).

3   Frequency Tuning Algorithm for Parametric Optimization of Loudspeaker Driven TAR

The value of the resonance frequency depends on a number of parameters [3,4,5] that are caused by structural features of TAD and environment influences. It can be approximately calculated by (1). In particular, for research TAR the resonance frequency is:

$$f = \frac{a}{\lambda} = \frac{\sqrt{\gamma k T/m}}{4L} = \frac{340}{1.2} = 283 \text{ Hz} \tag{15}$$

It should be noted that the value obtained by (11) is significantly different from the measured frequency $f_{res} \approx 229$ Hz. Thus, it is necessary to develop an algorithm for the resonant frequency identification.

Let's consider the approach of resonant frequency identification by the search of acoustic pressure maximum value using the extreme methods of optimization [1,2,10].

Extreme class of automatic control systems is widespread due to a simple technical realization. Among the various methods of extreme control [8] the most flexible is the method of active adaptation, which performs step-by-step tuning operations based on system responses to generated input signal. In developed algorithm such methods are applied to find the optimal operating frequency of a research TAR.

The starting point of the algorithm (Fig. 4) is assumed to be a frequency for a quarter wavelength resonator, which is calculated for research TAR [3] according to (15). This approach allows to avoid the problem of finding a local maximum of the function (8), because the next extreme value (half wavelength resonator value) is at a sufficient distance from the target point.

To improve the performance of synthesized algorithm the search operations are realized using the gradient ascending method [1,2] at which the current step change (17) in frequency is proportional to pressure gradient (16) on the previous iteration of the algorithm:

$$f_i = f_{i-1} - k\nabla p = f_{i-1} - k\frac{p_i - p_{i-1}}{\Delta f_{i-1}} \qquad (16)$$

$$\Delta f_i = f_i - f_{i-1} \qquad (17)$$

where Δf_i Δf_{i-1} – i and i-1 frequencies of acoustic signal; p_i, p_{i-1} – i and i-1 acoustic pressure values; k – iterative coefficient; i – number of current iteration.

The optimum frequency considered to be found if the value of acoustic pressure gradient becomes less than the allowable value: $k\frac{p_i - p_{i-1}}{\Delta f_{i-1}} < 0.01$.

Fig. 4 shows the results of designed algorithm simulation in case of resonance frequency search the with the iterative coefficient $k = 50$.

For further analysis it should be noted that the occurring thermoacoustic processes are result of sound waves interaction with the surface of the stack [7,11,13]. Thus, changes in temperatures T_H, T_C (Fig. 5) near the stack directly reflects the nature of processes and therefore these temperature fluctuations are significant and their optimization can improve TAR efficiency.

Analyzing the results of the experiment (Fig. 4) we can confirm that the founded with developed algorithm value of the resonance frequency ($f_{res} \approx 229$ Hz) allows for greater temperature difference ($\Delta T = T_H - T_C$) at the ends of the research TAR stack (23.7 °C to 12.8 °C) than the calculated (1) value of 283 Hz, and therefore it can be concluded that implementation of proposed frequency tuning algorithm can be used for parametric optimization of thermoacoustic devices and allows to increase the efficiency of the system.

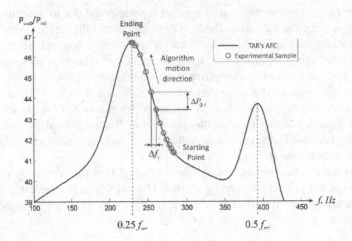

Fig. 4. Simulation results of the frequency tuning algorithm for parametric optimization of experimental TAR

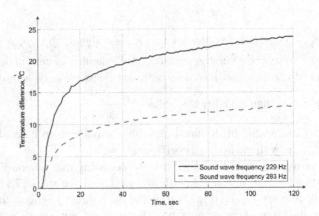

Fig. 5. Temperature difference $\Delta T = T_H - T_C$ at the research TAR stack ends

4   Conclusions

The efficiency of thermoacoustic devices highly depends on the values of acoustic pressure and sound wave frequency inside the TAD's resonator. Authors proposed the method of experimental measurement of acoustic pressure dependency $P_{out0}(\omega)$ on sound frequency $\omega = 2\pi f$.

Using the data obtained TAR's acoustic pressure nonlinear mathematical models were synthesized by means of regression analysis least squares method. Performed statistical analysis of obtained models adequacy (12) – (14) showed that the best outcome is reached by the mathematical model which approximates the experimental sample $P_{out0}(\omega)$ by a number of Gaussian functions (8), (13).

The resulting model is characterized by clearly expressed extreme nature of acoustic pressure dependence on the sound wave frequency. Peak value of pressure is achieved by the resonance frequency value. Therefore, authors implemented the resonant frequency identification algorithm by the search of acoustic pressure maximum value using the extreme methods of optimization. Experimental comparison (Fig. 5) of TAR work efficiency on theoretical $f = 283$ Hz and obtained by developed algorithm $f_{res} \approx 229$ Hz sound waves frequencies showed that proposed algorithm allows to increase the temperature difference on the stack ends and therefore leads to the system efficiency increase.

References

1. Attetkov, A.V., Galkin, S.V., Zarubin, V.S.: Methods of Optimization. Bauman Moscow State Technical University Publishing, Moscow (2003) (in Russian)
2. Kiselova, O.M., Shevelev, A.E.: Numerical Optimization Methods. Dnipropetrovsk National University Publishing (2008) (in Ukrainian)
3. Kondratenko, Y., Korobko, V., Korobko, O.: Multisensor Data Acquisition System for Thermoacoustic Processes Analysis. In: 6th IEEE International Conference, IDAACS 2011, Prague, pp. 54–58 (2011)
4. Kondratenko, Y., Korobko, O.: Synthesis of Thermoacoustic Devices Nonlinear Mathematical Models. In: MPZIS 2011: Proceedings, pp. 135–136. Dnepropetrovsk, Ukraine (2011) (in Ukrainian)
5. Kondratenko, Y., Korobko, O.: Analysis of the Sound Wave Frequency Impact on the Efficiency of Thermoacoustic Processes. In: Automatics 2011: Proceedings, pp. 390–391. Lviv, Ukraine (2011) (in Ukrainian)
6. Korolyuk, V.S., Portenko, N.I., Skorokhod, A.V., Turbin, A.F.: Handbook of Probability Theory and Mathematical Statistics. Science (1985) (in Russian)
7. Penelet, G., Gusev, V., Lotton, P., Bruneau, M.: Experimental and Theoretical Study of Processes Leading to Steady-State Sound in Annular Thermoacoustic Engines. Physical Review (2005)
8. Pupkov, K.A., Ehupov, N.D.: Methods of Classical and Modern Automatic Control Theory. Bauman Moscow State Technical University Publishing, Moscow (2004) (in Russian)
9. Rott, N.: Thermoacoustics. Adv. Appl. Mech. 135 (1980)
10. Shapiro, J.F.: Mathematical Programming: Structures and Algorithms. Wiley-Interscience, New York (1979)
11. Spoelstra, S., Tijani, M.E.H.: Thermoacoustic Heat Pumps for Energy Savings. Grensoverschrijdende akoestiek, Nederlands Akoestisch Genootschap (2005)
12. Strett, J.W.: Theory of Sound. The State Publishing of Technical Theoretical Literature, Moscow (1958) (in Russian)
13. Tomonaga, T.: Fundamental Thermoacoustics. Uchida Rokakuno Publishing, Tokyo (1998)
14. Wheatley, J.C., Swift, G.W., Migliori, A.: The Natural Heat Engines. Los Alamos Science 14 (1986)
15. Yudaev, B.N.: Technical Thermodynamics. Heat Transfer. High School, Moscow (1988) (in Russian)
16. Zarembo, L.K., Krasilnikov, V.A.: Introduction to Nonlinear Acoustics, Moscow (1966) (in Russian)

Author Index